THE THAMES EMBANKMENT

TECHNOLOGY
AND THE
ENVIRONMENT

JEFFREY K. STINE AND
WILLIAM MCGUCKEN

SERIES EDITORS

DALE H. PORTER

THE THAMES EMBANKMENT

ENVIRONMENT, TECHNOLOGY, AND SOCIETY IN VICTORIAN LONDON

THE UNIVERSITY OF AKRON PRESS

AKRON, OHIO

Copyright © 1998

The University of Akron
Akron, Ohio 44325-1703
All rights reserved.

LIBRARY OF CONGRESS CATALOGING–IN–PUBLICATION DATA

Porter, Dale H.
   The Thames embankment : environment, technology, and society in Victorian
London /Dale H. Porter.
      p.   cm. — (Technology and the environment)
   Includes bibliographical references and index.
   ISBN 1-884836-28-3 (cloth). — ISBN 1-884836-29-1 (paper)
   1. Embankments—England—London—History.   2. Public works—Social
aspects—England—London—History—19th century.   3. Thames River (England)
—Regulation—History—19th century.   4. Sewerage—England—London—
History—19th century.   5. Social values—Great Britain—History—19th
century—Sources.   I. Title.   II. Series: Technology and the environment
(Akron, Ohio)
TC533.P67   1998
363′.0942′09034—dc21   97-34773
                                                                                CIP

In memory of F. Elbert Porter

Contents

Illustrations

Series Preface

This book series springs from public awareness of and concern about the effects of technology on the environment. Its purpose is to publish the most informative and provocative work emerging from research and reflection, work that will place these issues in an historical context, define the current nature of the debates, and anticipate the direction of future arguments about the complex relationships between technology and the environment.

The scope of the series is broad, as befits its subject. No single academic discipline embraces all of the knowledge needed to explore the manifold ways in which technology and the environment work with and against each other. Volumes in the series will examine the subject from multiple perspectives based in the natural sciences, the social sciences, and the humanities.

These studies are meant to stimulate, clarify, and influence the debates taking place in the classroom, on the floors of legislatures, and at international conferences. Addressed not only to scholars and policymakers, but also to a wider audience, the books in this series speak to a public that seeks to understand how its world will be changed, for ill and for good, by the impact of technology on the environment.

Preface

This book, like the Thames Embankment itself, was constructed in three phases over a period of twelve years. My initial research aimed to show that the process by which historians compose accounts of the past was analogous to the conception, design, and building of innovative technological artifacts. By highlighting their common elements of purposeful design in the face of contingency, I hoped to refute the notion, fashionable in the early 1980s, that narrative was a false and perhaps pernicious mode of representing the past. Using the Embankment as a focus was prompted not just by my desire to return to London but by a comment, in a book review by Asa Briggs, that someone should really study the Victorian London engineering community from a social perspective.

As often happens when historians engage in philosophical enterprises, the subject matter soon became more interesting than my theoretical arguments. A paper detailing my initial research on the London community of civil engineers was, by happy coincidence, accepted for a session at the 1987 meeting of the Society for the History of Technology at which the Dutch scholar Wiebe Bijker outlined a new model which he, Thomas Hughes, and Trevor Pinch had formulated for analyzing the social contexts of technological innovation. Further correspondence with Bijker led me to recast the manuscript as an exploration of the many different inter-est groups involved with the Embankment, their changing conceptions of its intended functions and design, and their competition for control over the project.

When William McGucken and Jeffrey Stine read the manuscript for The University of Akron Press, they saw in it an environmental dimension that I had rather taken for granted. Based on their suggestions, I pursued a new round of research and substantially revised the manuscript once more. It is now devoted to the historical function of technology (in the form of a public works project) as an interface between a community and its environment. As in the case of the Thames Embankment, however, fea-tures from earlier stages of construction may still be visible in the text.

Acknowledgments

Generous support for this project was provided by the Western Michigan University Research and Creative Activities Fund; the WMU Burnham-Macmillan Endowment Fund; and by the American Philosophical Society. A sabbatical leave recommended by my colleagues and approved by the WMU Board of Trustees made the initial research possible.

A network of colleagues and archival staff members can contribute more to successful research than the most extensive bibliography. For help with the Thames Embankment, I wish especially to thank Miss D. J. Bayley and her successor Michael Chrimes, librarians at the Institution of Civil Engineers; Dr. Christopher Hamlin of the University of Notre Dame; Carl Harrison, archivist at the Lewisham Record Office; Ralph Hyde, curator of maps at the Guildhall Library; Professor Denis Smith of Northeast London Polytechnic; Dr. Norman A. F. Smith of Imperial College London; Dr. Joel Tarr of Carnegie-Mellon University; and Lady Alexandra Wedgwood, architectural archivist at the House of Lords Record Office.

Archivists at the Greater London Record Office (now the London Metropolitan Archives) were especially helpful in locating records of the Metropolitan Commissioners of Sewers and Metropolitan Board of Works, as well as relevant maps and illustrations. The Greater London Council engineering department not only produced working drawings of the Embankment but also answered my questions at various work sites along the river. The staffs of the British Library, the Public Record Office, the University of London Institute of Historical Studies, the Civil Engineering Library in New York City, London Transport Department of Civil Engineering, the National Museum of Science and Industry, the Victoria and Albert Museum, the Royal Institute of British Architects, the Blackheath Preservation Trust, and the Business Archives Council offered helpful service at every stage.

For their continuing friendship and hospitality, as well as professional assistance, I owe a great deal to Stephanie Morland of Glastonbury, to Neil and Elizabeth Rhind of Blackheath, and to Dr. Gloria Clifton, who patiently coauthored an article on engineering and contracting at the Metropolitan Board of Works.

My graduate student Kevin Vichcales provided editorial and bibliographical assistance, especially with environmental concerns. Christine Porter and Fritz Seegers prepared maps and illustrations, and Andrea Harger Donovan took time from her dissertation research in London to photograph the Albert Embankment. My daughter Stephanie, then age fourteen, gave up a day of shopping on Oxford Street to serve her apprenticeship in the archives. Finally, of course, I have to thank my wife Betty Porter for encouraging the project even while enduring years of anecdotes about sewers, rivers, and concrete.

Kalamazoo, Michigan
January 1997

THE THAMES
EMBANKMENT

Constructing the Victorian Environment

During intermissions at the Royal Festival Hall, symphony orchestra patrons like to carry their wine and conversation outside to the upper terrace. There they look northward over the Thames River as it sweeps through a great arc from the southwest to the east, framing a panorama of metropolitan London at dusk. The great field of lights is anchored by Westminster on the left, then Whitehall, Charing Cross, Somerset House, and the Temple, with the dome of St. Paul's glowing softly on the far right. It is a magnificent reflection of the symphonic stage within.

On special Sundays during the summer, a viewer at Festival Hall equipped with binoculars might discern a flotilla of tiny plastic ducks or dolphins bobbing downriver on the ebb tide, cheered by spectators as they pass under the bridges. They are racing for charity, each one marked by a contributor's number.

The royal panorama and the dinky dolphin derbies are each made possible by the same environmental construction. The Thames from Westminster to St. Paul's flows between smooth, parallel walls, which keep most of the toys from straying out of the central channel, and support the great expanse of built-up shoreline where Westminster and the Temple sit in state. The cranes, barges, wharves, and warehouses that dominate the Docklands to the east are not evident here. Aside from the tour boats, few vessels crowd the river. Cleared of its commerce to make it a public amenity, it has become a major feature of the city's architectural identity.

The relationship of the Thames to London is not a simple contrast between nature and human civilization. Both the river and the metropolis have blended natural functions with organized human enterprises for over a thousand years. People shaped the city with their buildings and the surrounding countryside with their demands for food and raw materials. They shaped the river by their embankments and piers, their demand for fresh water, and their discharges of waste. These two systems of human ecology, the urban and the riparian, collided in the middle of the nineteenth century. That collision produced the Thames Embankment.

In the autumn of 1863, the engineering department of the Metropolitan Board of Works, representing the City Corporation and all the London borough councils, began erecting the first of three retaining walls along the banks of the Thames River (map 2). The three embankments extended five and a half miles from the western suburb of Chelsea down to Blackfriars Bridge near St. Paul's Cathedral. Made of brick and concrete, thirty feet high and ten feet thick at the base, they were faced with granite and topped with lighted parapets. They streamlined the previously irregular shore, improving navigation and providing new docking facilities for millions of steam ferry passengers. The walls enclosed fifty-two acres of formerly polluted shoreline. The embanked areas also provided room for a new boulevard between the City and Westminster, new gardens and pedestrian walkways, and a new underground railway line. Most importantly, the embankments incorporated the final section of the London Main Drainage system, a network of sewer lines carrying the city's tremendous burden of human, animal, and industrial waste far down the Thames valley.

The Main Drainage and the Thames Embankment transformed a notoriously polluted, smelly, and unsightly commercial riverfront into an architectural monument that still impresses visitors from around the world. They restored the Thames to London, and their construction gave Londoners a new vision of themselves and their community. Their history, consequently, opens a window into significant if unfamiliar aspects of Victorian society and its environment.

The contemporary importance of the Main Drainage and Embankment lay in their status as public works projects, rather than in their technical difficulties. The sewer lines had to be dug to unusual depths, of course, and the engineers had to integrate them with a labyrinth of old drains while avoiding building foundations, quicksand, and other dangers. The work, however, was done mostly with hand and animal labor, with the large oval sewers built on site of brick. The technical aspects of the Embankment were equally commonplace. Though designed to 1863 specifications, most of it could have been built a century earlier.[1] During construction, engineers experimented with a new type of retaining wall, introduced steam-powered cranes, and learned how to use poured concrete in place of brick. But, in general, the Embankment differed from harbor and canal installations elsewhere in England only because it was designed as a multipurpose facility and because it had to withstand strong tidal currents. Planned by committees and built by a nascent bureaucratic agency, it lacks the dramatic appeal of individual achievements celebrated by Samuel Smiles, such as Telford's Caledonian Canal or Brunel's Clifton Suspension Bridge. Nor can it compare in interest to the buccaneering triumphs and disasters of the railway age immediately preceding. Significantly, the most extensive account of its origins and construction appears in an administrative history of the Metropolitan Board of Works.[2]

For historians, the Main Drainage and Embankment may serve as unique icons for Victorian ideas and practices regarding their society, their technology, and their environment. The phases of design, organization, and construction reveal complex interactions between urban and riverine environments, attitudes toward dirt, disease, and death, the nuances of class distinctions, and an emerging public identity. They reflect that curious Victorian habit of repressing conversational references to human anatomy and functions while perfecting the flush toilet, devising schemes for using "human guano" as fertilizer, and establishing magnificent Gothic "conveniences" in their public gardens. Planning and construction pitted entrepreneurs against government agencies, liberal economists against sanitary engineers, and aristocratic residents against media-led public opinion. The projects required innovations in public finance during a period dominated by parsimonious governments and wild fluctuations in the capital markets. The interlocking communities of engineers, contractors, public officials, and working men who built the structures mark the ascendancy of Engineering and Sanitation in the Victorian pantheon. The Thames Embankment shaped no great politician's career, caused no change of ministry, occasioned no financial crisis. But its construction was affected by all of these phenomena, and so serves to illustrate, confirm, or modify what we know about Victorian life from other sources. As a cultural phenomenon, it is an appropriate symbol of what Asa Briggs called the Age of Improvement.[3]

The Embankment and Main Drainage belong not just to the Victorians, however. They have been extended and modernized but remain today basically as they were constructed 135 years ago. The sewer lines are still there—maintenance men use the original access tunnels, and the overflow penstocks along the banks still dump raw sewage into the river at periods of heavy rain and high tides. Almost every visitor to London walks the Embankment or rides one of the tourist buses that regularly park along its roadway, blocking the incessant automobile traffic. Millions cross over it daily on a dozen road and rail bridges, or ride deep inside it on the Circle and District lines of the London Underground Railway. Unlike many structures in "heritage England," the sewers and the Embankment are not mere relics of the past. Even before they were completed, they began constructing their own history, changing their environment and the habits of Londoners.

Their story is known to many people in outline; it is ritually recited whenever an addition is contemplated or an old part renovated. The Main Drainage was the brainchild of the Benthamite public health advocate Edwin Chadwick and a group of civil engineers connected with the Metropolitan Commissioners of Sewers (MCS). It was intended to clean up the polluted Thames River and remove what many regarded as a primary source of disease. Legislation in 1855 created a new public authority, the Metropolitan Board of Works (MBW), which hired the MCS engineers (but not Chadwick) and implemented their plans. The Thames Embankment, authorized by Parliament in 1862, was built to enclose the lowest-level intercepting sewer, protecting the river from further degradation. The sewers and the Embankment not only solved London's sanitation problem but also turned the riverfront from an eyesore into a useful and ornamental public thoroughfare. Its success established the reputation of the MBW; its chief engineer, J. W. Bazalgette, was duly knighted when the work was finished in 1874.

So much is well known, and often rehearsed. The deterioration of London's sanitation, the incessant debates over water pollution, the administrative politics of the Main Drainage project, and the emergence of an engineering solution to the perceived crisis have been explored in several publications, which are reviewed in chapter 3. The Thames Embankment, however, is another matter. Surprisingly little has been written about this massive, long-lived edifice. Capsule accounts of the Embankment's origins appear in most London guidebooks, and details of specific sites can be found in contemporary surveys such as John Timbs's invaluable *Curiosities of London* (1867).[4] Nostalgic picture collections, such as John Betjeman's *Victorian and Edwardian London from Old Photographs* (1969), show us the way the Embankment used to look, when the old Thames sailing barges plied their trade alongside. But a complete account of its origins and development has never appeared. It may be that the project has been so often referenced in abbreviated form that it has been reduced to a stereotype. Perhaps because it was so typical of its age, or because it still functions in its quiet, massive, unassuming way, it is taken for granted. Even the current metamorphosis of the nearby South Bank, the Surrey Docks, St. Katherine's Dock, and Billingsgate—high-rise/high-rent office blocks and condominiums emerging from the shells of old wharves and warehouses —seems not to have raised curiosity about the history of Thames-side development. It is intriguing to find an object so public and so constantly in use, whose history is so obscure.

A present-day observer might respond, "If the Embankment endures as it was planned and built, then it speaks for itself. What else can its history reveal?" The answer is that public works, by their very nature, attract the interest and contributions of a variety of institutions, groups, and individuals. Some of these manage to gain control of the development process, imposing their designs and interpretations on the whole, or negotiating compromises with other interests. Once the project is complete, builders and their associates tend to consolidate their collective memory of how it came into being. They forget the alternative possibilities considered early on in the process, the compromises and changes made during construction, and the actions and perspectives of those who failed to realize their ambitions. It is the business of historians to recover, so far as possible, such "might-have-beens," to represent the experience of losers as well as winners. If we investigate why people thought the Embankment was needed, how they conceived and designed and financed it, who built it and who used it, we can discover a whole culture in microcosm: civil engineers striving for public recognition of their professional and social status; tough-minded wharf and barge owners ready to litigate any interference with their trade; a fringe of genteel dependents on Crown property threatened by public access; tightfisted public officials with vested interests; moralistic social reformers, crackpots, and thieves.

Public works projects also interact with the natural environment in complex ways. The Main Drainage was obviously a response to environmental degradation caused by previous urban growth and well-defined technical and political practices. But it was a particular *kind* of response, generated partly by the natural configuration of the Thames valley and partly by the special political and technological framework of the metropolis. The Embankment, however, embodied a much more complex set of goals and assumptions, promoted by a variety of groups over a lengthy period. Some type of embankment was foreseen as early as 1824, when engineers planning the replacement of old London Bridge realized that it would change the topography and flow of the river as well. They were concerned primarily with navigation, and the later Thames Embankment did address that problem. But it was also designed to reduce flooding on the south side of the river, to improve vehicular and steamboat traffic from east to west through the metropolis, and to provide open space conducive to the health and morals of urban residents.

The Thames Embankment as a public works project shows how technology mediates between cultural values, social groups, and institutions on the one hand, and the natural environment (as perceived and modified by humans) on the other. I have, therefore, organized this study of London society and its environment around the processes of design and construction. As a kind of interface, of course, technology itself is subject to environmental conditions and to the attitudes and practices of the society which utilizes it. It is full of contingencies. The Embankment was "constructed" socially as well as physically, and was modified in response to unexpected weather, high tides, quicksand, and the quality of raw materials available in the region.

The concept of "social construction" is not new to the history of technology. It was a natural derivation of the social studies of science pioneered by Thomas Kuhn in the 1960s. The late Melvin Kransberg embedded elements of it in the journal *Technology and Culture* during his long editorial tenure, and scholars like Thomas Hughes, Wiebe Bijker, and John Law have developed it into a useful, if sometimes controversial, model for studies of invention and public policy.[5] The present study extends its application to the history of public works. I assume that the Thames Embankment resulted from a multifaceted, competitive negotiation among a variety of groups espousing different ideas about embankments. The meaning of "The Embankment" stabilized when one set of groups imposed its interpretation(s) upon the community. Paradoxically, stabilization opened up possibilities for further development imagined by new as well as older parties. Analysis of the competing groups, the meanings they attributed to the Embankment, and the process through which those meanings were negotiated, stabilized, and elaborated provides a running commentary on the metamorphosis of the project. It tells us as much about the society that built it as it does about the engineering techniques they used.

Most people would not think of a five-mile concrete wall, backfilled with a million cubic yards of dirt, as a social or cultural informant. But, in fact, the voluminous files of technical reports and related construction correspondence tell a great many stories. The Embankment Gardens near Hungerford Bridge, for example, have the shape they do because the engineers who designed the site ran up against a crusty and powerful old aristocrat who refused to leave his family mansion. The engineers had to abandon the elaborate terraced gardens and shopping arcade that appears in contemporary illustrations. They left the old York Water Gate crumbling at the back of the property instead of mounting it on the new Embankment wall. At Pimlico, the waterfront has an entirely different appearance from the rest of the Embankment because Thomas Cubitt, the contractor who had his immense workshop and storeyards there, got fed up waiting for the government to act. He built a river wall and roadway at his own expense. Why are there no cafés or restaurants along the broad, tree-lined Embankment boulevard? Because Victorians associated such things with dissolute Paris society.

Comparing the Embankment as it stands today against original draw-ings, reports, speeches, and memoirs, we find that Victorians had many different ideas about what it was and what it meant. We have to treat the documentary evidence they left behind in the same way we treat the mate-rial objects they built—as clues to perception and experience, rather than as pieces of an objective reality. We can no longer think of an artifact like the Thames Embankment as the result of purely technical processes. It did not emerge automatically from a set of technical or environmental needs, and, however concrete it may be today, its construction was a matter of competing interpretations and countless contingencies. Its very success as a piece of engineering changed the criteria by which people evaluated it. So there were as many "Embankments" as there were institutions, groups, and individuals involved with it.

This social construction of technology, however, will only take us so far. The Embankment is, after all, a very substantial chunk of real estate which has endured for over a century. It was built in response to both long-term and short-term demographic, institutional, and ecological changes in the metropolis, some of which were accurately predicted and planned for. Specific techniques of surveying, excavating, pouring concrete, laying pipe, and pumping water from the work site were hardly matters of social interpretation. Materials, land, and labor had to be obtained, organized, and paid for. Once the Embankment was in place, the social and physical environments had to accommodate it. An adequate account of its history has to represent these objective conditions and processes as well as subjec-tive experience, has to give due weight to the structural as well as the cul-tural forces that informed its construction.

The environment, too, must be treated both as a relatively autonomous source of conditions and forces acting upon society and its physical sur-roundings, and as a product of sustained human intervention guided by historically circumscribed concepts. For example, the topography of the

Thames valley could not be altered by the Victorian scale of construction; rainwater and sewage were bound to flow towards the river. The sewage could, however, be diverted downstream through gravity. Londoners also had to accept the fact that the Thames was a tidal river, though they changed the impact of the tidal flow by their bridges and river walls. For historians, the term "environment" must be considered an abstraction from practical experience and knowledge of specific natural or human-built features: rivers, quarries, rain, manure, crowds. It exists in a particular time in relation to some phenomenological counterpart, such as a city, which it surrounds. It can never be directly, "naturally," experienced. Our ways of perceiving and interacting with it are always mediated by intellectual, religious, political, economic, class, and gendered filters.

In nineteenth century England, moreover, there were very few locations where nature could be said to remain in a primeval state. Deforestation had proceeded apace since Roman times. Sheep grazed everywhere in England, cattle invaded the Scottish Highlands, and slag heaps from mines grew into recognized topographical features. Even the famous grouse moors of the eastern uplands in northern England and Scotland, which are now designated historical landscapes and conserved as examples of natural wilderness, were actively cultivated by estate managers in the nineteenth century to produce heavier populations of grouse for increasing numbers of middle-class weekend hunters. Except for some sand dunes, cliff faces, salt marshes, and the odd undrained fen, natural ecosystems had all but vanished.[6]

By 1800, the London metropolitan region was well past the stage where one could discuss the impact of urbanization upon the natural environment as if the latter had been interrupted in its normal path of progression. It had become what I. G. Simmons terms a "cultural ecosystem"— that is, an ecosystem dominated by human diversions, destruction, domestication, and unique concentrations of raw materials.[7] The population of the built-up area amounted to nearly 750,000 on the Middlesex side and another 210,000 or so on the Kent and Surrey side, which had been opened up by the building of Westminster and Blackfriars Bridges in the mid-eighteenth century. Within a half century, the metropolis held 2.4 million people spread over 117 square miles, with satellite communities extending into five counties. In 1851, population density reached 100,000 per square mile in a dozen districts, where less than half the residents were native-born to London.[8]

The countryside for at least thirty miles around the residential core had been actively reshaped by market gardens, landscape architects, railways, and suburban real estate developers. Farmers converted commons and open fields into grasslands to supply hay for the city's horses and cattle and grain for its bread and beer. Over 8,000 dairy cows lived on the outskirts of the capital, mostly in Hackney and Islington. Before going to market, Irish pigs fattened on South London brewery wastes and Welsh cattle thrived on the chalk farms of Wiltshire. In return, the livestock yielded thousands of tons of manure every year for the "high farming" methods introduced by application of chemical science to agriculture.[9]

Londoners' intensive use of coal for hearth fires (they felt the more efficient Continental stove made rooms "close" and cheerless) helped raise the ambient temperature of the city two to three degrees Fahrenheit above that of the surrounding countryside, with a corresponding rise in humidity. Gas lights, introduced in the 1830s, produced another four to five degrees difference and added their fumes to the atmosphere. The smoke, mixed with caustic exhausts from coal gas works and the miasmas emanating from polluted watercourses, caused chronic respiratory ailments among its less fortunate denizens and blocked out perhaps three-quarters of the sunshine normally enjoyed by country towns. This "heat island," as Thomas Glick has called it, severely limited the species of animals, trees, and flowers (as well as parasites), favoring those which could flourish symbiotically with humans. Rats monopolized the city's sewers, and wrens proliferated amid the piles of manure in city streets. Pigeons, of course, adapted well to a regime of garbage and building ledges. Long before Darwin, they selectively bred for dark coloring as camouflage against the soot-covered buildings.[10]

When we examine the environment of Victorian London from the perspective of a public works project like the Thames Embankment, therefore, we have to remain aware of their long-developing relationship of mutual definition. This awareness was not part of the stock of concepts and values familiar to most Victorians, and it was not part of the story of the

Main Drainage and Embankment until very recently. The history of urban ecology is a relatively new branch of scholarship. People have always written about cities of the past but usually from political, institutional, or antiquarian perspectives. British subjects who noticed their cities crumbling in the late nineteenth century introduced the idea of "conservation," but until the 1950s that usually meant the preservation of ancient monuments and the upkeep of public facilities. The term "ecology," coined in 1866, denoted the distribution and succession of vegetation in a particular habitat. The incorporation of animals into ecological science in the late 1920s added the concepts of the food chain and the ecological niche. These concepts assumed that the dynamics of a habitat were self-regulating and tended toward increased complexity and ultimate stabilization around a dominant species. Ironically, people were seldom considered in such habitat studies, even though cultural anthropologists were using similar models of dynamic equilibrium to study the social interactions of relatively isolated tribes. In the standard ecology model, humans were regarded as rather nasty outsiders whose (often technological) activities impacted other species, and resources like air and water, in mostly deleterious ways. The publication of Rachel Carson's *Silent Spring* (1962) reinforced this view. More recently, ecologists have acknowledged that, even in human-free ecosystems, "disturbance and readjustment are continual and the outcomes of change are very poorly predictable."[11] Evidence of natural discontinuities and crises has led to models that include oil spills, soil depletion, and irrigation dams as roughly equivalent to volcanic eruptions or sudden climatic changes.[12]

In spite of the explosion of ecological and energy studies since the 1960s, British historians have paid little attention to the dynamic interactions between environmental change and human institutional and cultural patterns. Bill Luckin remarked, in 1986, that historical writing on these subjects remained undeveloped; almost a decade later, B. W. Clapp reiterated that it is "not yet widely practised in Britain." Lawrence Breeze's 1993 study of British river pollution lists only one article and no books devoted to Victorian environmental concepts or history.[13] The Urban History Group established by H. J. Dyos has studied the environmental impact of cities through innovative interdisciplinary methods, but their followers are often regarded as outsiders by those whose background and inclination are toward rural habitats, wilderness, and landscapes. The same estrangement has occurred in American scholarship.[14]

The major foci for studies of Victorian ecology have been landscapes and water pollution. Landscape history is a venerable practice in Britain, whose devotees typically describe how the built environment of particular regions responded to geological and geographical features.[15] More recent studies have applied social constructionist models to the work of the eighteenth-century landscape architects and, in doing so, have pointed out their influence on the perception of public spaces in Victorian cities. Research on pollution and disease began with conventional institutional histories of the sanitation and public health movement associated with Edwin Chadwick. New approaches to epidemiology, demography, and the social history of science and medicine have enriched our understanding of how different groups of Victorians experienced and defined disease and sanitation.[16]

Curiously, the same selective emphasis on landscape and pollution affected public consideration of the Thames Embankment and the London Main Drainage. Some people thought about these projects in terms of protecting the "natural" regime of the Thames River and its associated landscape from the ravages of urban expansion, and responded to the creation of a new shoreline with visions of health-restoring public parks. Others saw the sewers and river walls as means of controlling the threat of social disorder arising from epidemic disease and degenerate slum dwellers. It is worth looking at these attitudes more closely, because they directly influenced engineering designs, financing techniques, and political decisions.

When Robert Darnton explored eighteenth-century French society, he asked Levi-Strauss's question, "What things are good to think with?" Darnton's subjects thought with folktales, jokes, police reports, letters, public processions, and tree-shaped diagrams.[17] The people of Victorian England thought with (among other things) sermons, poems, property rights, separate gender spheres, newspaper reports, litigation, patronage and deference, engineering projects, gardens and parks, steam engines, and romances. Like their French counterparts, bourgeois Victorians suffered sudden and often painful deaths from epidemic disease. They developed complex funeral rites to deal with such tragedies, while the poor staged elaborate wakes. Survivors had to cope with their own physical debilitation, with repeated bouts of grief and loss, and with the threat of family disintegration.[18] Many people, and not just the elite, understood these experiences as divinely inspired tests of faith and character. They imagined themselves as characters in a novel or narrative poem replete with suffering and redemption (*Pilgrim's Progress* was still a staple of lower-middle-class homes). Faced with pollution, disease, and urban blight, they habitually responded with projects for moral as well as environmental improvement.

The Thames Embankment was also a typical product of Victorian thinking, a construction project with specific dates to begin and end, planned phases of development (very much like a plot), a designated cast of characters, and the intended goal of *improvement*. By the time the Embankment was authorized, civil and mechanical engineers had compiled an outstanding record of successful projects: canals, steam engines, railways, bridges, housing estates. As a group, engineers had earned the status of middle-class professionals, consulted by public officials, testifying frequently before Parliament, representing British interests all over the world. Construction projects were advertised regularly in the newspapers as investment opportunities. Progress on the great dockyards, the railways, the Main Drainage system, and the Embankment was followed avidly by the general public both in print and in person. Gentlemen sitting in their clubs wagered on where all the soil from the Main Drainage trenches would be dumped (it became the foundation for Wembley Stadium). Party drunks crawled up the unfinished sewer lines on a dare. Families took steamboat excursions along the half-built Embankment site. For millions of investors, laborers, officials, engineers, journalists, members of Parliament, and readers of popular periodicals such as the *Illustrated London News*, "projects" were a familiar, engaging way to think about the world.

A more specific idea entertained by Londoners during the nineteenth century was the notion of the public interest as it related to property. Both concepts have a long tradition in British custom and law, of course, but they came together in London partly as a result of the success of the Thames Embankment. In the Middle Ages, the public interest had been defined in terms of the King's peace, a domain of law in which the ruler claimed the right to intervene on behalf of the good of the realm and its subjects. Gradually, the Crown was subjected to the rule of the law as interpreted by the courts, and, after 1688, Parliament took over much of the royal claim. In the nineteenth century, three developments precipitated new questions about the public interest. Rapid urbanization, coupled with the invention of mass circulation newspapers, generated a sense of community with often radical overtones. Unlike the eighteenth-century mob agitation that supported John Wilkes and Henry Hunt, however, this new community image was identified with local parishes, boroughs, neighborhoods—in other words, urban territory. Second, turnpike and railway legislation and the various reform acts passed by Parliament (especially the Factory Acts, the Poor Law administration, and the Municipal Corporations Act of 1835) highlighted tensions between local authorities and royal government. Finally, public construction projects brought subjects and rulers face to face with the question, whose property is public property? Political authorities obtained the legal right to confiscate real estate and buildings that stood in the way of approved projects, so long as the owners were properly compensated. But who had a vested interest in the new landscape and structures that replaced them? The local vestry or borough council? The central government? An autonomous "improvement commission"? Residents? In the case of the Thames Embankment, some fifty-two acres of prime real estate were literally dredged out of the river in the heart of London. They were authorized by Parliament and the Treasury, claimed by both the Crown and the City Corporation of London under ancient law, built under the auspices of the recently created Metropolitan Board of Works, defended by aristocratic residents—and requisitioned, in the end, by the people of London for their own use. This new concept of public property was to have important consequences for future environmental issues. To cite a single instance, by the end of its institutional life, in 1889, the Metropolitan Board of Works came to manage over 2,600 acres of parks and commons, most of them taken over from private landowners as the metropolis swallowed up the manors and villages on its periphery, and opened to the public under pressure from the Commons Preservation Society.[19]

The emergence of an articulate concern for public space has sometimes been ascribed to a residual nostalgia for the countryside held by vast numbers of migrants to the city in the late eighteenth and early nineteenth centuries. But we have to ask, "which countryside?" and "which migrants?" Scientific agricultural practices, including enclosures, had changed the face of the agricultural countryside in the eighteenth century. The old open fields with their haywains and cottagers were rendered picturesque, along with the newly accessible "wilderness." They acquired aesthetic and sentimental value for aristocrats who associated them philosophically with Enlightenment concepts of "divine" Nature and more practically with the social and political privileges of country house living, hunting, and quarter sessions. Nigel Everett has associated such values with early nineteenth-century Toryism as articulated by Robert Southey and William Cobbett and later expressed by the Young Englanders, including Lord John Manners and Benjamin Disraeli. Everett argues that Tory paternalism, traditionalism, and sense of stewardship for the environment challenged the emerging Whig-Liberal view of Nature as a storehouse of commodities to be exploited for individual economic gain. Whether the evidence can sustain such a neat dichotomy, especially in the "obstinately non-industrial" social and economic structures of London, is questionable.[20] Great numbers of new city bourgeoisie proved eager to escape urban congestion, sample the romantic scenery of Rob Roy's brambled glens, and emulate aristocratic traditions. If they could not afford to rent a secluded estate for weekend hunting parties, they could at least enjoy nature walks, fishing, and boating. Aesthetic appreciation was made easier by the migration of disorderly and disheveled rural laborers to the city.[21]

By the end of the eighteenth century, "Capability" Brown and other landscape architects had established a new aesthetic of the managed landscape. For thirty miles around London, travellers and Sunday tourists could view leisure parks, artificial cascades, and carefully sculpted hillocks topped by Greek temples, "medieval" ruins, or artful clusters of trees.[22] Humphrey Repton and John Nash brought that aesthetic into London. They enjoyed the patronage of the Prince Regent, but their designs began to accommodate the greater mobility and less hierarchical class structure of the urban population.[23] The small green squares of the "garden" sub-urbs remained private pieces of artificial countryside to be shared among local residents. But both Regent's Park and St. James Park were opened to the public, and their functions were redefined. In 1833, after the first cholera epidemic dramatized the perils of crowded living, a Commons Se-lect Committee on Public Walks advocated additional parks and gardens for the laboring population. Its call was seconded by the Health of Towns Association and partially answered by Pennethorne's East End Victoria Park in 1841.[24] The campaign grew slowly before midcentury, but, even in its early phase, it carried moral and hygienic overtones. Walkways and gar-dens were intended to divert workers and their families from public hous-es and illicit pleasures, as well as from dirt and squalor. By mixing with higher ranks in society, they would imbibe more civilized habits.[25]

These early efforts to ameliorate urban congestion proved inadequate. In the 1830s and '40s, as cities rapidly grew unmanageably crowded and polluted, the countryside emerged, in contrast, as an area of relative salubrity, a haven of fresh air and breezes, of sunshine undimmed by pu-trid fog, of room for exercise. Factory operatives and dockworkers, who had no cause to feel sentimental about the rural degradation from which they had been expelled, nonetheless sought open spaces for conversation and companionship. The railway allowed them to visit the country on Sundays. It also enabled city merchants to commute from modest country houses as far away as the Chilterns, where they took a proprietary interest in the natural environment. The countryside thus became an amenity rather than just a pretty sight. Like the new urban parks, it took on moral and hygienic qualities. Enjoying it provided an opportunity for personal enrichment.[26]

As an amenity, nature fit rather awkwardly into the liberal economic as-sumptions of the day. On the positive side, it could be regarded as a re-source with specific productivity values. Fresh air and water made workers healthier, women more fruitful, and managers restored to competitive competence. Air and water were essential for many new manufacturing processes and were still needed by farmers and fishermen. On the negative side, precisely because air and water were "natural" and available over an extended area, their quality was only minimally regulated by the voluminous body of property law that distinguished and defined so much of British society. That body of law protected rights to exploit the countryside by mining raw materials and polluting streams, but the law said little about nature's amenity value for weekend wanderers and sentimental residents. As Garret Hardin pointed out long ago, it may have been to everyone's advantage that the supply of air and water be kept clean, but it was to no individual's advantage to be held responsible for that condition.[27]

A resident of the upper Thames told a Royal Commission, in 1866, that "nature has given to all running water the faculty of purifying itself."[28] The organic matter in sewage, for example, would undoubtedly be oxidized after a dozen miles in the Thames current. This popular belief in the natural purity and restorative powers of rivers and streams was challenged by city dwellers' growing perception of environmental treachery, especially after cholera invaded the country in 1832. Fouled by industrial waste and urban sewage, rivers literally generated life-threatening diseases which medical science was unable to decipher. A city could be held hostage by its own environment. Civility itself became problematic. As one result, the concept of "pollution" acquired multiple meanings. If rivers were naturally pure, then sewage (once considered any form of drainage, but, since 1827, denoting concentrations of human waste) polluted them. If cities were meant to be orderly, then the disorder spread by epidemic disease was also a form of pollution. If the incidence of disease fell most violently upon the crowded, chaotic, and often disorderly lower classes, then pollution might be a social condition as well. Given the moral framework of Victorian discourse, it was perhaps inevitable that "an intemperate and ill-disciplined way of life" became one of the factors disposing individuals to cholera, and "personal moral strength and resilience in adversity" became the key determinants of who would survive an epidemic.[29]

It slowly dawned on the British that their cities were not mere victims of pollution but agents as well. Morality was not enough; the urban infrastructure had to be changed, and the concept of landscape brought in from the countryside. Pollution and public space thus became the province of politicians, local government administrators, and engineers. Public works contractors and their financial backers, advocates of urban transport systems, and architects primed with elaborate designs joined the discussion. Newspapers responded with advice and criticism. Residents and their government representatives organized to protect their interests.

Because the Thames Embankment embodied or denied the aspirations of so many of these groups, its history cannot easily be shaped into a unified narrative (although that would have pleased my Victorian subjects). The chronology offered in the appendix, which might be taken as the skeleton for a narrative, will suggest the difficulties. It makes more sense to use each stage of the Embankment project, from the earliest notions to its twentieth-century extensions, as a window into specific communities, their perceptions and priorities.

For similar reasons, the social, technical, and environmental aspects of the project are not treated in separate chapters. Rather, the engineering technology of the Embankment acts as an *interface* between society and the environment within each phase of construction. When that construction is examined closely, it can be seen as a set of discontinuous, overlapping processes. The most visible components are the three massive walls built on either side of the river, at slightly different periods and for slightly different reasons. The sewer lines, underground railway, and utility tunnels within the walls originated separately, as did the topside boulevard, gardens, and ferry docks. The overall project design was debated for three decades until a consensus emerged, but it was radically changed at the last moment. Construction took a dozen years, filled with delays, modifications, failures, and substantial additions to the original plans. Royalty celebrated its completion in 1870, and again in 1874. Although the Embankment was considered successful in its own right, it immediately evoked proposals for extensions in every direction, becoming both an inspiration and a prototype for future development. As the extensions gradually materialized, the term "Embankment" came to refer both to the original sections and to the whole connected series. Thus, like the term "construction," it signifies both an object and a process.

The Embankment appears as an *object* in relation to its human-built environment. If an historically minded tourist follows the course of the Thames from the Cotswolds down through London, he can compare the riverscape of today with what would have been perceived in the early nineteenth century. A geological perspective shows the Embankment as part of a *process*: the dynamics of the Thames basin led to the original placement and eventual (1830) rebuilding of London Bridge, which required reconstruction of the river's banks. At the same time, a shorter-term process of unregulated urban growth polluted the Thames with sewage. Professional engineers "solved" this problem by constructing the London Main Drainage system, which also required an embankment of the river. Other objects and processes contributed to the genealogy of the Thames Embankment, leading to the design approved by Parliament in 1862. After that decision was reached, new institutional, legal, and financial arrangements had to be created, or old ones adapted for use. Each stage attracted a different array of individuals and groups with a stake in the project. The actual construction process extended over a dozen years and weathered a host of social and environmental contingencies. It stimulated technological innovations, the emergence of a civic identity, and the professionalization of London's civil engineering community. The historical tourist, reviewing both object and process from the perspective of the present, can discern the Embankment's impact on Londoners' image of their city and their river, and trace its role as an enduring paradigm for riverside development.

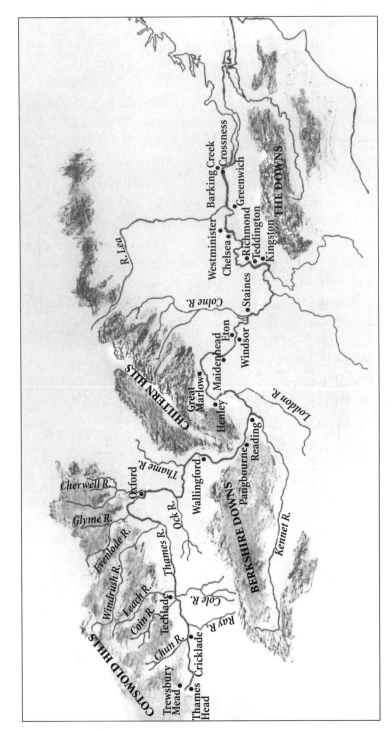

Map 1. The Thames Valley.

CHAPTER II

The Structural Background

I. The Riverscape

The Thames River and the Thames Embankment define each other in a dynamic relationship that starts long before the river reaches London (see map 1). Rising in Gloucestershire, on the southeastern slopes of the Cotswolds some two hundred miles from the North Sea, the Thames flows northeastward along a broad escarpment, accepting half a dozen tributaries, until it cuts through the ridge and abruptly turns south a few miles above Oxford. In this region of chalk, clay, gravel, and rock, with few human settlements to negotiate, the spring-fed river ran bright and clear in the early nineteenth century. Small boats could navigate as far upstream as Lechlade, but above that village a countryside tourist would have to walk through grassy banks or ride one of the increasingly infrequent barges on the Thames and Severn Canal that paralleled the stream. Even in this remote countryside, where William Morris found his bucolic Utopia, mill foundations, bridges, manor house parks, and towpaths molded the river's edge. At the juncture of the Thames with Colne Brook, the westward finger of the Colne River, the Thames Conservancy rebuilt old locks and dredged the riverbed in a vain attempt to improve navigation. Other locks and weirs functioned farther downstream.[1]

When the Thames reached Oxford, its waters deteriorated sharply. Oxford, in 1850, was the largest town on the upper Thames, with a population of about 25,000. Already the railway and gasworks along the river were spoiling the view of ancient Oxford Castle. The town's residential water supply was piped, by a private company, from a nearby artificial lake to many of the wealthier citizens. Workers and the indigent tended to move around too much to keep up water service. They lived mostly in the lower, marshy ground where the River Cherwell ran down to the Thames from the north, and where factories and workshops were situated. (The word "slum," first used in the 1820s, derives from the older "slump," meaning a wet mire.) Like most cities of its time, Oxford was only partly sewered. Working-class districts were usually difficult to drain. In any case, the poor were not trusted to maintain the water closets, which were company owned. They made do with neighborhood privies and leaking cesspits that contaminated the surrounding soil. But the household sewage from higher up was discharged through outfalls, some within the municipal boundaries, and factories poured their wastes directly into the river. Mill owners who required clean water for processing fabrics joined the chorus of complaints raised by university dons and landowners with fishing rights. Scientists testified that Oxford's sewage ought to be fully diluted and oxidized before it reached London, but this hypothesis, even if correct, was no comfort to the 800,000 inhabitants of the towns in between.[2]

Leaving Oxford, the Thames then meandered southeast through fairly flat countryside to Wallingford. Its banks rose more steeply as it passed between the Chilterns and the Berkshire Downs to its junction with the Kennet River at Reading. Like the headwaters of the Thames, the Kennet had once been noted for its purity, but, by 1850, the effluent from tanyards and paper mills had thoroughly wasted it. Coal tar residue flushed into the river from several gas works sank into the river bed, releasing noxious chemicals that poisoned the fish. Residents in the poorer sections of town who took their drinking water from the Thames suffered frequent epidemics of fever. Reading's public officials, like those in most other riverside towns, were eager to build a sewer system, but they feared legal action against discharging sewage into the rivers and knew of no other way to cope with it.

Below Reading, the Thames makes a broad loop to the north along the slopes of the Chilterns past Henley and Great Marlow, then again cuts through a low escarpment to run southeast. It flows between rising banks near Maidenhead and then around the north side of the great outcropping of ancient bedrock at Windsor. Windsor Castle, adopted as a residence by George III and his successors, had been restored and remodelled in the early nineteenth century and was lavishly furnished by Queen Victoria. Its drains, however, were poorly designed and prone to back up under the pressure of high water and wind. When that happened, the grassy parklands sloping down to the river flooded with sewage, which the royal groundskeepers had to sweep down when the waters receded.

Below Windsor, the Thames is joined by the branched mouth of the Colne, running down from the Chilterns, and then reaches Staines, which marked the upper limit of medieval London's authority over Thames navigation. Just above Kingston, the river again turns northward and begins to snake toward London. Kingston had already become a railway suburb by 1850, and its population was growing quickly past 15,000. Its waterfront was rapidly developing, but it had no sewer system until the late 1850s. This fact bothered Londoners, because the intake pipes for several of their water companies had been relocated from polluted areas within the metropolis to points just downriver from Kingston. The Kingston local authorities tried to connect to London's new Main Drainage system but, having been refused, built instead a wholly inadequate system of intercepting sewers. They were sensible enough to position the main outfalls below the water companies' intake pipes, but a layer of "black slime" spread over the river, unnerving visitors from London who knew where it was heading.[3]

The weir and locks at Teddington, built four and a half miles below Kingston in 1811, mark the beginning of tidal action. The tides flow about twice daily from the North Sea, carrying a wall of saltwater up the river. Incoming tides impede the river's flow, and flotsam on the surface moves upstream with it—one can watch plastic cups and half-submerged logs move westward, against the river's own current, or swirl around in eddies by the shore. At ebb tide, the effect is reversed: the river flows faster below Teddington than it does upstream, scouring its channel and the shoreline on the outer edge of each bend. If the tides washed down all the way to the North Sea, the river would be cleaned twice daily. But the movement is only incremental, like a ratchet; what goes downriver tends to come back up, at least part way. When the tide changes, the water comes to a standstill for about an hour, allowing silt, sewage, and flotsam of all sorts to precipitate among the shallows. As will be explained below, tidal action between Teddington and London had been exacerbated, after 1830, by the rebuilding of London Bridge, by the removal of millions of gallons just upstream by the London water companies, and by a program of channel dredging undertaken by one of several navigation authorities, the Thames Conservancy. The new regime affected even the aristocratic pleasure gardens of Richmond and fashionable Twickenham just downriver, where mud banks began to widen at low tide and attract increasing amounts of odoriferous muck. In the 1840s, pleasure boats began to avoid the area.[4]

At high tide, however, the river could still be charming. The riverfront at Twickenham was enhanced by a low embankment, and, farther downstream, a walkway still fronts the old palace grounds at Richmond. The Thames, already civilized by its progress through broad valleys, was bordered but not constrained by its suburban towns. From a boat, one discerned workshops, wharves, and warehouses, but also public houses, meadows, parks, the Royal Botanical Gardens at Kew, and the sports grounds at Chiswick. Today, gliding past Richmond Deer Park and around the bend of Kew Gardens, the river eases against a low inclined bank faced with fieldstone and overgrown with grasses and bushes, while Syon House and other buildings opposite are screened by verdant islands. Just downstream, the boathouses of London rowing clubs line the south bank, facing the green lawns of once fashionable country houses. One might be hundreds of miles from the largest city in England. The worst of the Victorian industrial workshops are now gone, and the Thames water is attractive, if not actually hygienic. It still carries the miscellaneous contributions of its many urban neighbors, the runoff from modern agricultural fertilizers, and a variety of recreational effluvia. Some of these are filtered at the huge waterworks opposite Hammersmith.

From Richmond, the Thames begins winding like an enormous snake north and east through the London metropolitan area. After a loop down through Fulham and Wandsworth, the river curves up to Chelsea, where it finally meets the Thames Embankment. Chelsea Reach was once the

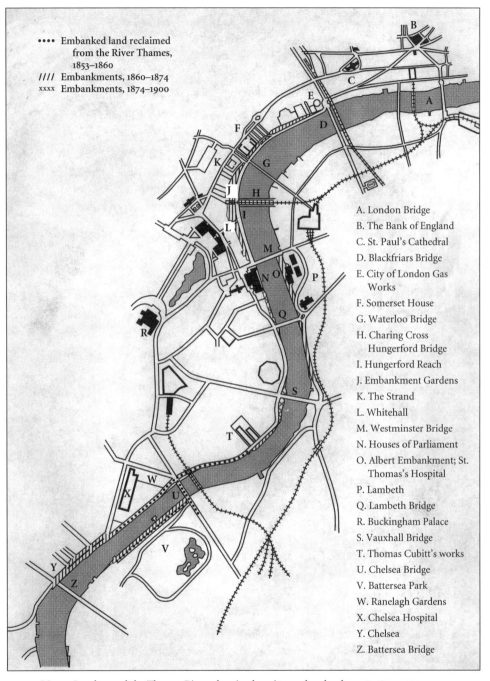

Legend:

•••• Embanked land reclaimed
from the River Thames,
1853–1860
//// Embankments, 1860–1874
xxxx Embankments, 1874–1900

A. London Bridge
B. The Bank of England
C. St. Paul's Cathedral
D. Blackfriars Bridge
E. City of London Gas
Works
F. Somerset House
G. Waterloo Bridge
H. Charing Cross
Hungerford Bridge
I. Hungerford Reach
J. Embankment Gardens
K. The Strand
L. Whitehall
M. Westminster Bridge
N. Houses of Parliament
O. Albert Embankment; St.
Thomas's Hospital
P. Lambeth
Q. Lambeth Bridge
R. Buckingham Palace
S. Vauxhall Bridge
T. Thomas Cubitt's works
U. Chelsea Bridge
V. Battersea Park
W. Ranelagh Gardens
X. Chelsea Hospital
Y. Chelsea
Z. Battersea Bridge

Map 2. London and the Thames River, showing key sites and embankments 1830–1900

widest part of the river west of London. It had a wonderful view of the Thames valley and was convenient to Westminster (by boat, of course— the bankside road was an almost impassable set of tracks through muddy clay). Thomas More built his house here, and the manor of Chelsea was given to Catherine Parr by Henry VIII. In the early nineteenth century, the manor was developed into housing estates by Baron Cadogan of Oakley, and had already become a fashionable suburb for artists and writers, although it contained slums as well.[5] It was fronted by Cheyne Walk, a dirt lane supported by a low rock wall, which turned into Cheyne Row east of Battersea Bridge (map 2). The connection between the walk and the bridge was obscured by a maze of rambling wooden ramps used by steamboat passengers to cross over the muddy shoreline. Considered picturesque, though inconvenient and often noisome, Cheyne Walk was incorporated, in 1874, into the last section of the Thames Embankment to be built. There are still those who regret the "improvement."[6]

The Chelsea section of the Embankment is basically a solid concrete wall rising about four feet above the present road and walkway, capped with marble and a row of lampposts. With trees set into the sidewalk, it gives the appearance of a boulevard, a feature that owes something to the contemporary efforts of Baron Haussmann in Paris. The wall itself, made mostly of poured concrete faced with granite, extends down into the riverbed some twenty feet below low water and gradually broadens to a twelve-foot-thick base. Unlike the more famous Victoria and Albert sections of the Embankment, the Chelsea wall contains few tunnels, arches, or utility lines—just a length of the low-level sewer serving the London Main Drainage system. In fact, if it were not located in such a fashionable part of London, it would be quite unremarkable. Just west of Chelsea Bridge, near the Chelsea Military Hospital, the "new" Chelsea Embankment merges with an older section, built in the 1850s as part of a vast project that ultimately failed. The Chelsea Hospital Garden on the north side of the roadway was, until 1802, the famous Ranelagh Gardens, where notables and nobodies promenaded around a huge rotunda to the strains of an orchestra. Now it features a different sort of extravaganza—the annual Chelsea Flower Show.

Before the Embankment, this side of the river was shored up with wooden retaining walls and fieldstones laid into packed earth. But on the west side of the hospital, between Tite Street and Swan Walk, present-day excavators have found layer upon layer of crushed granite, laid down each spring to facilitate approaches to the old Chelsea Free Dock, from which passengers and their horses might cross over to Battersea.*

The older river wall continues past the hospital along Grosvenor Road to Pimlico where it joins a short embankment built by the Victorian contractor Thomas Cubitt. Cubitt's immense contracting works, a major landmark on this section of the river, have disappeared. The embankment encloses a restaurant, the Westminster Boat Club pier, and Pimlico Gardens, across the road from St. George's Square. The short run up Grosvenor Road to Vauxhall Bridge, presided over by the ancient "Spread Eagle" tavern, includes one of the last unimproved sections of the riverbank in the metropolis, a line of semiderelict warehouses now used as storage sheds and car parks. Next to the bridge, the twenty-seven-acre luxury apartment complex called Crown Reach (motto: "Becoming more exclusive every day") signals the future of the river as a setting for cosmopolitan lifestyles. It features a new granite-faced river wall and exact replicas of the original Embankment lampposts.[7]

The Chelsea Embankment is paralleled on the south side of the Thames by the low terrace walk of Battersea Park, completed along with Chelsea Bridge in 1858. The park grounds were raised out of a low marsh, using the soil excavated for the West India Docks far downriver. They incorporated a well-known tea house and shooting club called the Red Horse Tavern, originally reached by ferry from the opposite shore. Farther on, the park gives way to the derelict Battersea power station and the Nine Elms Gas Works.

At Vauxhall Bridge, the river turns due north, throwing its current against the low Lambeth shore. Before the Albert Embankment was built, flooding was so common here that wharves and warehouses were equipped with sluice gates to let the Thames flow unimpeded through the

*In 1985, the author came upon a work crew trying to drill through these layers, now about six feet thick, to reach one of the Main Drainage junctions that needed repairs. Apparently, the Victorian sewer builders tunnelled underneath it to lay their pipes. An inspection of the original Embankment plans and some contemporary local maps revealed the source of the granite.

premises to wasteland at the rear. The Albert, the second of the three main sections of Embankment, begins just above Lambeth where the roads from Camberwell, Clapham, and Battersea converge on Vauxhall Bridge. The bridge takes its name from Vauxhall Gardens, a pleasure grounds built in 1661 and famous for fireworks until 1859.[8] The Embankment stopped short of the gardens, allowing its roadway to swing inland to meet the foot of the bridge. The garden site, occupied by commercial facilities for many years and then left derelict, is now being developed as an office and residential complex, and the river wall is being extended as it was originally meant to be. Farther north, near the fire brigade headquarters at Black Prince Road, a square opening under the Embankment marks the one remaining draw dock in the area; approached from the other side of the roadway, it is now used only for the occasional regatta landing.

Lambeth Bridge divides the Albert Embankment. Opened in 1862, when the Embankment was still in the planning stage, and rebuilt in 1932, the bridge crosses the river near the site of an ancient horse ferry, owned by the Archbishop of Canterbury as lord of Lambeth Palace. James II is said to have dropped the Great Seal into the Thames while crossing via the ferry on his escape from England in 1688. After Lambeth Bridge, the roadway is carried up to Westminster Bridge on the east side of St. Thomas's Hospital, relocated here in 1869 after the Embankment was built. Doulton's pottery works were the other attraction along this section, but they are gone now. A pedestrian walkway fronts the Embankment with a splendid view of the Houses of Parliament across the river.

To get to Westminster from Vauxhall Bridge on the Middlesex side, one must skirt the office blocks that decorate either side of the bridge footing and drive along the Grosvenor Road, also known as Millbank, a low embankment built by the Earl Grosvenor in the 1830s as part of his estate development. The Tate Gallery (1897) replaced the notorious Millbank Penitentiary, which ran on Benthamite principles of solitude and hard work, and served as a waystation for hardened criminals sentenced to transportation. One of the great mooring piles used for convict ships bound for Australia is displayed on the Embankment walkway there, with an appropriate memorial plaque.

Approaching the Horseferry Road, opposite Lambeth Palace, the Embankment leaves room for a narrow triangle of park, complete with public toilets, at the south foot of Lambeth Bridge. This triangle forms a tip-end to Victoria Tower Gardens, which run up to Parliament's Victoria Tower. The gardens were formed on the site of a tangle of tenements, wharves, and stonecutters crowding up to Parliament until the turn of the century. From Tower Gardens, the Embankment wall fuses with the terrace of the House of Commons and extends to Westminster Bridge.

Many parts of the river frontage from Chelsea to Westminster on the northern side of the Thames are of earlier or later construction than the Embankment of the 1860s, but they have a fairly consistent appearance and thus unify the whole stretch of riverside. This is not because river-wall construction is inherently simple and restricted in design. Many proposed additions, both in Victorian and in recent times, have been elaborate and fanciful. The consistency came about largely because the Embankment was envisioned as a whole from the start, and the earliest sections were built with an overall plan in mind. The design of the three great sections that serve as the focus of my account—the Victoria, Albert, and Chelsea Embankments—was based partly on the earlier effort and, in turn, dictated the look of the smaller pieces filled in later.

The histories of many public works projects have a reflexive character. They grow out of a vision that is articulated in specific designs, but if they are originals, or the first of their kind in a particular location, people aren't really sure what they are building until they are well under way. From the design alone, it is difficult to tell whether everything will function as intended and what the project will mean to the community when it is completed. If successful, the bridge or harbor, road or embankment will almost immediately redefine its public environment. Recreational patterns, traffic flow, utility lines, wind patterns, and river currents may all be affected. But the completed artifact will establish itself as a reference for future construction. This was the case with the Thames Embankment. It was originally conceived as part of a comprehensive riverside development for the entire metropolitan region, which proved impracticable in the political, social, and economic conditions of the 1840s. The scaled-down version built in the 1860s, however, immediately caught the imagination of the public, so that every addition thereafter was judged by reference to it.

The Victoria Embankment was the "star" in this process. It begins at the foot of Westminster Bridge (itself a brand-new structure in 1860) as a seventy-foot roadway, with a wide pedestrian walkway on the riverside decorated with plane trees and ironwork benches. The Westminster ferry dock, reached by a flight of stairs and hinged to float on the tides, is the main launching point for Thames river tours. Underneath the roadway are pedestrian subways linking the two sides of the street with the Westminster underground railway station, built between the river wall and the original shoreline. Parallel to the railway, buried next to the thick Embankment wall itself, is the low-level intercepting sewer of the metropolitan drainage system, the original impetus for the whole project. Above the egg-shaped sewer pipes, a reinforcing archway carries gas, electric, and other utility lines. At important junctions with secondary drains that carry the city's sewage down the side of the valley to the Thames, there are huge underground catchment and overflow chambers called penstocks. If a visitor leans over the parapet in the right place, at low tide, he or she can see the penstock outlet covers mounted in the Embankment's granite facing. They were designed to open at ebb tide to release diluted sewage when rainstorms flooded the system, and they still do from time to time. Easier to see, and certainly more picturesque, are the lion-headed mooring rings mounted high up on the wall.

From Westminster to Blackfriars Bridge, the Thames curves eastward almost at right angles, throwing its weight against the Middlesex shore along a stretch called Hungerford Reach. It parallels Whitehall, the Strand, and Fleet Street. In medieval times, the Thames reached almost to the Strand, which was only a dirt lane leading from the City gate known as Temple Bar to the crossroads (Charing Cross) just west of the hamlet of St. Giles. The south side of the Strand once sloped directly down to the river, but it was gradually filled in to support mansions, gardens, and terraces, with water stairs down to the Thames.[9] Hungerford Reach was a wide, shallow stretch of river subject to curious changes of depth. At high tide, it spread out into a large bay that lapped the stairs and gardens of riverside residents. At low tide, this "embayment" revealed some twenty-seven acres of mud, crossed by open sewer ditches. By the 1840s, it reeked with the accumulated effluvia of London's exploding population. Nevertheless,

Hungerford Reach remained an exclusive residential area. Whitehall Mansions then occupied the site of the present Ministry of Defense buildings, next to Horseguards Avenue. Well-connected residents enjoyed government-subsidized apartments there, with an unobstructed view of the river. The present Whitehall Gardens facing the river were then strictly private, having originally been the King's Privy Garden. They were bordered by the mansions of the Dukes of Richmond, Montague-Buccleuch, and Portland, among others. Montague House, in fact, had just been rebuilt, in 1862, at a cost of £5 million. Just to the northeast, the Duke of Northumberland's palace occupied the middle of what is now Northumberland Avenue.[10]

The Reach was divided by Brunel's Hungerford Bridge (1845), rebuilt in 1862 as the Charing Cross Railway bridge but still known by its original name. The new railway station at Charing Cross replaced the Hungerford market, which had operated there since 1680 and had grown into a two-story arcade offering art exhibits, coffee shops, an orchestra, and cheap tourist items. Its memory is sustained by the clutch of fruit and flower stalls now decorating the back of Charing Cross Station. The Hungerford Bridge abutment on the river's edge was a major rendezvous for passenger ferries. Over two million people embarked and landed here annually until the Embankment's underground railway line undercut the ferry business.[11]

The Thames between Hungerford/Charing Cross Bridge and Waterloo Bridge spread out over what is now Embankment Gardens. At high tide, the water reached the ancient York Gate, still preserved at the rear of the Gardens on the end of Buckingham Street. Most of the ancient palaces, such as York House, had terraces and river walls fitted with boat landing stairs, but there was no continuity to them and no common shoreline, even by the nineteenth century. The famous Adelphi Terrace was built there by the Adam brothers, in 1770, and graced the site until 1936, when it was replaced by the Adelphi office building. The Terrace was built over the Thames foreshore upon vaulted arches, a common method of gaining storage space. Over the years, the Adelphi vaults became a "little subterranean city" whose inhabitants included draft horses used for local work, milk cows who never saw daylight, vagabonds, ragpickers, and thieves.[12] Next to the Adelphi stood Salisbury House, built by descendants of the first Earl of Salisbury, Robert Cecil, Lord High Treasurer to James I. The seventh Earl was living there at the time of the Embankment and actively resisted plans to develop approaches to Embankment Gardens through his property. After his death, Salisbury House was replaced by the Hotel Cecil, considered the most magnificent hotel in Europe at the end of the century, and then, in 1930, by the Shell-Mex building with its monster clock tower. The final building in this row, the Savoy Hotel, was not built until 1889. It was named after the Savoy Palace once held by Edward the Black Prince but burned and blown up by Wat Tyler's rebels in 1381. At the time the Embankment was constructed, the site was occupied by a number of churches built by Charles II for the use of foreign communities. One of them remains tucked away on Savoy Street as it runs down from the Strand.[13]

Visitors today approach the Thames Embankment from its backside, so to speak, walking down Villiers Street from Charing Cross, down Savoy Street from the Strand, or up from the underground at Westminster or Temple Stations. The edifice really faces the river, which it borders and channels. The Victorians who built it were used to thinking of the river as an east-west thoroughfare, so they planned it as a kind of complementary quay. Traffic connections to the inland metropolitan area were added almost as an afterthought, expedient and inadequate. The best way to view the Embankment as it was originally conceived is to take the tourist boats from Hampton Court downriver to Westminster and thence to the Tower and Greenwich, trips which the Victorians themselves enjoyed.

The Embankment landscape along Hungerford Reach is a busy thoroughfare of automobile and tour bus traffic, with brown-bagging office workers competing for park benches with the homeless by day, and crowds rushing across Hungerford Bridge to Festival Hall and Waterloo Station on the South Bank in the evening. At the foot of Northumberland Avenue, in the shadow of Hungerford Bridge, the Embankment wall displays a memorial to Sir Joseph Bazalgette, chief engineer for the Thames Embankment. On the other side of the bridge, past Charing Cross Pier, the wall sports a narrow obelisk, "Cleopatra's Needle," brought to England from Egypt in 1878, in a fit of imperial bravura, and mounted on the Embankment wall, flanked by stone sphinxes.

The road and pedestrian way swing eastward under Waterloo Bridge and cut across the lower arches of Somerset House, repository of wills and other records, whose eight-hundred-foot facade once opened directly onto the Thames across a low plaza. Fifty feet above the plaza ran a spacious terrace in Palladian style, overlooking a central watergate surmounted by a colossal mask of "Father Thames."[14] The upper terrace now affords a view of the rest of the Victoria Embankment as it sweeps past Temple Gardens and then curves up to meet Blackfriars Bridge. Halfway between Somerset House and Temple Gardens, one can walk up from the Embankment to Temple Station, an original part of the underground railway system. The flat "roof" of this station, which remains level with the street behind it, was originally designed to support a shopping arcade giving onto the Strand at the rear and overlooking public gardens along the river.

East of Temple Gardens, medieval religious settlements, including the Whitefriars, once provided sanctuary for miscreants, and from the time of James I the area was known as the thieves' quarter of "Alsatia." In the nineteenth century, the City of London Gas Works was established here. Its rights to river access were preserved despite the efforts of Embankment engineers, so the roadway that now rises up to Blackfriars Bridge was originally carried on open arches to allow coal barges into the Gas Works wharf. Ironically, the Gas Works was bought out and moved to Beckton soon after the Embankment was completed. The area remained undeveloped until Sion College and the City of London School moved into new quarters there in the 1880s.

Blackfriars Bridge marks the end of the Victoria Embankment. The underground sewer and gas pipes, and the Metropolitan District Railway lines, curve diagonally away from the river up Queen Victoria Street toward the Mansion House in the City. When the first Blackfriars Bridge was built in 1767, the Thames downriver from Blackfriars to London Bridge was an active commercial frontage. As new docks were built to the east and railway lines brought goods into the city, the area declined. Heavily bombed during World War II, the old docks and warehouses in this section have been replaced by rows of office blocks and hotels with renovated river frontage. A new portion of Embankment roadway now passes under Blackfriars Bridge, connecting with Upper Thames Street and thence along the shore to the Tower. The shore below London Bridge, where

Billingsgate and the Custom House stand, was the official port of London. As both the river and its shipping were blocked by the massive stone arches of London Bridge, this area formed the Pool, terminus of most cargoes. The north side was embanked following the Great Fire, part of an elaborate rebuilding plan by Sir Christopher Wren. Drawings of the site in 1751 and 1753 show a spacious customs quay extending in a straight line from the bridge to the Tower.[15] Billingsgate, the old fishmonger's hall, has recently been reconstructed as a set of luxury flats, and the whole area is being gentrified. When renovations are finished at Cannon Street Station, Londoners will see Wren's vision made manifest.

After the eighteenth century imperial wars with France and Spain, the port of London became an entrepôt for world commerce. Ships from every nation came upriver as far as London Bridge, where their cargoes were transferred to riverside wharves by a flotilla of lighters. The Pool just below the bridge grew so congested by 1800 that Parliament underwrote construction of the West India Docks across the neck of the Isle of Dogs, where the Thames loops to the south and back. The Royal Victoria and other docks followed, taking the bulk of long-distance shipping transfers downstream. By midcentury, the waterfront between London Bridge and Chelsea supported only local industries, coal gas companies, and wharves supplying timber, hay, coal, and cement to the immediate urban locality. This shift in trade would eventually make it possible to conceive of the upriver shoreline in terms of public architecture rather than commercial utility.

Those familiar with London will realize I have not included the whole south bank of the Thames from Westminster eastward. It was, as we shall see, also part of an early design for embanking the river. But construction was put off, and then abandoned, for many of the same reasons that had delayed its northern counterpart. A souvenir of its commercial importance is the statue of a lion at the foot of Westminster Bridge by County Hall, guarding the entrance to the Queen's Walk. The lion is one of a pair that adorned the roof of the Lion Brewery in that section of the south bank considered, for a long time, commercially too valuable to be replaced by a public embankment. By the 1950s, its value had sunk, so the Royal Festival Hall was built there to revive it. The only section to be developed prior to that was County Hall, at the corner of the bridge. It was authorized in 1911 but not finished until 1933. Headquarters for the London County Council and its successor, the Greater London Council (GLC), it came to be seen by Mrs. Thatcher as a politically inconvenient and often obstreperous metropolitan authority. Labour militants in the GLC liked to hang huge banners across the front of County Hall that could be seen from Parliament's terrace across the river. Mrs. Thatcher dismantled the council under the Local Government Act of 1985 and privatized its headquarters.

Extending eastward from County Hall, a broad pedestrian walkway leads past Jubilee Gardens, developed for the Festival of Britain in 1951. The walk passes under the Hungerford Bridge to the South Bank Arts Centre, anchored by Royal Festival Hall on one end and the National Theatre on the other. Although the view from Festival Hall is magnificent, the South Bank arts complex itself is harder to appreciate, both because it is wrapped around a bend of the Thames bisected by Waterloo Bridge, and because its several buildings were put up over a twenty-five-year period when the prevailing architectural style leaned heavily toward reinforced concrete slab construction. Before 1962, the site was adorned with a 140-foot shot tower, clearly marked on many maps and panoramas of London, that some people think was aesthetically superior to the Centre.

As late as 1987, the pedestrian walkway and the modern embankment ended at the National Theatre. To reach Blackfriars Bridge, one would have to skirt London Weekend Television, several commercial buildings, and another power station along a clutch of shabby alleyways. But with the development of Bankside and the rest of the South Bank down to London Bridge and beyond, the whole Thames riverfront has been opened up for public access. The old warehouses and wharves have been transmogrified into office suites, luxury or public housing, boutique galleries, and tourist attractions. The irony here is that twentieth-century developers have succeeded in blending public gardens and walkways with commercial, rate-paying enterprises, a combination that the Victorians could never finally accept, let alone conceive. The key to this change has been to abandon the use of the Embankment as a thoroughfare for trains or automobile traffic from city to suburb, making it possible to rediscover the charm of the waterfront and the unique perspective afforded by the broad expanse of the

Thames. The original Thames Embankment was designed to parallel the river—to channel its flow, carry sewage along its gradual decline toward the sea, move carriage and omnibus traffic east and west, and redefine its architecture as befit an imperial capital. Subsequent embankments have been oriented perpendicular to the shoreline, conceived of as river*front* rather than river*side*. They thus generate problems of access from the interior and force designers to consider their relationship to streets, buildings, railway stations, and parking structures behind them. Nevertheless, the new South Bank office blocks, with their decorative piers, have occasioned a modest rebirth of passenger ferry traffic up and down the river.

Although it is useful to contrast the parallel orientation of the Victorian embankments with the perpendicular orientation of recent additions, one should remember that no such comparison guided the original design. The Victorian planners had to deal with a set of immediate problems in a rapidly deteriorating situation without benefit of modern concepts of city planning—in fact, without a real metropolitan authority to organize thinking in such a fashion. We must distinguish between the historical riverscape that we can interpret today with benefit of hindsight, and the project the Victorians were able to envision, given the conditions of their time.

II. London Bridge is Falling Down

My description of the upper Thames focused on its general geographical orientation and on the riverfront features a boat passenger might see around 1850. Approaching the metropolitan region, I incorporated the embankments constructed in the nineteenth century and modifications to the shoreline since that time, primarily on the Surrey and Kent side. This humanized riverscape was built up over many centuries in sometimes overlapping, sometimes discontinuous developments. Very few lengths of it could be assigned historically to a single century, let alone a single decade. Overall, the pace of environmental change was gradual. Only in the two decades before 1850 had communities grown to a size and organized their local authorities with sufficient means to impact the Thames beyond a basic human scale. The weirs, locks, and bridges slowed and sometimes diverted the current but did not significantly interrupt its flow.

Most of the workshops and industries that discharged wastes into the river were still individual enterprises or partnerships, although the scale of industrial and civic intervention was visibly expanding.

This modest framework of human ecology was radically altered after 1830, particularly in the tidal reaches, by a unique intersection of geological, topographical, and technological changes. In outline, a glacially slow declination or downward tilting of the Thames valley toward the North Sea brought more intense tidal action upriver to erode the foundations of London Bridge. The bridge itself had shaped the river's regime for over seven hundred years by its sheer bulk; its gradual replacement, over a period of seventy years, dramatically rerouted the currents and the contours of the Thames channel. The river's instability disrupted navigation, a problem which required embankments to remedy. Concurrently, the accelerating growth of London, Oxford, Reading, and other Thames communities put an immense strain on the river's ability to absorb human and industrial wastes. Water and related atmospheric quality deteriorated from "undesirable" to "intolerable" within a few years. Pollution drove plans for the London Main Drainage. The Thames Embankment was, at least initially, intended to complement the rebuilding of the bridge.

Fernand Braudel advised historians "to imagine and locate the correlations between the rhythms of material life and the other diverse fluctuations of human existence."[16] A Braudelian might arrange the changes outlined above in order of temporal duration, with the geological *longue durée* of the Thames valley overlaid by midterm riparian responses which, in turn, impacted the relatively sudden reconstruction of London Bridge. But as I remarked in the previous chapter, we have learned to appreciate dramatic interruptions to the apparently incremental pace of environmental evolution, such as volcanic explosions and tropical storms. We have also realized that human structures can have an immediate and sometimes profound impact on their surroundings. Thus the interplay of technology and ecology proceeds in polychronic fashion, sometimes anticipated but more often not.

The most general feature of the Thames basin was its gradual sinking in relation to the North Sea.[17] The low hills that border the Thames valley— the Berkshire Downs and the Chilterns to the north, the Hampshire

Downs, the Hog's Back, and the North Downs to the south—represent folds in a layer of chalk that extends right across southern England, rising from the southeast in a series of inclines punctuated by escarpments (see map 1). The Thames runs parallel to one such escarpment west of Oxford, then cuts through it and turns south to enter its lowland basin through the Goring Gap between the Berkshire Downs and the Chilterns. Below Reading, where it is joined by the considerable drainage of the Kennet, the river runs through alternating levels of sand and clay deposited during the Tertiary period. Occasionally, as at Windsor, the underlying chalk rises to the surface. London clay is the most widespread and uniform covering, producing low, heavy-soiled country from Regent's Park northward and also south of the river, where flooding was frequent in the eighteenth and nineteenth centuries. After the original oak and elm forests had been cleared (except for St. John's Wood, Enfield Chase, and Wormwood Scrubs), the London clay region was used for pastureland. Described as "stiff, tough, numb, dumb, and impervious" by area farmers,[18] it was generally too wet to build on, and did not provide the wells and springs associated with layers of sand and gravel. In the neighborhood of London, the clay areas were largely uninhabited until population pressure in the early nineteenth century led to the development of cheap housing estates. Even then, drainage remained a serious problem until the main sewer system was completed.[19] This same impermeability, however, made the bed of the Thames a dependable base for the pilings and foundations of large-scale embankments.

On a geological time scale, the Thames has gradually worn through a hundred feet or more of rock and has deposited its own alluvium in the valley during periods when the sea receded. During the last Ice Age, the valley, along with the rest of southeast England, was covered by a succession of giant ice sheets that left vast deposits when they receded. The freezing of so much water lowered the level of the North Sea, so that the land was relatively much higher than it is today. As the Thames wore its way down through layers of glacial and marine deposits in successive periods, it created a terraced terrain. Remains of these terraces make up the wide, flat ledge extending from Paddington to Holborn (including Hyde Park) and from St. Paul's to Bethnal Green. Higher terrace outcroppings appear at Clapham Common and at Hampstead Heath, among other landmarks.[20]

In historic times, the Thames has busied itself widening its valley rather than cutting through the rock. As its tremendous burden of ice melted, the land surface of southeast England actually rose, but the level of the North Sea rose as well, so the Thames developed a fairly gentle gradient and began meandering through its basin. Much of its flood plain below London was marshy wasteland, so travellers would have to go upriver to find a suitable place to cross. At that time, the tides reached about forty-five miles up from the Nore to a series of small gravel-capped hills that remained from the river's last terracing. The location provided access by land and by boat. London Bridge was built there, its thick arches anchored in the gravel. Chelsea, Putney, and Battersea, their Saxon endings meaning "island," are situated on other hills, as was Westminster, originally called Thorney.[21]

The marshlands east and south of London presented a constant challenge to the Romans, Angles, Saxons, Jutes, and Normans. They realized that proper drainage could produce good farmland, but drainage required raised embankments to keep the water from returning. Their primitive dikes gradually became part of the local geography, and from the time of Edward I there is documentary evidence of royal concern for their upkeep.[22] Charges were placed on local rates, rents, and tithes along the river to ensure their continuing repair. One such charge is immortalized in the old epithet "Tenterden's steeple is the cause of the breach in Goodwyn Sands." It is credited to an old Thames-side resident who advised a village commission investigating the collapse of local dikes in Kent. The man insisted that the dikes had held firm until a new steeple was built on Tenterden church. His theory was dismissed as addle-brained superstition until it was revealed that the Bishop of Rochester, believing or hoping that floods had ceased to occur, began siphoning off dike repair funds to finance the new steeple. Despite the ultimate accuracy of the old man's diagnosis, his initial words have come down to us as a lesson in the fallacy of causal explanation, *post hoc, ergo propter hoc*. The story is more appropriate, perhaps, as a lesson about the interaction of geographic and human scales of change. We could say, by analogy, "London Bridge was the cause of the Thames Embankment."[23]

The dikes constructed along the lower Thames had constantly to be raised because the drained peat and clay behind them tended to shrink as they dried, so the level of the land, relative to the river, actually dropped, as it did even more dramatically in the Fenlands to the north.[24] But the real problem for medieval Thames-side dwellers was that, after about 1000 A.D., the landmass of southeast England began once again to subside relative to the North Sea, at a rate of about one foot per century. Tides gradually flowed much farther up the river, increasing the frequency and severity of flooding. This gradual shift led to construction of the new weir and locks at Teddington ("Tide-end town"), in 1811, and the subsidence continues today. By itself, it accounts for the need to build almost all the embankments recorded since Norman times.[25]

The fact that the Thames is a tidal river magnifies its structural propensity to flood. The tides run in progressive cycles, averaging two per day. The flood tide takes two hours for the forty kilometers from the Nore to London Bridge, and another two hours for the twenty kilometers from London Bridge up to Teddington. The ebb tides run faster, with the regular drainage of the river basin added to them. The variation in water levels is dramatic: by the nineteenth century, the average tidal change at London was twenty-six feet.[26] Whenever the incoming tide encounters low ground, as it used to do at Hungerford Reach, it spreads out sideways to the main channel, unless embankments hold it in. The tide first impedes the regular flow of the river toward the sea and then draws the river down with it at high velocity (fast enough to race toy fish from Westminster to Waterloo Bridge for charity drives). Before 1860, the changes of water level, combined with irregular currents and eddies caused by man-made protrusions, formed a patchwork of shifting gravel shoals and sand bars along the meandering riverbed, clogging wharves and undermining dikes and buildings. Walking along the Embankment today, one can see such mud banks still forming and shifting despite the massive effort to eliminate them. A fairly sizable shoal rests against the Albert Embankment opposite the Tate Gallery, a favorite fishing site for desperate anglers. Others form in response to the Festival Hall pier, the tour boat docks, and bridge abutments. From the time of James II, the City of London Navigation Committee was responsible for dredging the river along the length of its tidal flow to reduce or remove these shoals, to keep the river in its channel, to prevent flooding, and to improve navigation.

Figure 1. Old London Bridge in 1756. The bridge's thick foundations raised the upstream level of the Thames River by as much as five feet. (From G. Walter Thornbury, *Old and New London*, 1881.)

The tides interfere with the river's natural tendency to carry solid material—soil runoff, sewage flotsam and jetsam—downstream in a gradual cleansing action. Noxious substances thrown into the Thames are carried back and forth a number of times before disappearing below the city. Anyone who observes the same stretch of river for several days in a row can track the odd-shaped piece of wood or plastic container through this movement.

A bridge over the Thames at London had existed since the Conquest. Early construction was of wood, periodically rebuilt. According to a history by the royal architect Nicholas Hawksmoor, the stone London Bridge was started in 1176 and finished by 1209.[27] Until 1729, it was the only bridge over the Thames, so the city grew densely at either end of it. Its nine thick arches rested on boat-shaped footings called "starlings" (see the view of London Bridge in figure 1). Whereas the cross-sectional area of the river near that point was roughly 20,000 square feet, the bridge covered only 7,360 square feet.[28] The difference was made up by the ramps approaching the bridge from either side. The massive starlings, together with the approaches, dammed up the regular flow of the river. They kept the flood tides from advancing upstream as easily as they might, and impeded the ebb tide as well as the downstream current of the river. In the period 1580–99, water mills were constructed within the first five arches, interrupting the flow by another 25 percent. The current "above bridge" became torpid enough to allow the river to freeze occasionally along King's Reach. When this happened (in 1684, 1716, 1788, and 1814), Londoners organized elaborate ice fairs with skating, games, dancing bears, vendors of all sorts, and even a printing press that published souvenir newspapers. At ebb tide, the water below bridge receded faster than that above it, which therefore remained some five feet higher than below. Most of the current then flowed through the two or three open arches at the center of the bridge, creating a set of torrential waterfalls. Boats coming upriver had to transfer their goods to lighters to be taken upstream, so that the waterfront from the bridge up to Westminster developed differently than the Pool below it. Small craft travelling downstream could try their luck at "shooting the bridge" through the central arches, a pastime enjoyed by rowdies and young swains to impress their girlfriends. The arches were a favorite spot for suicides, too: laden with rocks, people jumped overboard and were drowned in the great rush of water.[29]

By the eighteenth century, the old bridge was deteriorating seriously. The roadway had been repaved so many times it grew eight and a half feet thick over the arches, creating enormous pressure on the foundations. Buildings that had grown up along the parapet, a feature of many old prints, were torn down to lighten its weight. In 1759, architect George Dance removed the last houses, took out the water wheels, and combined the center two arches into one.[30] Dance's modifications increased the flow of the river, along with its "scouring" action. The channel narrowed and deepened, lowering the average water level upriver from the bridge about a foot (although the upcoming tide now rose about four inches higher than before). The narrower channel, in turn, exposed a wider area of mud banks, which were generally covered with sewage and traversed by ferry passengers. The increased scour also began to undermine the foundations of London Bridge itself.

The old bridge had required periodic repairs and replacements right from the start (prompting the famous nursery rhyme), but, by 1799, it was truly in danger of falling down. Parliament felt compelled to take action. A Commons select committee of 1800 consulted architect Robert Mylne, who said London Bridge should be replaced rather than repaired. This opinion seems to have been shared by most of the London engineers, but it is difficult to tell whether it was based on an accurate assessment of the bridge or a wish to contract some new construction.[31] In any case, the committee agreed. At first, they favored an advanced iron design by Thomas Telford and James Douglas, with a single sixty-five-foot arch, but the banks of the river were found inadequate to support the required abutments, and a five-arch stone design by an engineer named R. Dodd gained approval.[32]

Apparently, the resumption of war with France cut short the project because no further reports are evident until 1814, when plans were submitted to the City of London Navigation Committee, which had been appointed Conservators of the Thames by James II. The delay and the new authority meant a recapitulation of the original review process and a battle between James Walker, the new City engineer, and the respected engineering consultant John Rennie over the appropriate site for construction. Finally, in May 1821, a new Commons select committee recommended a revised version of Dodd's original design. Noting that 9,600 tons of stone and rubble had been dumped around the old bridge's starlings since 1800 to keep them from washing away, they urged that the new bridge be built "without delay."[33]

When a new bridge had first been proposed in the 1790s, engineers assumed that any design would open the river to seagoing ship traffic as far up as Blackfriars Bridge, built in 1769 by Robert Mylne. Mylne had enhanced the bridge approach by building an embankment westward to Temple Gardens. It was natural to extend his improvement eastward to the Tower along the line of the well-known "forty-foot way" planned by Sir Christopher Wren after the Great Fire. Officials also envisioned this area as an extension of the "legal quays" of the port established below London Bridge in the time of Queen Elizabeth, now crowded and difficult to police. They therefore planned to renovate the wharves and warehouses and provide a smooth wall to improve the river channel and aid navigation. Telford and Douglas's proposal for an iron London Bridge had been accompanied by plans for an embankment about thirty yards wide on both sides of the river, filled with dredging from the riverbed. Engineer Dodd had provided sketches of a similar embankment, with four-storey fireproof warehouses resting on a continuous Romanesque archway supported by iron pillars set into the existing riverbank. The revenue generated by rents on reclaimed land and new buildings would pay for both the embankments and the bridge: Telford and Douglas figured net profits of £945,918 over twenty years.[34] These embankment plans had to be shelved when Dodd's stonework design was adopted, because, without a much wider arch, significant ship traffic could not pass upriver. Also, larger docks were being contemplated farther east, near the Isle of Dogs. The movement of commercial shipping was steadily in that direction. To fund the new bridge, trustees tapped the ancient Bridge House Estates and Approaches Fund, which originally provided for maintenance expenses, including compensation for widows and orphans of workmen killed during bridge construction and repair.[35]

The contract for the new London Bridge went to John Rennie, who died in 1821 after producing some preliminary sketches. It was finished by his sons John Jr. (later Sir John) and George. The first stone was laid in June 1825; the last arch was completed in 1828; and the bridge was opened in August 1831.[36] The City Navigation Committee insisted on having the new span built just above (west of) the old one in order to keep traffic flowing. Only a few of the old starlings were removed to permit construction of a cofferdam (retaining wall) for the new bridge, and the last of the old pile was not gone until 1834. The younger John Rennie recalled that building upstream presented a real problem: the river currents cascading against the mass of the old bridge had gouged a "deep hole" into the bed of the channel.[37]

New London Bridge almost doubled the span of its predecessor, pushing back the approach ramps and opening its arches to a greater volume of water. The impediments that had shaped the river's behavior for seven hundred years were thus removed. The Thames engineers were quite aware that this would lead to drastic changes in tidal action, water levels, currents, and erosion. Already, in 1821, James Walker and Stephen Leach had been hired by a select committee of Commons to work out the probable effects of replacement.[38]

In 1831, the City Navigation Committee hired Thomas Telford, Sir John Rennie, and Alexander Mylne (son of Robert Mylne) to write a second report. Before that time, and again in 1843, the Crown's Office of Woods and Forests surveyed cross sections of the river, and the City of London surveyor made a complete set of tidal observations.[39] Unlike historians, these engineers were able to apply known laws and hypotheses to a finite set of conditions and make fairly accurate predictions. The removal of the old arches allowed the flood tide to reach much farther upriver, and a greater volume of water to flow downriver at ebb tide. The velocity of the tide-enhanced currents above the bridge increased by as much as 50 percent, scouring the channel twelve to eighteen inches deeper. The average low-water level between Blackfriars and Westminster dropped about three and a half feet, while the average high-water mark rose about three feet. At the same time, the average difference in water level on either side of London Bridge was reduced from five feet to three inches.[40] The engineers had also anticipated some erosion of the foundations of Thames-side buildings and shifts in the mud shoals along the riverbed. They did not realize that the more vigorous scouring action would undermine the pilings of the upriver bridges as well. Westminster Bridge had to be rebuilt in 1860, Blackfriars in 1869. Waterloo Bridge, which had been built in 1817 with shallow piers, was shored up with 17,000 tons of crushed stone.[41]

In their 1831 report to the City of London Navigation Committee, Rennie and Mylne wrote that the changes wrought by the replacement of London Bridge were compounded by the uneven width of the Thames as it ran through the metropolis. Their survey corroborated the testimony of engineer James Walker that the river was 600 feet wide near the Penitentiary (where the Tate Gallery is now); 1,050 feet opposite Millbank (just southwest of Parliament); 1,480 feet at the Hungerford footbridge; 720 feet at Southwark Bridge; and 690 feet at London Bridge.[42] These variations, together with the tides, created an endlessly shifting series of shoals that impeded navigation and the offloading of barges. Walker had predicted that, after rebuilding London Bridge, the increased action of the river would make the shoals worse. The only remedy in the long run was to make the width of the river uniform by a set of embankments; in the short run, until the new regimen of the river worked itself out, a schedule of dredging would have to be maintained. This view was shared by Rennie and Mylne and by a team of City surveyors: Benjamin Whinnell Scott, principal clerk to the City Chamberlain, and James Francis Firth, principal clerk for records concerning wharf frontages.[43]

Although a solid embankment was not immediately recommended, all the engineers agreed that the City Navigation Committee, in cooperation with H. M. Commissioners of Works, should establish a line on either side of the river to which all future wharf construction should conform, so that the combined frontages of all the wharves, docks, warehouses, and terraces would gradually function as a kind of rough embankment and would be in the proper place when a permanent embankment was built. Dwarf pilings would be used to secure the river banks where frontages did not extend, and the river would be dredged uniformly between the two lines.[44] James Walker, then president of the Institution of Civil Engineers, was commissioned to survey the two lines, one on the north side from Millbank to Waterloo Bridge, and one on the south side from Vauxhall nearly to Blackfriars Bridge. Walker presented his survey to the Office of Works, in March 1836, and to the City Navigation Committee in April. The committee agreed to fund the project in their section of the river, provided government would support a bill officially establishing Walker's line and giving the committee full authority to enforce it. They proposed to recoup their expenses by collecting a quitrent of 4d. per foot on embanked property. In August 1838, a deputation from the committee conferred with the Works Commissioners in Whitehall to organize a detailed plan, estimates, and the necessary legislation. The two agencies agreed to split the costs involved. The Office of Works proposed Walker for the job, and the committee added Stephen Leach, their clerk of works, to the team.[45] Walker reviewed the survey made by Thomas Telford in 1823, conferred with Leach, and came up with a revised plan in April 1839. The Navigation Committee adopted this plan and promised to enforce it "subject to such breaks and recesses as may be required for private accomodation."[46]

Walker's line became a basic part of every subsequent plan for embanking the Thames and was incorporated as the guide line for the Embankment eventually built. However, it did not fulfill its interim function of regulating frontage construction. The City's plan to recoup its expenses by charging quitrents for embanked property discouraged frontage owners from cooperating. Owners also found it impossible to get the approval of neighbors to extend their premises to Walker's line, because it would block out the neighbors' access.[47] In one area, however, Walker did achieve the desired effect. He was appointed engineer to construct an embankment for the new Houses of Parliament, and he made sure that it conformed to his established line.[48]

The concerted effort to replace London Bridge and to predict and mediate the effects of that project on the river environment illustrates the interaction of short-term human endeavors with long-term aggregate patterns of change. Since we now are beginning to understand the complexity of human (and urban) ecology, it seems useless to insist upon the primacy of structural and infrastructural developments as "underlying" realities, relegating social and technological decisions within a local culture to a category of ephemeral disturbances. We should, rather, study how these several levels of change enhance, impede, interrupt, reflect, or contrast with one another. It is revealing, for example, to see how the growing skills and status of civil and mechanical engineers in the late eighteenth and early nineteenth centuries coincided with the decay of London Bridge, encouraging attempts to streamline its arches before the dynamics of river and tidal currents were fully understood. In fact, had Telford's iron span been built when he proposed it, around 1800, the changes in water level, tidal action, and channel scouring in the river would have been much more sudden, and more destructive, than they were in the 1830s. The engineers would have been ignorant of much of the information gained through the surveys of the 1820s. That the replacement of London Bridge took a half century to accomplish was just possibly a very good thing.

CHAPTER III **The London Main Drainage**

In most surveys of Victorian London, the Thames Embankment is described as the last and most visible part of the London Main Drainage scheme. Its other features are treated as appendages to that essential function. Of course, the Embankment had its own history, predating the Main Drainage by many centuries, and improvements to navigation, transport, and commerce were as important, if not more important, to its evolution than sewers. However, because the construction of the Main Drainage prefigures in many ways the construction of the Embankment, and because the completion of the Drainage required an immediate decision about the Thames waterfront, it is worth examining separately.

Sewers and sewage have become such an integral part of modern urban infrastructures that it is difficult to imagine alternative ways of thinking about the disposal of excrement, industrial waste, and rainwater runoff. Modern industrial nations are just beginning to separate toxic substances from effluent and landfill wastes and provide recycling facilities on an effective scale. Although the technology for in-house human waste disposal is available, its development is stifled by the immense social, institutional, and financial investment in sewer systems. It is instructive, therefore, to go back to the time when such systems were introduced. One can argue that WCs and sewers were neither inevitable nor even the best possible solution to the problem of waste disposal. They required, first, a redefinition of sewage from "inconvenient runoff" to "dangerous pollution," and a redefinition of sewers from "open streams" to "underground waste removal pipes." The growing community of professional civil engineers and contractors then overlaid the new environmental and cultural perceptions with a technical one: sewage became a hazardous by-product of metropolitan life that was to be flushed underground and out of the city. The engineers designed a system of pipes that, in effect, reproduced and relocated the polluted streams that had once constituted the natural drainage of the Thames valley. Finally, the method of disposal and political control over the system was held hostage to feuds between government agencies and competition from commercial entrepreneurs. Thus the London Main Drainage was a product of the interaction of social, technical, political, and environmental processes. The fact that it was almost universally acclaimed upon its completion is due less to its success as an engineering project (which was considerable) than to the elimination of alternative conceptions of the problem and its possible solutions.

During the half century leading to the replacement of London Bridge, the British experienced demographic and technological changes whose cumulative and synergistic impact shocked contemporaries. Between 1801 and 1851, Birmingham tripled its population (from 71,000 to 233,000), Manchester quadrupled (75,000 to 303,000), and Bradford multiplied eight times (13,000 to 104,000).[1] During the same period, the population of London (though it is difficult to calculate because of shifting boundaries) grew from about one million to over two million. By 1861, it was 2.4 million, more than three times the size of Birmingham, Manchester, and Bradford combined, more populous than the nine largest American cities put together, greater than the combined populations of Australia and New Zealand.[2] This explosion was partly due to the sheer growth of population in England and Wales and the shift from water power to steam power, which brought manufactures from the countryside to the towns. But it was also due to the increasing concentration of political, financial, and cultural leadership in the capital city of a new imperial power. These changes, which Kitson Clark called "blind forces" in contrast to the intentional movements for political reform and religious toleration, may rightly be seen as structural and basic. However, being basic does not make them overriding, in the sense of being most important in all cases. We have to evaluate them as they intersect with other types of change.

The particular feature of population growth and industrialization that most directly affected plans for the Thames Embankment was the pollution of the river by sewage. The excrement of a million and half humans, over 100,000 horses, and perhaps 10,000 cattle and other animals mixed with rainwater runoff, manufacturing waste, tanning and slaughtering byproducts, soil erosion, and shipping spills. Much of the human waste was held in cesspits beneath the houses, but the rest was washed into street drains or into the half-dozen streams that found their way down to the river. Tree limbs, rotting animal hides, barrels, and enough corpses to warrant an official annual count floated back and forth on the tides.[3] The sheer size of the population doubled the amount of water flowing through residential areas and down to the river in two generations. Anthony Wohl has pulled together the multitude of sources on pollution and health problems in Victorian Britain in his book *Endangered Lives*, and there is no need to survey again what the Victorians themselves publicly investigated and deplored. Henry Robinson, a regular observer of the Thames, lamented, in 1856, that "this object of so much tender solicitude and legislative care, having seen its trade carried off by railways, and its beauty marred by sewers, having, as regards that part traversing the Metropolis, ceased to some extent, to be either useful, or ornamental, is now again before the public as a shame and a disgrace, and is exalted to the position of the crying nuisance of the day."[4]

London, of course, was neither the only nor even the worst site of contamination. Rapid industrialization in northern towns such as Manchester and Leeds produced ghastly scenes of contamination immortalized by Engels, Dickens, and Gaskell. In Yorkshire, the region drained by the Aire and Calder Rivers supported 440 woolen and worsted mills and over 100 mills and factories of other types. Manufacturing discharges raised riverbeds, causing increasing seasonal flooding. In Bradford, which nearly doubled its population in the decade of the 1840s, the industrial canal was so polluted that teenagers set fire to it on hot summer days.[5] Yorkshire and Lancashire industrialists tended to be laissez-faire liberals advocating economic expansion without restraints on "private" property. They were less likely

Figure 2. The Thames shoreline before embankment at Lambeth, 1850. At left is Doulton's pottery works, later replaced by St. Thomas's Hospital (see figure 9). (Photo courtesy of the Victoria and Albert Museum.)

than London landlords to install water closets in working-class houses, and, if they built sewers at all, it was to dump their wastes downstream for others to deal with. They were pitted against farmers, fishermen, and landed aristocrats who defended the rural landscape and ecology in ways that echoed the struggle against political reform. In these towns, pollution and its remedies came to be defined largely in terms of industrial waste.

London, by comparison, had a mixed economy but far greater residential population. Manufacturing tended to be on a smaller scale and was interspersed with commercial, professional, and political interests. Londoners experienced somewhat different kinds of environmental degradation than other urban areas and, as a result, took different approaches to it. Pollution came to be defined largely in terms of human and animal wastes, rather than industrial effluents. Moreover, suburban housing developments and regional market gardening softened the contrast between city and country that struck visitors to the North.[6] Despite its gigantic size, therefore, London may have been better prepared, as a community, to deal with its sanitation problems.

Before the 1830s, "sewage" was defined as any surface runoff involving liquids. It was perceived as a source of nuisances but not particularly dangerous, and, even during the 1840s and 1850s, experts could not agree what to do about it. There were several reasons for this. First, neither sewage nor agencies for regulating its removal were anything new in London. What was new was the scale of the operation. Cesspits were common in the main areas of London from very early on, and the poorer residents simply dumped their waste in the streets to be washed downhill by rain or water wagons. "Night soil" and horse manure was also collected and carted out of the city by local contractors. Seven district Sewer Commissions were established for metropolitan boroughs by Act of Parliament in 1532, but their authority was strictly local, and their interest was primarily in surface water. One of their charges was to prevent households from clogging up the open street sewers, which were intended to carry rainwater and liquid waste. Although common sense led most people to suspect filthy water, "sewage" was not defined as pollution until 1834, when a parliamentary select committee on metropolitan improvements used it to refer specifically to household cesspool discharges and street runoff. Until then, a "sewer" was considered more like a drainage ditch than an underground pipe.[7]

Second, the masses of people immigrating to London brought with them rural perceptions of sewage and its disposal, which took time (and the shock of cholera) to change. It is not necessarily true that cottagers and agricultural labourers all lived in filthy hovels and held no concepts of sanitation. Rather, the countryside was less densely populated and could tolerate both the relatively low levels of waste produced there and the more informal methods of disposal. When cottagers moved into city tenements, they found rows of outhouse toilets packed into cramped courtyards, uncertain and inadequate water supplies, and refuse from thousands of workshops. London was not much better for the upper classes. The necessities for cleanliness among the aristocracy and gentry—a water supply and a means of removing human excrement and the byproducts of industry—required, before the nineteenth century, a class of servants. And, though the number of servants increased almost as fast as other groups in the nineteenth-century urban population mix, the duties they might have performed in a country manor house were subverted in the city by a lack of sanitary facilities.[8]

Third, cleanliness is in the eye of the beholder, and there is a difference between the dismal levels of sanitation and public health that historians perceive in hindsight and how most contemporaries perceived them. Late Georgian London was generally considered clean and healthy by most people. Donald Olsen cites evidence that visitors to London before 1830 were quite pleased by the healthy climate and sanitation of the capital. He goes on to speculate that London's early Victorian reputation for filthiness stemmed most directly from the totally unexpected and shockingly swift cholera epidemic of 1832. The epidemic threw the medical profession into a panic, for no one could explain what cholera was, let alone how to prevent it. Its horrifying symptoms and devastating mortality led the public and the media to overdramatize its actual impact. Although cholera returned in 1848, in 1853, and in 1866, each time prompting cries for pollution control, it actually killed fewer people than probably any other epidemic infection.[9] John Snow's famous demonstration of the waterborne nature of the disease was not accepted until after the last of these epidemics. Prevailing medical theory linked disease to impurities in the air ("miasma"), usually thought to derive from decaying organic matter. Thus, Londoners were more apt to worry about the pollution of the atmosphere than the actual quality of the water. Since the air could also be corrupted by dirt, feces, and even perspiration, moralists pointed to squalid, undisciplined slum neighborhoods as the probable source of the disease.[10]

The miasma theory of disease led to a conviction among both experts and laymen that human and animal waste could be rendered innocuous by immersing it in water. The water closet, patented by Joseph Bramah in 1778, became popular among the upper middle classes in the 1820s and gradually spread throughout the metropolis in the following decades.* The WC was supposed to eliminate the need for carting human waste

---

*An ambitious scheme for embanking the Thames, promoted in the 1820s by Colonel Sir Frederick Trench, included provisions for collecting sewage in large locks or basins built inside the proposed river wall. At high tide, the locks would be closed and allowed to fill up; at low tide, the contents would be released into the river on the analogy of flushing a toilet. (Barker and Hyde, *London As It Might Have Been*, 85.) Several wharves and docks along the Thames were fitted with large gates to form tidal locks; flushing the foreshore at low tide kept it clear for boats and barges.

through the streets and to render it harmless by dilution. The result, however, was to fill up the residential cesspits faster than ever before and pollute the surrounding soil. Flushing all those WCs would have been bad enough, but, until the 1870s, many had leaky valves, and some householders just tied the chain down and let water flow continuously through the pipes.[11]

Edwin Chadwick (figure 5), leader of the public health movement, was convinced by his associate Dr. Southwood Smith that cholera was borne by "epidemic atmosphere" bred in piles of filth and cesspools. His solution, embodied in the Building Act of 1844 and enforced by the new Board of Health in 1848–49, was to "flush out the city" by providing water service to all residents, hooking up house drains to the public sewers, and abolishing cesspits and cesspools altogether. He intended to collect all the runoff into an integrated sewer system, but he failed to gain the necessarily legislative authority for this vital part of the system (see below). Household sewage, therefore, flowed down into the Thames.[12] By the late 1850s, the river was receiving some 260 tons per day through seventy-one main sewer outlets. The result, in David Owen's eloquently understated phrase, was "a grave deterioration in the state of the river."[13]

The tidal action of the Thames made matters considerably worse. Most people expect a river to flow downhill and take with it whatever sewage may be dumped into it. Certainly the towns above Teddington operated on that principle. But, in the tidal region of the Thames, a wall of saltwater carried raw sewage back upstream roughly twice a day, right through the center of London, only gradually shifting it downstream. At high tide, the water rose higher than the mouths of the main sewer outlets, blocking and backing up the sewer discharge for hours, sometimes into nearby basements, and sending streams of sewer gas out through the street openings.[14] When the tide stopped, there was about a twenty-minute period before the ebb began, when solid matter precipitated near the shore. The irregularity of the shoreline encouraged deposits. At low tide, this precipitate mixed with the sewage being discharged onto the exposed mud banks from drains to form a ribbon of black putrescence that stretched near each shore from Putney down to Woolwich. Around and beyond these black ribbons, the water was yellow and gave off a "urinous smell." Exposed to the sun, the sewage on the river banks decomposed rapidly and gave off hydrogen sulphide, a potentially lethal gas that nauseated passersby and killed at least eight sewer cleaners.[15] On cool days, a dense, eye-stinging mist rose from the riverside.[16] Ironically, due to the widespread belief in the miasma theory of disease, this noxious effluvia prompted stronger demands for sanitary reform than the sewage which generated it.

Given the quality of London's atmosphere at the time, the Thames mist must have been truly caustic for people to regard it as a special nuisance. By the early 1840s, industrial and residential coal fires regularly created a column of smoke extending twenty to thirty miles, which hovered over the city day and night. Twenty years later, still more coal fires, plus the body heat from almost two million people and animals, were said to raise the mean temperature of London two or three degrees above the surrounding countryside during the night. After the general deployment of gas lights, the nighttime temperature rose another four to five degrees. Coal gas manufacturing companies proliferated along the river's edge, spewing acid, soot, and ash into the neighboring air. Dust loaded with fecal matter, hot air, sewer gases, and smoke metamorphosed into the famous London fog, sometimes of a bottle-green color, sometimes pea-soup yellow. "Fuliginous matter" in the mixture, including ammonium sulphate, often crystallized on window panes in tree-like patterns. Many species of plants could no longer grow in London; one reason the later Thames Embankment boulevard was lined with a variety of tree called "London plane" was its hardy adaptation to the local atmosphere.[17]

The rebuilding of London Bridge compounded the pollution problem by altering the river's regime upstream through King's Reach, Hungerford Reach, and Westminster. The faster currents cut deeper channels, exposing ever more of the mud banks at low tide (see figure 2). Hungerford Reach was especially vulnerable because the Thames, curving eastward at this point, spread out into a wide, shallow "embayment" fronting the government offices and aristocratic residences of Whitehall. At low tide, the sewer outlets were left far from the water, dumping raw waste along the flat shoreline for the duration. Perhaps it is no coincidence that the Duke of Buccleuch, whose city mansion lay hard upon Hungerford Reach, chaired the 1844 Royal Commission that expanded and tried to implement Chadwick's 1842 *Report on the Sanitary Condition of the Labouring Population of Great Britain.*

Chadwick's *Report* skillfully wove statistics, examples, and lurid descriptions into a warrant for government intervention into the urban environment. He dramatized the need for pollution abatement, waste removal, and clean water supply.[18] His report inspired a national Health of Towns Association promoting "the sanitary idea" in the manner of the anti-Corn Law and abolition crusades. Chadwick himself wrote legislation to consolidate the eight district London Sewer Commissions, to bring the water supply companies under government supervision, and to create a national Board of Health which would oversee sanitary reform on the Benthamite model used for his Poor Law system in 1834. These initiatives evoked passionate hostility from the vestries (parish or neighborhood councils) and paving commissioners, from the private water companies, and from the City Corporation, always jealous of its liberties. They aroused Londoners' antediluvian instinct for local autonomy and piecemeal action, while threatening a myriad of literally vested interests. Each parish vestry sponsored paving, lighting, sanitation, and water boards, which were only gradually being opened to ratepayer participation. The district Sewer Commissions, originally appointed to cut through vestry parochialism, wielded almost unlimited powers within their boundaries. Each had its peculiar methods of doing business, its own staff of engineers and surveyors and clerks, and even its own specifications for drain pipes, rates of incline, methods of construction, and cost accounting. As G. M. Young observed, "hardly any one had yet thought of the county as an administrative unit."[19] The City Corporation, governing the original square mile of London with its medieval array of guilds and aldermen, had used its Parliamentary representation to escape inclusion in the Municipal Corporations Act of 1835 and to create its own Poor Law and Sewer Commissions. Now, like many other towns throughout England and Wales, it refused to join the new Board of Health, hastily producing a bill that set up its own board under a medical health officer. The *Times* commented later, "we may really say that there is no such place as London at all, the huge city passing under this title being rent into an infinity of division, districts, and areas. . . . Within the metropolitan limits the local administration is carried on by no fewer than 300 different bodies, deriving powers from about 250 local Acts, independent of general Acts. The number of commissioners employed, though not precisely ascertainable, [is estimated] at about 15,000."[20]

It should be said here that most other large towns faced the same problem of overlapping or competing jurisdictions. Because the borough councils set up under the Municipal Corporations Act of 1835 were supposed to end the corruption of ancient corporations, they were often excluded from managing just those services that expanding cities needed: paving, lighting, water, and sewerage. These remained in the hands of local boards or were assigned to "improvement" commissions, which established themselves as a new layer of politicized bureaucracy. A significant proportion of their members were drawn from the class of manufacturers who created much of the pollution in the first place.[21]

In 1847, Chadwick succeeded in taking control of a new Metropolitan Sanitary Commission set up by Parliament for London improvements. Lacking engineering training himself, he hired consultants and held a competition for design ideas. He gradually evolved an integrated scheme that linked new sources of fresh water to a matrix of sewer lines that would intercept all the old drains and concentrate the sewage at a few major outfalls. There it could be collected for treatment and recycling as agricultural fertilizer. In 1848–49, he gained parliamentary approval to consolidate the Metropolitan Commissioners of Sewers and create a Board of Health, both under the authority of H. M. Commissioner of Woods and Forests. The latter was a Cabinet member whose traditional role was to oversee Crown properties. This was not such an anomaly as one might think, because "Woods and Forests" also included an Office of Works which (in the absence of any other authority) had undertaken several public improvement projects in London neighborhoods. The MCS, the Board of Health, and the Office of Works thus constituted the elements of a potential centralized sanitation system for the metropolis.[22] Chadwick made sure he and his associates were appointed to each of the three agencies.

The Office of Works was a permanent establishment, but the MCS and the Board of Health were commissioned to run for only six years, until 1854. During that period, the Office of Works disintegrated into a bureaucratic shambles, and the other two agencies were rendered ineffective. This was partly due to pamphlet warfare waged by anticentralizing local officials and partly to Chadwick's quixotic style, which insulted sanitation experts and invited personal attacks. According to Christopher Hamlin's detailed study of the controversies, a failed commercial venture in partnership with a professional sanitary engineer left Chadwick feeling that such men were hypocritical jobbers who would take any position or project for a fee. He voiced such criticism openly during meetings of the Sanitary Commission in 1847 and brought his negative attitude to the new MCS the following year. During the cholera epidemic of 1848–49, however, Chadwick delayed reconstruction of London's drainage until he could obtain accurate topographical data from the ongoing ordnance survey. As a result, 14,000 people died of the disease, and Chadwick was relieved of his post at the MCS.[23] Chadwick shifted his campaign to the Board of Health, generating further controversy. He instructed Board of Health inspectors to warn local authorities against professional sewer engineers, to sabotage their contract negotiations, and even to threaten prosecution if conventional arrangements were adopted.[24]

Chadwick's sewer design was based on the theory that water pressure could be increased almost indefinitely in a gravity-fed system by running it through incrementally smaller pipes. Economical small-bore clay pipes could produce a high-pressure discharge to flush out any sediments and other obstructions that might otherwise build up. Most professional engineers followed Continental theory in advocating much larger, stronger, egg-shaped brick sewers, which could be cleaned and repaired from inside. They were willing to consider the merits of small-bore pipes but rejected them as a panacea for all situations and clients. As pragmatic consultants, they reserved judgment until concrete performance data were available.

In the meantime, they grew increasingly furious at Chadwick's attack on their professional integrity. Like the majority of urban clients they served, the engineers preferred local, piecemeal projects rather than grand designs, as Christopher Hamlin has noted in his account of the sewer-pipe "wars":

In opposing Chadwick's system . . . these engineers espoused principles of decentralization reminiscent of modern appeals for "appropriate" technology. Yet they did not see themselves as endorsing a philosophy of technology but only as advocating good engineering, which included designing sewers with local needs in mind and making allowances for human error, heavy storms, or town growth. Such an outlook was a professional necessity, for few towns wanted integrated systems; they wanted lengths of sewer here or there or partial waterworks. They expected engineers to respect their concerns and often asked them to change their designs in all sorts of ways and for all sorts of reasons. Chadwick had invited engineers to discard carefully constructed relationships and reputations for the security of a hierarchical bureaucracy, but, with a few exceptions, engineers rejected the risk.[25]

The irony of Hamlin's observation is that the engineers who resisted centralized power when proposed by Chadwick were the same engineers who later designed and built the large-scale, integrated Main Drainage system for the Metropolitan Board of Works, and then built the Thames Embankment to complete what was seen, in retrospect at least, as a "grand design."

Civil engineers, however, were rapidly consolidating their position in the upper middle classes, transforming themselves from employees into consultants and gaining respect as expert witnesses before Parliament. Their stronghold was the Institution of Civil Engineers (ICE) in Great George Street, a few hundred feet from Parliament, where they had many friends. One of the ICE's more reputable members (he became president in 1871) was Thomas Hawksley, the engineer with whom Chadwick had quarrelled in their joint commercial venture. In a series of debates sponsored by the ICE, Hawksley and his colleagues attacked every aspect of Chadwick's design, pointing out his ignorance of basic engineering principles and challenging his data. Thomas Page, another ICE member with the title of "Thames Embankment Engineer" at H. M. Office of Works, produced a lengthy government report, in 1852, condemning the clay pipe design in principle and castigating the Board of Health for its unethical operations. Finally, the assistant and then chief engineer at the MCS, Joseph W. Bazalgette (figure 7), conducted several unauthorized surveys of pipe sewers around London, in 1852–53, reporting high rates of accumulated deposits and breakage.[26] Although his data were suspect, Bazalgette was able to convince other engineers at the MCS that the large, egg-shaped brick sewers favored by him and his predecessor, Frank Forster, and by his counterpart at the City of London engineering department, William Haywood, were the only viable option for large-scale drainage. However, he was forced to agree with many other engineers and public officials that the only way to prevent the further pollution of the Thames was by constructing an integrated set of sewer lines, running roughly parallel to the river, that would intercept all the local drains as they wound their way down the valley slopes and carry the entire burden out of the city to the east. Piecemeal amendments would no longer work. Thus Chadwick, whose strategic planning was always better than his tactical maneuvers, lost the technical and political battles but won the war.[27]

Taking command of the MCS, Bazalgette and his assistants John Grant and Thomas Lovick drafted a set of plans for what was henceforth known as the Main Drainage. In 1854, Bazalgette and William Haywood prepared project specifications for the northern side of the Thames. The following year, he tackled the difficult conditions along the marshy lowlands on the southern side. But, though the MCS engineering department was enterprising enough, the commissioners themselves had too little authority and too few resources to carry out such a far-reaching scheme. They could legally raise rates for public works or pledge future rate receipts with private commercial firms for long-term loans; but few of them would countenance rate increases, and the insurance companies would not risk the loans. The Treasury provided scant funding, arguing (despite the prevalence of Crown lands and government offices in the area) that London's sewerage was a local, not a national, expense.[28]

The statutory authority of the MCS and the Board of Health came to an end in 1854. The parliamentary representatives from the London area, led by the member for Marlyebone, Sir Benjamin Hall, had no intention of letting Chadwick's centralizing agencies continue in operation, even temporarily. When Sir Benjamin was appointed the new president of the Board of Health, Chadwick's career was over. London clearly needed some kind of sanitation system, but the vestries would accept only a system with local control.

Once again, nature, in the form of a cholera epidemic, supplied the necessary motivation for reform. The summer of 1854 brought reports of staggering sewer backups and riverside pollution in the boroughs, whereas, in the City itself, Dr. John Simon, the new medical officer of health, had organized extra water supplies and enforced sanitation codes with a vengeance. The City suffered 211 deaths from cholera; the rest of the metropolis suffered 10,527. Sir Benjamin Hall, a practical businessman (see figure 5), went down into the slums to see the conditions for himself. He came away converted to a policy of vigorous action by government authorities. He instituted an aggressive course of sanitation reforms at the Board of Health. In March 1855, as soon as he was promoted to First Commissioner of Works in Lord Palmerston's first ministry, he introduced a bill to create a strong new sanitation authority for metropolitan London.[29]

The Metropolis Local Management Act of 1855 is a prime example of politics as the art of the possible. Hall knew that the vestries would refuse to cooperate with a central authority so long as it looked like a traditional government. For that matter, both the City Corporation and Parliament would look askance at a new political body established in their midst. So Hall proposed a Board of Works, which seemed like a sewer commission enhanced by paving, lighting, and other municipal utilities. Its governing board would be elected indirectly from the vestries themselves (with some of the smaller vestries being grouped into roughly equal electoral districts), and the City would have its own delegates. Like the MCS, the MBW operations would be financed through the local rates. The local management bill sailed through Parliament with little fanfare. Although the new board took a long time to organize and tended to favor lengthy public meetings (drawing the epithet "the Board of Words" from the newspapers), Bazalgette and the other engineers transferred from the MCS got down to work immediately on plans for the new sewer system.[30]

The basic principle of Bazalgette's London Main Drainage (see map 3) was to intercept the discharge of the several "natural" sewers (the Westbourne, Ranelagh, and Tyburn to the west, the Fleet and Walbrook at the center, Hackney Brook and the River Lea to the north and east, and the Falcon, Effra, and Ravensbourne on the south side) and of the hundreds of large and small street and residence sewers snaking down the valley toward the river. The sewage would be diluted by rainwater and carried to downriver outfalls on either side of the Thames. Pumping stations of the kind designed by Forster would supplement gravity where needed. The principle of interception had been introduced by an engineer named Creasy about 1774 and was generally accepted by 1850, there being a dozen or so cases of such sewer lines in operation. Intercepting sewers were, in fact, proposed for London as early as 1832 by a painter and amateur architect named John Martin. Martin's plans (frequently modified and republished) were endorsed by the Royal Institute of British Architects, many members of Parliament, and Fellows of the Royal Society. Many of the designs proposed in the 1850s, including Bazalgette's, were based on his general concept.[31]

The question was, what should be done with the collected wastewater?

The answer depended on one's attitude toward sewage as the source of pollution and disease. As we have seen, many if not most people thought disease came from airborne organic matter, produced only in part by sewage water. One option, therefore, was simply to dump the sewage, treated or untreated, into the Thames. Special tidal locks would restrict the discharge to the early hours of ebb tide so that it attained maximum dilution and downstream velocity. Embankment walls would streamline the river channel and eliminate the shallow mud banks. John Martin's plans, for instance, envisaged discharges at Hungerford Bridge, the Temple, and Blackfriars. These sites were effectively eliminated by the growing pollution of the river, which defied any calculations about its carrying capacity. A curious variation was revived, in 1858, by the ventilation officer of Parliament, Goldsworthy Gurney, who proposed deodorizing and detoxifying the sewage by burning off its miasmas through a set of steam-jet furnaces of his own design. The solid residue, now theoretically harmless, could be safely dumped into the river. Gurney proposed lining the riverbanks with a uniform gravel coating that would let the sewage slide easily into the water, and then dredging deep channels near either shore that would guide the sewage downstream without mixing into the tidal flows above it.[32]

It is significant that Gurney was trained as a surgeon-apothecary, had given public lectures on the elements of chemistry, and had written pamphlets on the nature and treatment of cholera as a public health problem. As a medical man who prided himself on keeping up with the latest research in Britain and the Continent, he subscribed to a version of the miasmatic theory of disease: he believed that microscopic "animalculae" from decaying organic matter were transmitted to human bodies through fetid air. These, he proposed, could be destroyed by steam heat.[33] He undoubtedly was aware of John Snow's now-legendary study correlating polluted water sources in Broad Street with the incidence of cholera in 1854, but, like many scientists of his time, he thought it an inconclusive replication of similar research by Chadwick and others. The fact was, despite the success of Dr. John Simon's sanitation efforts in London, medical science was not able to give definitive advice about pollution and disease to the authorities who so desperately needed it. As Christopher Hamlin has shown in his detailed study of the period, science, and especially chemical science, was heralded in the early nineteenth century as an objective technical authority to which questions of industrial processes and public sanitation could appeal. But most scientists did not have expert positions at that time (the title "chemist" was not used professionally until the 1840s) and so built careers by lecturing, consulting, and testifying before various commissions on behalf of paying clients. Their success depended on "selling authority," that is, convincing decision makers of the accuracy and reliability of their evidence. Thus science became not a touchstone for truth but a mode of argument, a set of terms used for discussion. People with opposing views could summon "experts" to legitimize their proposals with assessments and predictions based on reams of data and compelling examples (just as they do today). When Parliament sought the hypothetical "correct" answer to the question of water quality, therefore, it received advice based on the interests of commercial waterworks, competing government agencies, ambitious consultants, or local ratepayers. "Science was a rich and expressive medium in that conflict," comments Hamlin, "one characterized by the ideal that there was a best answer, a natural truth, for any question, and yet possessing vast flexibility, being capable indeed of giving expression to nearly any argument one wished to advance."[34]

The water quality controversy rose to its highest pitch during the years 1849–54, between the two cholera epidemics, when Chadwick was trying to centralize and reform the water supply companies. Both the government and the water companies hired medical chemists to conduct studies and testify on their behalf, but neither side could prove its case. No one was sure whether bacterial, mineral, or crystalline analysis yielded the right data, or whether the analysis itself contaminated the samples. As a result, commercial and political factors decided the issue. The Metropolitan Water Act of 1852 forced the water companies to move their intake pipes upstream, filter the water, cover their reservoirs, and give a more constant supply. It did little to regulate them or contest their monopoly of service.[35]

In spite of, or perhaps *because* of, the experts' inability to define and measure pollution in water, the question of whether to allow sewage to flow into the Thames was answered by the time-honored use of the senses. The river smelled horrible, tasted poisonous (if one dared to drink it at all), looked ghastly, and felt slimy. Almost everyone agreed that the sewage must go elsewhere. But where?

Bazalgette and the MBW engineers proposed to divert all of London's northside sewers to a single outfall at Barking Creek, five and a half miles east of the River Lea, and the southern sewers to an outfall at Crossness, seven miles east of Greenwich. Other engineers doubted whether this was far enough downriver to prevent sewage from washing back up on the tides, and the more foresighted predicted the outfalls would soon enough be overtaken by metropolitan expansion, but the MBW had to balance these concerns with the need to contain costs. They argued, reasonably enough, that the outfalls could always be moved downriver without requiring reconstruction of the basic Main Drainage.[36]

The engineering solution offered by the MBW fit the prevailing hypotheses about pollution and was straightforward enough to be depicted through maps in the popular press. It drew upon the heroic image of engineers and the general respect for large-scale technology. As the *Times* would plead a few years later, "This is preeminently an iron age . . . So we beg to suggest that a hearing be given to those engineers who proposed to deal with this matter in the spirit of an iron age. As things are these days, this is an affair of machinery from first to last."[37]

The MBW plan still allowed for discharging metropolitan sewage into the Thames. This feature remained its weak point for years to come. Meanwhile several groups of entrepreneurs, their scientific advisors, and parliamentary supporters promoted a popular alternative: recycling. They argued that sewage, being composed mostly of human waste, could be thought of as a type of manure, rather like horse manure, which in fact was one of its other ingredients. Chemically detoxified and sufficiently diluted, it could in theory be reconstituted as agricultural fertilizer. For that purpose, they proposed to construct a set of sewage collection and treatment plants at the main sewer junctions, rather than simply discharge waste into the river. These plants could be built within the metropolis, because they would deodorize and sanitize the sewage for transport to the surrounding agricultural regions.*

The concept of recycling domestic sewage had been touted for years before urban pollution reached crisis proportions. In 1835, a "Thames Improvement Company" was formed for the purpose but failed after a few years.[38] Edwin Chadwick, as mentioned above, tried to start a private company for sewage reclamation with the engineer Thomas Hawksley in the early 1840s. Research and experimentation accelerated in those years. The House of Commons initiated several inquiries into the prospects for agricultural application, beginning with a select committee in 1846. The select committee investigated and finally supported the claims of William Higgs, an agricultural chemist, whose Metropolitan Sewage and Manure Company proposed to convert the outflow of the Ranelagh and King's Scholars' Pond sewers, southwest of Westminster, into fertilizer. Higgs's initial plan to build settling tanks near residential districts drew fire from angry neighbors, but the select committee encouraged him to go forward with a modified arrangement. A more ambitious scheme for draining the whole of London into an intercepting sewer system was offered at the same time by engineer Thomas Wicksteed on behalf of the London Sewage Company. Wicksteed, an aggressive consultant for a dozen London-area water companies, originated or borrowed Frank Forster's idea of using pumping stations to maintain a steady flow within the low-level sewer along the Thames shore. The sewage was to be filtered through huge treatment beds on the east side of the metropolis and the solid matter converted to agricultural fertilizer.[39] The select committee of 1846 thought the scheme too grandiose for the present state of practical knowledge in the field, and the uncertain state of financial markets kept its promoters from pressing their

*Londoners do not appear to have copied the Parisians, who applied manure from the 96,000 public transport horses to some 19,000 acres of municipal gardens. Straw mats and cloches increased the process of fermentation, allowing gardeners to grow three to six crops of vegetables per season. The manuring increased the temperature of the urban atmosphere but also supplied 2.4 percent of the Parisians' protein. (Simmons, *Environmental History*, 34.)

case. But the select committee cited experiments in Essex, Edinburgh, Belgium, and the London suburb of Croydon to show that commercial development was an entirely feasible alternative to publicly financed works.[40]

By midcentury, in fact, nearly a dozen facilities were irrigating large tracts of farmland with treated or filtered sewage diluted in water. A Royal Commission on Rivers Pollution in 1866 found many of these operations effective and published practical information for town officials wishing to build their own. There were problems, of course: many farmers (and their neighbors) remained cautious about the danger of odors and disease; larger towns such as Manchester and Nottingham could not find sufficient acreage to absorb their enormous output; smaller sewer authorities could not always guarantee a large or constant enough supply to the countryside. In most schemes, the projected costs of treatment and transport usually outweighed estimated profits, and few towns broke even in the long run. Nevertheless, over eighty towns would adopt sewage irrigation by the 1880s.[41]

In London, during the 1850s, two new ventures challenged the MBW's plan for discharging sewage into the Thames: the London and Provincial Towns Sewage Irrigation Company and the Metropolis Sewage and Essex Reclamation Company. These firms sponsored scientific research on the relative merits of solid and liquefied sewage in comparison with artificial fertilizers, and on the types of soils most responsive to fertilizing. Their engineers made comprehensive, detailed plans for the construction of reservoirs, pumps, and pipes, while their directors calculated costs, profitability, and capital resources. Pilot projects were set up, including an experimental farm near Rugby. Some developed into regular operations. When Bazalgette submitted preliminary plans for the Main Drainage in 1853, the possibility of converting the outflow to fertilizer was seriously discussed.[42] By 1862, when the MBW was seeking final approval to complete the low-level intercepting sewer, the two companies were able to present impressive evidence for commercial alternatives to the House of Commons.[43] Even after the MBW began construction of its downriver outfall facilities, which would dump relatively untreated sewage into the Thames, the Metropolis Sewer and Essex Reclamation Company gained a select committee's support for substituting its own scheme.[44]

The MBW was at first willing to consider contracting for disposal of its sewage from the outfall reservoirs. It advertised for tenders, both in 1860 and in 1863, in response to the select committee reports from Parliament. Over the objections of City delegates, who thought the raw sewage should be sold at a profit to the projectors, the board entered into agreements with the speculative partnership of Napier and Hope to process the contents of the northern outfall reservoirs, and with a second entrepreneur, Thomas Ellis, for the southern. Both ventures fell victim to a collapse of capital investment markets in 1866. Although recycling continued to attract adherents in other towns around Britain, MBW officials began to suspect that, while small-scale experiments might work well enough, a commercial operation of sufficient size to handle the London outflow was still technically and financially unfeasible.[45]

Parliament's continuing eagerness to give sewage reclamation promoters a hearing was linked to a widespread conviction, shared by many engineers and contractors (including Edwin Chadwick), that public works, even those of the magnitude of the London Main Drainage, could best be carried out by private enterprise. The great canal, railway, and harbor works had been done this way, and prevailing economic theory advocated free competition with a minimum of government supervision.[46] The first efforts of the MBW did not dispel such convictions. Because the Metropolis Management Act of 1855 had carried over the inadequate financing arrangements from the MCS to the MBW (even the MCS debt was charged to the new agency), Bazalgette had been unable to make much progress on the Main Drainage scheme. As a result, there were frequent proposals to amend the Main Drainage clauses of the act in favor of private contractors. Bazalgette, however, was determined that his department should have sole control over the project. In this, he was supported by the City's chief engineer William Haywood, with whom he had forged a strong working alliance, and by the MBW's stolid but relentless chairman, John Thwaites (figure 5). The two engineers spent much of 1855 undermining the credibility of Wicksteed's London Sewage Company. Bazalgette, who possessed the grasp of detail, stern demeanor, and self-confidence required for dominating parliamentary committees, devoted long hours to defending the MBW plans against commercial competitors. He gained a respite, in 1857, when the disruptions of the Crimean War and the Indian Mutiny (among other factors) caused a financial crisis in the London money markets, leaving private ventures strapped for capital. But the question of sewage disposal remained the weak point in his design.

Sir Benjamin Hall had inserted into the Metropolis Management Act of 1855 a clause that required the MBW to submit for his review and approval any scheme estimated at over £50,000, and to submit schemes of over £100,000 directly to Parliament. This proved a great hindrance. Sir Benjamin prided himself on blunt, detailed critiques of departmental operations. When Bazalgette submitted plans for the Main Drainage in 1856, Hall requested numerous changes. He was especially adamant that the main outfalls be removed to a point much farther downstream, regardless of the added expense (for which he was not liable). A feud broke out between the two agencies, with proposals and counterproposals, delegations and letters angrily exchanged and the newspapers blaming the whole mess on the board. In December 1856, Hall appointed an independent review committee, which took seven months to produce a thick, detailed, and generally useless study. It was immediately challenged by Bazalgette and two MBW consulting engineers, Thomas Hawksley and George Parker Bidder.[47] Hall then appointed Parliament's ventilation officer, Goldsworthy Gurney, to make a new report on pollution in the Thames. Gurney reiterated the popular concept of miasmatic "animalculae" as the source of pestilence and offered once again to burn off London's sewer gases in his steam-jet furnaces, leaving the sanitized residue to be swept away along a reconfigured Thames riverbed. A Commons committee wasted more months investigating Gurney's claims and, in July 1858, authorized a trial of his furnace in the complex of sewers around the Houses of Parliament. Gurney did not realize he was but a diversionary tactic in a bureaucratic war until Bazalgette, Hawksley, and Haywood showed up to conduct the official test of his arrangements. They quickly pronounced the furnace totally ineffective both in principle and in practice. Despite evidence of sabotage on their part, Gurney's invention suffered the same fate as Edwin Chadwick's small-bore sewer pipes.[48]

In February 1858, the MBW won its fight against Sir Benjamin Hall by default when the Whig ministry fell from power. The new Chancellor of the Exchequer, Benjamin Disraeli, trying desperately to revive the Tories' popularity under the reluctant leadership of Lord Derby, was eager to satisfy the public outcry for action against London's pollution. He appointed one of his "Young England" associates from the 1840s, the astute Lord John Manners, as First Commissioner of Works. The two prepared legislation freeing the MBW from the oversight of the Office of Works and provided a Treasury guarantee for loans up to £3 million, based on projected receipts from a special sewer rate of threepence.[49]

The board was still suspect in the eyes of many, but the summer of 1858 dispelled all doubts about the need for immediate action. Drought and heat combined at record levels. By early July the temperature was in the 90s Fahrenheit, and the volume of water in the river, already lowered by the demands of the water companies, dropped to dangerous levels. The concentration of sewage in the Thames rose to 20 percent, and much of it lay spread out over the widening mud banks, literally boiling in the sun. The "Great Stink" became the stuff of legend. Men working near the river experienced nausea, cramps, and sore throats, dizziness and temporary blindness. Patients on board the hospital ship *Dreadnought* suffered miserably from the fumes, and diarrhea spread through the shore neighborhoods.[50] Fish died by the thousands. River steamers lost business. Travellers went out of their way to avoid crossing Westminster Bridge or used smelling salts to survive the trip.[51] Cholera, the "shock disease" of Victorian Britain, was said to be approaching England.[52]

Members of Parliament, who would normally have adjourned for the summer, stayed at Westminster to deal with the urgent issues of the Indian Mutiny. The "New Palace" had an overflowing sewer line laid right through its foundations and a large outfall just outside its new terrace. Noxious fumes invaded the committee rooms and took "a firm, repulsive grip on the Library."[53] Goldsworthy Gurney covered the windows with lime-soaked cloths and then shut them up entirely. On June 25, he advised the Speaker that he could no longer be responsible for the health of the House.[54] Chloride of lime, a disinfecting bleach powder obtained as a byproduct from the coke-fired gasworks along the river, was spread by the ton along the shores of the Thames.[55] Although the *Times* claimed the old men in Parliament had "the constitution of water rats,"[56] the condition of the river forced them to act decisively. Disraeli's MBW legislation passed on the second of August with little discussion.

Ten days later, the board voted to accept the plans of Bazalgette, Hawksley, and Bidder for the London Main Drainage. They were an amalgam of features proposed over a score of years by amateurs, officials, and commercial projectors and generally known to the community of professional engineers. Bazalgette himself acknowledged this fact, but as David Owen commented, he had the difficult task of extracting the useful elements out of the mass of conflicting proposals, blending them into a workable strategy, and applying that strategy to the special needs of each district.[57] In 1865, reporting on the finished project, he bristled at the suggestion that he was taking credit for other men's ideas: "I find that there are now a great number of persons who claim to be the 'pioneers' of my design, and it is a source of regret to me, that I had not been aware before of the existence of many of the schemes which had been referred to, for it would have saved me many anxious days and sleepless nights, in working out for myself, that which it is now alleged had all been done and prepared before." He promised to acknowledge any designs submitted in proof of such claims. In any case, he added, "it is satisfactory to me to find, that they approved of what had been done, and it would also have been very satisfactory to me, supposing the works had failed, to have found so many gentlemen ready to come forward and say, 'Well, never mind, we designed the same thing, and are ready to share the opprobrium with you!'"[58]

The northern high-level sewer was to run nine miles from Hampstead to the River Lea. The middle-level line would stretch from Kensington through the heart of the metropolis to Hackney Brook, and the low-level line was to be built along the riverbank from Chelsea up through the City to Stepney before curving up to the Lea. All three would be connected to the northern outfall sewer, which would carry the combined effluvient through its immense, nine-foot-diameter brick tunnels five miles down to Barking Creek. For the less populous and lower regions south of the Thames, only two main sewers were planned. These joined near Deptford Creek to form the southern outfall sewer, which discharged seven and a half miles downriver at Crossness (map 3). The system as a whole would cover a little over eighty-two miles and have a capacity of 400 million gal-

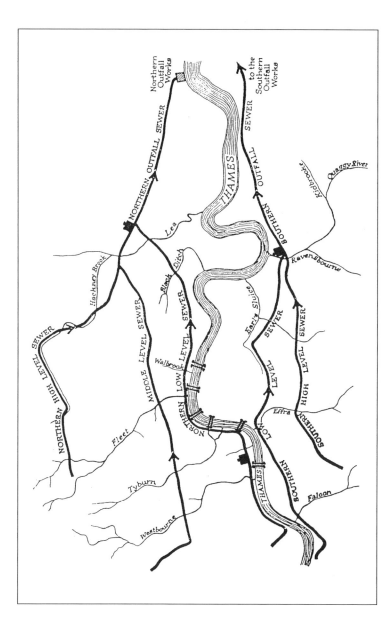

Map 3. The London Main Drainage, designed by J. W. Bazalgette, 1858. High-, middle-, and low-level sewer lines intercepted all the local drains and carried their sewage down the Thames Valley to outfall works at Barking Creek (north) and Crossness (south). (D.H. Porter, based on Institution of Civil Engineers *Minutes of Proceedings*, vol. 24 (1858): 314.)

lons. It would divert fifty-two million gallons of sewage from the river within London to a point fourteen miles from London Bridge where, released on the ebb tide, it would travel another twelve miles downriver.[59]

Details of construction on the Main Drainage have been ably recounted by David Owen.[60] Work began under the first of the twenty-six contracts in January 1859 and, despite labor stoppages and contractor bankruptcies, was sufficiently advanced by mid-1861 to impress the critics. In late 1864, Bazalgette deemed the system completed enough to warrant a formal opening. The following April, the Prince of Wales, watched by five hundred distinguished guests, switched on the pumping stations to set the lines in operation. The Board of Works, once an object of widespread opprobrium, could now bask in the appreciation of newspapers and politicians alike.

One crucial part of the system remained problematic. As soon as Bazalgette published his plans for the Main Drainage in 1856, it was clear that the route for the northern low-level sewer, designed to intercept and carry away the contents of the shoreward drains, would require extensive construction along the commercially built-up banks of the river from Westminster down to Blackfriars Bridge. Although the sewer line was not itself an expensive item (Bazalgette calculated the cost at about £18,000), getting it into the ground would be difficult, expensive, and time-consuming.

There were four ways to proceed. One, surveyed by the MBW engineers in 1856, was to lay the sewer along the foreshore of the Thames outside the line of wharves and docks. This would entail building a £200,000 cofferdam to keep the river back during construction, a cost ratio of 10 to 1 that Bazalgette deemed unacceptable. A second option was to lay the sewer without a cofferdam, plowing through or under the long row of warehouses, wharves, coal gas refineries, and small factories from Westminster to the Tower. The owners would demand compensation, and the work would be very slow and expensive. In any case, the omission of an embankment would leave the noxious and unsightly waterfront uncovered. A third option was to carry the sewer line under the Strand and Fleet Street and out through the City. Bazalgette presented this option to the board very reluctantly. Construction would completely disrupt the main east-west thoroughfare for a period of almost two years. In addition, quicksand was known to underlie parts of the Strand all the way down to Waterloo Bridge, and there was evidence that sewer excavations might undermine or damage the foundations of St. Paul's Cathedral.[61] In view of these problems, Bazalgette and Haywood recommended a fourth plan, namely, that the low-level sewer be built within an embankment from Westminster to Blackfriars Bridge, and then underneath a new street leading to Ludgate and the City. The embankment would follow the frontage line laid down, in 1840, by James Walker to improve the river channel, but it would also accommodate barge and lighter traffic to existing commercial wharves. After Parliamentary committees were formed in 1860 and 1862 to study the four options, the fourth gained a clear preference. A Royal Commission, appointed in 1861 to review various embankment plans, included an enclosed low-level sewer as a basic design element, rejecting all proposals without one.

Thus two patterns of human ecology intersected. The first pattern, centered on London Bridge, arose from the interaction of long-term geological and riparian changes with the gradual domestication of the shoreline by embankments and the relatively swift decay of a man-made structure. Rebuilding the bridge then triggered further (and faster) changes in the river's regime, which necessitated plans for streamlining and embanking its shoreline through the metropolis. The second pattern, leading to the London Main Drainage, grew out of the effects of urban population growth and prevailing sewage technology on the condition of the river water. Human reaction to the resulting stench and perceived menace to health prompted a reorganization of metropolitan government, with the specific aim of removing the menace. These patterns illustrate what is now a familiar theme—the ability of concentrated human activity to quickly devastate environmental features that have evolved over millennia. However, they also indicate that sudden environmental upheavals, such as floods, droughts, and cholera epidemics, can destroy or redefine years of human achievement and stimulate new human enterprises.

We must beware of isolating the environment, technology, or society as an autonomous "ground" for explaining what happens in the other two arenas. Their interactions are too complex for that. Moreover, each of them is a mixture of objective phenomena and human perceptions, a mixture that influences both human initiatives and environmental response. The concentrations of excrement and the water pollution, miasmas, and disease that led Londoners to build a sewer system and consider using its contents for agricultural fertilizer would not have existed without a previous history of human intervention in the natural environment based on specific ideas about rivers, sanitation, and engineering.

The London Main Drainage was designed by the Metropolitan Board of Works in response to conditions defined as a "pollution problem" according to a particular mix of institutional, technical, and environmental criteria. In order to undertake this enterprise, engineers and scientists had first to translate the public's growing fear and disgust into problem-defining and problem-solving language. They employed this language to compete for authority with political agencies led by technically naive politicians, who were partly influenced by public opinion as expressed through the popular press. Whereas the scientific community was unable to reach consensus on the nature of the environmental problem, the professional engineers succeeded in defining sewage and sewage disposal as a technical matter. The MCS and MBW engineers had the advantage of institutional continuity over commercial entrepreneurs who were more vulnerable to fluctuations in the capital markets. The *perceived* degradation of the metropolitan environment took many decades to develop, but at key moments (1848, 1854, 1858) the impact of cholera and climate provoked decisive action by public authorities, which the engineers were in a position to exploit. We also have to acknowledge the importance of timely initiatives by highly skilled, ambitious individuals. Everyone who has researched the origins of the Main Drainage and Thames Embankment must agree with David Owen that without Sir Benjamin Hall's conversion to a policy of government intervention the creation of the MBW would have been unlikely, and without the leadership of its chairman John Thwaites and its chief engineer, J. W. Bazalgette, the MBW would have failed in its mission. Yet it is well to remember, as we turn to the history of the Thames Embankment, that the almost universal acclaim enjoyed by these individuals and institutions upon the completion of their project came after the disparagement and elimination of alternative conceptions of the problem and its potential solutions.

CHAPTER IV   # Before Its Time
The First Thames Embankment

Although it is useful to treat technology as an interface between the natural environment of the Thames valley and the social organization and values of its human communities, this approach tends to discount the extent to which technology itself developed over time. The rebuilding of London Bridge and the emergence of "the sanitary idea" occurred during the height of the industrial revolution and the consequent rise of engineering as a socially valued profession. The expanding scale of operations inherent in Lancashire mills, regional railways, and industrial cities demanded more sophisticated types of engineering organization and management. These are aspects of the social environment that cultural historians such as Robert Darnton include in the *mentalité* of a period. Technology theorists like Wiebe Bijker consider them elements of "sociotechnological frames."[1] To understand what they mean in practice, we have to compare the discourse and practice of individuals and institutions over a relevant period of time. For example, the sophisticated organization of the Thames Embankment project by the Metropolitan Board of Works and the professional style of its engineering staff emerge more clearly if we examine the origins and misfortunes of an earlier embankment project, the agency which tried to manage it, and the professional style of the engineer who worked on it.

Why do some institutions prove adept at exploiting opportunities for constructive growth, while others languish? How do individual officials and engineers respond to the opportunities open to them? We have already seen that, in the campaign for sanitation reform, Edwin Chadwick had a vision of an integrated, centrally directed system of water supply and sewage disposal, whereas his opponents believed in local control and piecemeal construction. But Chadwick also had an idiosyncratic style of advocacy that generally polarized his colleagues within the MCS and Board of Health and threatened town officials, finally sabotaging the realization of his vision.

During the 1840s and 1850s, when the "sewer wars" were raging, one of Chadwick's critics, Thomas Page, held the title of "Thames Embankment Engineer" under Her Majesty's First Commissioner of Works. From the perspective of the great edifice that later bore that name, such a title seems presumptuous and a bit mysterious. Although his work was obviously visible to the public, it excited little comment in Parliament or in the newspapers of the day. Subsequent accounts, if they bother to distinguish Page's section from the later Chelsea Embankment, characterize it as elementary. It lacked the sewer line, landing steps, pedestrian walkway, and tree-lined boulevard that made the Victoria Embankment such a showpiece.[2]

Yet Page's project was the remnant of an ambitious plan to embank both sides of the Thames from Chelsea and Battersea all the way to the Tower, a plan actually approved by the government in the 1840s. Page received the title of "Thames Embankment Engineer" after winning a public design competition against more experienced rivals. He not only designed but supervised construction of Chelsea Bridge and its related embankments, as well as the new Westminster Bridge (1860–62). He was in the process of completing this work when the "Victoria" section of the Thames Embankment was being discussed and authorized in Parliament. The Royal Commission of 1861, which examined dozens of Embankment plans, called him to testify as an expert witness. Then he disappeared altogether from the official record.

The question immediately arises: if there was already a designated "Thames Embankment Engineer" engaged in building a previously authorized section that was part of the overall improvement of the river, and if that engineer worked for a government official with a Cabinet appointment, why didn't Page and H. M. Office of Works go on to build the Victoria, Albert, and Chelsea Embankments? Alternatively, if the Metropolitan Board of Works claimed jurisdiction over the Embankment project as an extension of the London Main Drainage, why didn't they engage Page as one of their chief engineers?

A brief answer is that Thomas Page practiced a style of engineering which was typical and appropriate for the 1840s but which grew less and less appropriate for the scale of operations needed for the new metropolitan system envisaged by 1860. He was also employed by a dysfunctional government agency, H. M. Office of Works, marked by weak leadership and limited vision. These characteristics were thrown into relief by the far more professional managerial style developed by J. W. Bazalgette and the vision articulated by his superior at the MBW, Chairman John Thwaites, as they struggled with the environmental, technological, and institutional complexities involved in the London Main Drainage and the Embankment. Page and the Office of Works might have done the same, but they did not. In fact, the First Commissioners' experience with their mercurial and often negligent chief engineer, over the period 1842–60, made unthinkable the prospect of pursuing a large-scale project under his supervision.

Her Majesty's Office of Works had its origins under the late medieval monarchy, with the gradual transfer of responsibility for royal buildings from sheriffs or local "clerks of works" to a small group of officers at Westminster. A clerk and/or comptroller of the king's works was first appointed in 1378, along with a master mason and a master carpenter. By the seventeenth century, the head clerk was usually an eminent architect styled "surveyor general of works" (Inigo Jones served from 1615 to 1643). The office was dissolved during the Civil War, but, in 1660, Charles II reestablished it under the direction of Christopher Wren, who brought in Grinling Gibbons, Vanbrugh, and Hawksmoor. In 1814, as the Crown became more involved in the planning and construction of public buildings and streets, the office was placed under Treasury control. The surveyor general's staff was increased to four or five resident clerks, six "clerks of works" who supervised building sites, three mechanics and four carpenters, a master bricklayer and master plumber, and about twenty laborers.

Three professional architects were "attached" to the office on a consulting basis.[3]

In 1832, the Office of Works was consolidated with H. M. Office of Woods and Forests as an economy measure, although it still received a separate parliamentary appropriation. Woods and Forests dealt with Crown lands, which by ancient treaty included the foreshores of all English rivers. The combined office was to be directed by a board of three commissioners. The first or "chief" commissioner sat in Parliament as a government appointee, almost but not quite in the Cabinet. The junior commissioners, who had technical training, continued the operations of the former offices as separate departments. This arrangement proved impracticable. Each of the three commissioners had his own account at the Bank of England, and the relationship of engineering to administration and finance was quite arbitrary. For every local street and sewer project approved by Parliament, the commissioners were legally incorporated as a separate "Improvement Commission" to buy land, dispose of property, and let contracts. Keeping the accounts, contracts, and personnel straight was impossible. After Peel's government fell in 1846, the Whigs under Lord John Russell began to ferret out what they regarded as extraneous government expenditures, looking especially hard at Office of Works' contracts for building the new Houses of Parliament. Two select committees scrutinized the Works administration and revealed a history of confusion, abuses of power, and outright fraud.[4]

Despite its poor reputation, the Office of Works managed to build or commission most of the new government offices, parks, and monuments in the metropolitan area, together with the Houses of Parliament, the British Museum, the National Gallery, the Law Courts, the (old) Treasury, the Foreign Office, and the Home Office in Whitehall.[5] The real problem came when the Treasury decided to assign it responsibility for planning and building a whole network of new streets for the expanding population of London. Beginning in 1836, a series of Metropolitan Improvement Commissions recommended ambitious projects for which the Works staff was completely inadequate, and for which the increasingly parsimonious government was unwilling to pay. The Office of Works was, as Robert Peel charged during the inquiries of 1848, a convenient dumping place for miscellaneous jobs no one else wanted, "the common Sewer of all the flotsam and jetsam" of every other department in government. Reforms prohibiting Works architects and surveyors from taking on private commissions forced the Office to hire outside contractors for its own projects. Yet Parliament would not expand its staff or budget. Even Prince Albert proposed enlarging the office, but to no avail. According to the official *History of the King's Works*, the office's confusion and apparent inefficiency was the result of too much work rather than the cause of too little.[6]

It is instructive to look ahead at this point to see that, after 1860, the Metropolitan Board of Works also became a repository of jobs no one else wanted. New streets, additional sewer lines, public gardens and lavatories, the management of former village commons swallowed up by urban growth, building regulation, the fire department, orphanages, flood control—all were charged to the MBW. It was transformed, willy-nilly, from a Board of Works into a metropolitan government, without a sophisticated political structure to support it. Eventually, it, too, succumbed to charges of mismanagement and corruption.[7]

The MBW, however, had the advantage of starting with a few large, well-defined engineering projects authorized by Parliament and supported by specific revenues. These projects required and justified the expansion of the MBW's technical and administrative operations and gave it jurisdiction over a wide but specific territory. H. M. Office of Works, by comparison, had grown through incremental additions to its traditional responsibilities, which were heterogeneous to begin with. It also had no territorial focus or constituency because its work sites were scattered far beyond the metropolis.

The Office of Works might have responded more vigorously to the demands of metropolitan growth had its leadership been more ambitious and more directly interested in its mission. But the commissioners who served from 1832 to 1860 were not an impressive lot. For most of that period, the appointment was a convenient sinecure for aristocratic men, politically useful in the Cabinet or Privy Council, who were not considered among the top rank of ministers. Most were capable amateurs, but few of them devoted much time to the office, looking instead to parliamentary or family obligations, hoping to move on to different appointments. Early

Reform politics ensured frequent Cabinet reshuffling or changes of ministry. Day-to-day business and continuity of policy were left to the two junior commissioners and the experienced head clerk, T. W. Phillips.[8]

Seven men filled the office of commissioner between 1832 and 1851, but only three of them served more than six months. John William Ponsonby, Lord Duncannon headed both Works and Woods and Forests during the 1830s. Considered a man of high principle, accuracy and industry, his common sense and easy manners made him a useful negotiator of political and personal differences among his Whig colleagues. He guided selection of designs for the new Houses of Parliament and "took an active part in the improvement of the metropolis."[9]

Lord Lincoln (1811–1864), a young Peelite, succeeded Lord Duncannon in September 1841. The respect given to Peel's masterful budget and the high calibre of his second administration do not seem to have extended to the Office of Woods and Forests. Heir to the Duke of Newcastle, Lord Lincoln was deeply involved in county affairs. Family and marriage quarrels kept him from attending to the Metropolitan Improvement Commission of 1842, which took up the question of the Thames Embankment. A priggish, egotistical, and headstrong man who fancied himself an amateur architect, he quickly alienated his junior commissioners and staff. Although he was a capable speaker and represented the Office creditably in Parliament, his tenure is marked by an inability to get projects funded by the Treasury or accepted by the affected landowners.[10]

When the Whigs returned under Lord John Russell in 1846, the cultivated Lord Morpeth, future Earl of Carlisle, became First Commissioner. A man of "endearing gentleness" but "without commanding abilities or great strength of will," Morpeth shared the government's uneasiness about Edwin Chadwick's aggressive reforms. He reluctantly agreed to create a Metropolitan Commission of Sewers and a General Board of Health during the cholera epidemic of 1848, but he soon lost control of their operations to the dynamic Chadwickians. In the autumn of 1848, the prime minister charged him to clean up the growing administrative mess created by the sanitation reformers, but, when he succeeded to the Earldom and moved to the House of Lords, he neglected the matter entirely. Gladstone and the other fiscal hawks among the Whigs began scrutinizing the Office of Works' accounts, and, in March 1850, Morpeth was shunted off to an even less visible Cabinet appointment.[11]

Chadwick and his associate Southwood Smith showed that it was possible for energetic subordinates to push the First Commissioner into ambitious projects, but weak leadership could not protect them from attacks by vested interests and professional rivals. Nor could they count on support from the Office of Works staff. The two junior commissioners who supervised operations during the 1840s were career officials of limited imagination and talent. Charles Alexander Gore (1811–97), brother to the Earl of Arran, had been found a clerkship in the Pay Office at the age of eighteen, and then made private secretary to Lord John Russell in the 1830s. Having married Lord Duncannon's daughter, he was placed in the Office of Woods and Forests, in 1839, where he remained until he was seventy-four years old.[12] Public works, buildings, and metropolitan improvements were supervised by the architect Alexander Mylne, whose father had surveyed the Thames with John Rennie in connection with the rebuilding of London Bridge. Mylne was, no doubt, competent enough, but his office was swamped with the recommendations of various Metropolitan Improvement Commissions and projects shifted from other agencies. Since he was required to work through independent contractors for the larger projects and was only weakly supported by his superior, his control over the Office of Works was hardly secure.

After the Royal Commission audits of the later 1840s, the Office of Works was separated from Woods and Forests and placed under the nominal authority of a new set of commissioners. Only the senior or "First" Commissioner was actually involved with the office and sat in Parliament as a member of the Cabinet. He was assisted by a secretary (the former head clerk, T. W. Phillips) who supervised operations as he had before 1832. The other two commissioners were purely formal designations, usually given to junior Lords of the Treasury. Woods and Forests, which managed the royal estates, reverted to its former Crown orientation, and Charles Gore stayed on as the new First Commissioner.[13]

The Office of Works continued to suffer from weak leadership. Lord Seymour, who succeeded Morpeth in 1850, was known more for his beautiful wife and his Mediterranean yachting cruises than for management skills. His qualifications as First Commissioner seem to have been his consistent Liberalism and his aristocratic connections. In any case, he was not long in office before the Whig government fell.[14] The Tories appointed Lord John Manners, a "Young Englander" described as a "dextrous and resourceful debater" in the House. Unfortunately, "his wisdom in council was of greater value than his capacity for action."[15] He was replaced, in December 1852, by Sir William Molesworth, a philosophical radical more interested in colonial reform than local improvements, who died within three years.[16]

By this time, Works was in shambles, overloaded with responsibilities and short on both leadership and resources. The Metropolitan Commission of Sewers had proved incapable of carrying out its mandate to build a new main drainage system, and the Board of Health was thoroughly discredited. These two agencies were virulently attacked in Parliament by Sir Benjamin Hall (figure 5), who became president of the reformed board in 1854. The following July, he was appointed First Commissioner of Works in Palmerston's first ministry.[17]

Sir Benjamin later reported that, when he took over the Office of Works, he discovered that the chief clerk had been out sick for the past two years, while the surveyor was "mentally incompetent" and absent most days. Staff meetings were haphazardly attended, and the clerks of works, who supervised on-site construction, met infrequently.[18] Hall made a habit of denigrating the operations of his predecessors and of claiming credit for dramatic reform, but, in this case, his charge that debility was virtually a staff tradition is supported by plenty of other evidence. He set about tightening work schedules and assigned the office solicitor to straighten out the tangled project accounts.

It was against this background that Sir Benjamin proposed to reconstitute the MCS as a largely autonomous "Board of Works" run by delegates from the local vestries. He had never liked the idea of having metropolitan improvements controlled by an agency of the central government, and it was clear that the MCS had failed in its mission. Moreover, his cabinet colleagues preferred to levy the expense of building a giant sewer system upon the residents of London, rather than on the government's own budget. But Hall was understandably cautious about entrusting a new, untried agency with the power and resources required for the Main Drainage. Thus the Metropolitan Management Act of 1855 required his approval for all projects costing more than £50,000, a provision which led to the struggle between Hall and the MBW recounted in the previous chapter. In February 1858, Lord John Manners returned as First Commissioner of Works, swiftly expediting the legislation required to get the Main Drainage underway, and initiating talks between his office, the MBW, and the City of London's Thames Navigation Committee regarding construction of the Embankment.

When the Whigs (now calling themselves Liberals) returned under Palmerston in June 1859, the matter of the low-level sewer and the Embankment became an urgent priority. Fortunately, an able First Commissioner, William Cowper (1811–1888), came into office in February 1860 (figure 5). The second son of the fifth Earl Cowper, William was as well connected as a Whig could be. His mother was Melbourne's sister, and, when she was widowed in 1839, she married Palmerston and supported his career with a popular salon. William entered politics at the age of twenty-four as MP for Hertford and private secretary to his uncle, Melbourne. After sponsoring legislation for medical and sanitary reforms and holding several minor posts, he was made president of the Board of Health, in 1855, when Sir Benjamin Hall moved up to the Office of Works. In 1859, he became vice-president of the Board of Trade (one of the junior commissioners of Works) and then First Commissioner. Edward Barry, who was completing his father's New Palace at Westminster, considered him "a man who was quite without official airs ... considerate and accessible."[19] A friend of Ruskin, he was interested in architecture and the preservation of public spaces. In the 1860s, with John Stuart Mill and G. J. Shaw-Lefevre, he would sponsor the Metropolitan Commons Acts to protect village commons as London swallowed up the surrounding countryside, and he campaigned against enclosures elsewhere in England.[20] Thus he was more likely than previous commissioners to agree that the proposed Thames Embankment should be built as a public enterprise and that land reclaimed from the river should be reserved for public recreation.[21] In this, he had the enthusiastic support of John Thwaites, chairman of the Metropolitan Board of Works (figure 5).

Thwaites was a Yorkshireman who came to London as apprentice to a woolen-draper. After joining the business as partner, he became active as a parish overseer and Poor Law guardian in Newington, on the south side of the Thames. Having served as a metropolitan sewer commissioner, he was elected to the new Metropolitan Board of Works as a representative of both Southwark and Greenwich.[22] David Owen described him as "determined rather than imaginative ... solid and dependable, an excellent moderator and a sound judge among opposing points of view."[23] Elected chairman of the new MBW in 1855, he somehow managed to organize the heterogeneous, intensely parochial delegates into an agency for effective planning and change. Although the Works committee included virtually all the delegates, Thwaites managed to exclude the press and casual visitors from its meetings. Owen commends Thwaites's style of presiding:

> When in the chair at the Board, the expression of his eyes was heavy with gloom; his face reflected no emotion. Neither praise nor blame seemed to affect him, and his somber smile could never develop into a laugh. Yet nothing seemed to escape him. In that often rowdy "Senate of Sewers" he used the gavel when it was needed, sometimes rising in his seat to command silence. . . . While Thwaites was in command, the Board seemed to be in diligent, firm hands.[24]

Although he could be aggressively outspoken when defending MBW interests against competing agencies, Thwaites did not aspire to national prominence and, unlike the First Commissioners of Works, did not treat his office either as a gentlemanly preferment or as a way station to higher office. He gave the MBW a continuity of solid leadership that the Office of Works conspicuously lacked.

From the very beginning of the board's existence, Thwaites had campaigned for the right to build the Embankment in connection with the low-level sewer line. His plan was frustrated by the claims of rival agencies (a topic discussed in chapter 6) and then by Sir Benjamin Hall's interference. Even after the MBW gained independent authority and stronger financial backing in 1858, it was clear that Parliament would have to authorize the Embankment as a project separate from the Main Drainage, with its own source of funding. That posed a danger for the MBW, because the Embankment extended into territory legally claimed by the Crown and by the City of London. Moreover, Parliament and the Cabinet might award the project to an independent commission, as had been done for decades at the Office of Works. To avert this possibility Thwaites met with William Cowper, in March 1860. The two men quickly came to appreciate one another, and Cowper gave firm support to the MBW's claim. They arranged for Sir Joseph Paxton, a veteran of metropolitan improvement commissions, to bring in the necessary bill and chair the select committee of Commons.[25] Paxton's committee obligingly recommended an embankment to be built by the MBW along the river, to be paid for by some wholly anachronistic but very convenient tariffs still levied on coal and wine imports into London. The bill said little about the functions of the embankment, simply assuring the wharfingers along the riverside between Westminster and Blackfriars that they would have "improved facilities." No mention was made of railway lines, tidal gates for barge access, a roadway, or commercial buildings. The design of the whole project, in fact, was left up to the expertise of a proposed Royal Commission.[26]

Thwaites objected strenuously to this last provision. In a letter to Cowper of 8 October 1860, he argued that the Embankment, though requiring new financing, was really part of the Main Drainage scheme and thus a matter for deliberation between the MBW and its constituents, the local ratepayers. The London Main Drainage Act of 1858 had expressly abrogated the First Commissioner of Works' veto over MBW plans. Appointing a Royal Commission would reverse that legislation and add one more agency to those already arguing over the project. Besides, why waste time on all the preliminary plans when any final plan drawn up by the MBW would still require review by a select committee of Commons?[27]

Cowper, for his part, was mindful that the MBW had no direct representation in Parliament and would need a friend in the Cabinet to ensure a fair hearing. He convinced Thwaites that a Royal Commission was politically necessary and even persuaded him, with great difficulty, to accept appointment to it. Thwaites served with characteristic energy and insight, even though often at odds with the commission chair, William Cubitt. Since becoming head of the MBW, Thwaites had developed a vision of London as a modern imperial city, something more than a collection of neighborhoods, vestries, and ugly commercial districts. He convinced his colleagues on the Royal Commission that the wharves and warehouses between Westminster and Blackfriars, which had frustrated and complicated

Embankment plans for three decades, could be replaced by a monumental wall and public thoroughfare at reasonable expense. Despite this achievement, his worst fears were realized when the Royal Commission, during a meeting in July 1861 which Thwaites could not attend,* decided to recommend appointment of *another* commission to actually build the Embankment. This would, in effect, create a new legal entity for the purpose, as had been done for the Chelsea Bridge and other Works projects.[28] Given the history of competition among the agencies already involved with the Thames, and the tangled question of financing, such a body would constitute an administrative and political monstrosity.

At this point, Cowper's political acumen proved decisive. After further meetings with Thwaites, the First Commissioner simply ignored the recommendation of the Royal Commission, introducing a bill giving the MBW authority to execute the Embankment. Like Sir Benjamin Hall in 1855, he insisted that the MBW act under the aegis of the Office of Works as a kind of independent contractor, so that its plans were still subject to his review. However, after the basic design was approved, his oversight was only a formality. Cowper did refuse Bazalgette's proposal to embank the whole south side of the Thames, simply because there were insufficient funds for it. But he agreed with Thwaites that the northside Embankment (the "Victoria" section between Westminster and Blackfriars Bridges) should be developed as a public rather than commercial space, and he personally chaired the select committee which finally gave the MBW a mandate to clear out the existing wharves. When residents in the government-owned Whitehall Estates tried to prevent the Embankment roadway from crossing their property, Cowper agreed to a weak compromise in a Commons committee but then (with Palmerston's help) killed it on the floor of the House.[29] As he had promised Thwaites, his ability to represent the MBW in the Cabinet was probably more help than hindrance. Fortunately, he stayed in office for six years, long enough for the Victoria and Albert Embankments to get under way, and was succeeded by the equally sup-

---

*At least Thwaites contended he was absent and avowed, on his word as a Baptist preacher, that his colleagues had all agreed to give the project to the MBW. William Cubitt, who was Lord Mayor of London as well as the Commission chair, assured Parliament that Thwaites was indeed present. (Owen, *Government*, 77.)

portive Lord John Manners. Subsequent First Commissioners of Works reverted to type: the last to oversee construction on the Embankment, A. S. Ayrton (1869–73), was so zealous at cutting his own budget for public works that he finally had to be transferred to another post.[30]

The characters of the First Commissioners of Works and the workings of their office show why, despite their previous control over some metropolitan improvement projects, they were not prepared to undertake construction of the Thames Embankment. Except for Edwin Chadwick and Sir Benjamin Hall, few Works officials were ambitious enough to want such an expansion of their authority. They maintained only enough staff for general planning and strictly local design, construction, and repair. The office had been systematically starved of funds by successive governments and had a reputation, fostered by parliamentary inquiries and reports from its own commissioners, for bad management. The special "Improvement Commissions" through which it had operated were temporary legal expedients, clearly inadequate for the long-range project proposed in 1861. At that time, faced with the question of who would finally build the Embankment, William Cowper provided a traditional British solution. He delegated practical authority to a man whose judgment he respected and to a makeshift Board of Works that was more than just another commission but less than a government. He empowered the only agency around that had actually accomplished a sizable engineering project (the Main Drainage) but retained formal authority in his own office to provide the political leverage required for success.

That is one answer to the question of why the Office of Works did not build the Thames Embankment. But there is another answer, related to the Office's experience with its special Thames Embankment engineer, Thomas Page. Page practiced a buccaneering, hands-on style of engineering that flourished in the railway boom of the 1840s and in faraway countries where young British engineers learned their trade building bridges and factories on speculative contracts. That style could produce ingenious solutions to technical problems, so long as the scale of operations did not require delegation of authority. Comparing Page's approach to that of the superbly organized, politically astute J. W. Bazalgette, we will be able to see why the latter gained control of the project.

In 1838, a local group called the "Chelsea Commissioners" recommended erecting a river wall along the wide muddy banks of the Thames on either side of Battersea Bridge. This was to connect with low embankments running from the new Houses of Parliament up to Millbank Penitentiary and Vauxhall Bridge, and thence to a proposed new bridge near the Royal Hospital.[31] Not all residents favored this scheme. Chelsea, in the 1830s, was still separated from Westminster by broad fields, clay pits, and a wretched road along the river. Many artists and writers were attracted to its rural character, and the view across the Thames to Battersea was considered especially charming. But the river was steadily eroding its banks along Cheyne Walk, and mud shoals were beginning to impede navigation. Three large sewers discharged in the neighborhood, where several waterworks took their supplies from the river. Letters demanding action were written, in 1839–40, by local leaders such as the Marquess of Westminster (even then one of the wealthiest landowners in England), master builder Thomas Cubitt, and the Chelsea Water Works Company.[32] In 1842, a Royal Commission responded by recommending embankment of the river. The Office of Works apparently introduced a bill for the purpose, but without success.

The following year, the First Commissioner of Works, Lord Lincoln, convened a Royal Commission for "Improving the Metropolis and Providing Increased Facilities of Communication," which included Charles Gore, Sir Robert Smirke (the architect of Millbank river wall), Charles Barry, and local notables. Inquiries among property owners found a majority in favor of paying half the cost of a new embankment, the other half to be borne by the government. In their first report, dated January 1844, the commissioners announced that "Upon a careful review of the many subjects of improvement for which Plans had already been before the public, or were subsequently submitted to us, we considered AN EMBANKMENT OF THE RIVER THAMES to have the first claim to our attention."[33] However, they did not limit themselves to the Chelsea region. Between January and August 1844, they took testimony from a variety of witnesses and examined embankment plans submitted by the leading engineers and promoters of the time.[34] Their report advocated embanking both sides of the Thames from Chelsea all the way down to London

Bridge. The most feasible design for this majestic enterprise, they announced, was one submitted by Thomas Page.

Page was born in London in 1803, the eldest son of a solicitor. His youth was spent at Rowald Kirk on the Tees, where he was educated for naval service but also studied civil engineering. Though he remained interested in ships and gunnery all his life, his career developed on land. After a few years as a draughtsman and architect's apprentice, he moved back to London and was soon elected a member of the Institution of Civil Engineers. In 1835, Sir Marc Isambard Brunel chose him as assistant engineer for his long-delayed Thames Tunnel project and, the following year, made him acting engineer. He served in that capacity until the end of 1842 and was then hired by Lord Lincoln at the Office of Works to survey the river all the way from Battersea down to Woolwich. His embankment plan was based on this survey.[35]

Page proposed to disarm the opposition of wharf proprietors, who feared loss of river access, by building an arched roadway and terrace parallel to the shoreline, with drawbridges along the way allowing barge access, at high tide, into the side channels between roadway and wharves (see figure 4). Thus the wharfage facilities would be improved, and construction of the embankment need not disrupt barge traffic. The roadway, raised on a brick viaduct faced with granite, would be fifty feet wide from Blackfriars up to Whitehall, where it would join a new street swinging up to Charing Cross. It would be connected to the streets along the shore by a series of ironwork bridges. After Whitehall, the embankment would continue up to Westminster as a broad public promenade linking the existing shoreline to the gardens in front of Whitehall Estates. The river wall would make the width of the river more uniform, improving its channel for navigation.[36]

Page and the Office of Works architect James Pennethorne also surveyed Chelsea Reach and designed an embanked roadway to run along the shoreline from Battersea Bridge down to Vauxhall Bridge, crossing in front of Thomas Cubitt's huge contracting works, with a new street cut from Sloane Square down to the White Horse Ferry (where Lower Sloane Street now leads to Chelsea Bridge). This portion of embankment would not be separate from the shore, because there were few wharves or warehouses to be considered. The Royal Commissioners offered several amendments to Page's plan but, in May, recommended it to Parliament as the best option. The estimated cost was £302,000 to £367,000.[37]

Page was the first engineer to work out the concept of a detached river wall serving to streamline the river channel while still providing access to the wharves inside. This feature was incorporated into most later plans, even one proposed by Bazalgette as late as 1860. (See figure 4.) Bazalgette also incorporated a granite-faced brick retaining wall in his design. The terrace Page designed for fronting Whitehall Gardens was revived as a serious proposal by Whitehall residents in 1862. The advantages claimed for Page's embankment were echoed by other projectors in the 1860s: it would provide an alternative east-west thoroughfare to Fleet Street and the Strand; it would create a "noble and disembarrassed walk in the midst of the fresh air of the river," as well as safe stairs and easy passage for steamboat passengers; and it would open up "a magnificent picture of the finest aspect of the metropolis" in place of the unseemly line of wharves.[38]

Although Page's design guided thinking and discussion about embankments for over eighteen years, it failed to get off the drawing board in 1844. When Lord Lincoln introduced an authorizing bill in August of that year, Sir Robert Peel was trying to reduce deficits incurred by the Whigs during the recent depression, and Gladstone at the Board of Trade was already introducing reductions on tariffs. The ministry was not in favor of large capital outlays supported by anachronistic coal duties. Wharfingers along the proposed embankment route protested angrily against the business interruptions and costs they were likely to incur. Lord Lincoln himself was distracted by marital and family problems. Finally, the City of London Navigation Committee had already proposed an embankment of its own. Fearful of losing its jurisdiction over the river, it sued the government in chancery court. This lawsuit (described in chapter 7) dragged on until 1857, effectively preventing development along the Westminster-Blackfriars section of the Thames.[39] Lord Lincoln's bill was withdrawn in February 1845.

Thomas Page did not abandon his great project. In fact, he was still acting in an official capacity the following year. Lord Lincoln let him "approve" minor embankment projects and repairs in answer to notices sent by the City of London Navigation Committee, a tactic, no doubt, linked to their jurisdictional dispute over the riverbed. The Navigation Committee objected that their notices were mere formalities and that Page had no right to question them. Works secretary Charles Gore replied, "the whole point of having to give notice was to obtain approval or objection." Furthermore, he said, Mr. Page "was and had always been referred to as Engineer to the Board."[40]

Because the Office of Works was prohibited from contracting with its own employees, Page was hired as a consulting engineer, paid according to a schedule of fees rather than by salary. He was, therefore, free to accept other commissions. He drew up plans for a great railway terminus at Charing Cross, for a pneumatic railway (powered by a vacuum tube) that would pass through the Thames Tunnel, for harbor facilities used by the Irish mail service at Holyhead, and for docks at Swansea. Within a few years, he was also designing new roads and bridges. On the whole, he was to have a successful career, but the failure of the Embankment bill in 1844–45, as his memorialist wrote later, "deprived him of the opportunity of associating his name with the great work which has since become one of the chief ornaments of London."[41]

Because most of the opposition to an embankment came from wharfingers in the Westminster-Blackfriars area, Works officials decided to try building along Chelsea Reach, where commercial facilities were fewer and smaller and the proprietors apparently more willing to cooperate. Page revised his plans several times, adding a new toll bridge over the Thames and the development of public gardens across the river at Battersea Park. From 1846 through 1852, the Office of Works sought parliamentary approval and a line of credit from the Treasury, while soliciting contributions from local property owners and lessees to make up the required funds. However, only £26,000 was pledged through private agreements, far less than needed.[42] Riverside proprietors proved as stubborn as elsewhere, and many residential properties were tied up in litigation. Thomas Cubitt, impatient over Works' ineptitude, used his own men and £15,000 of his own money to build 3,200 feet of new frontage and a roadway between his Pimlico workshop and the river (now part of Grosvenor Road), as well as a public garden (still in use).[43]

Sir William Molesworth, who took over as First Commissioner of Works, in December 1852, was convinced that further negotiations would be successful. He accepted a new design by Page for a higher embankment with a wider road that would cost an estimated £92,579. Construction would begin on the west side of the Chelsea Hospital and proceed eastward (that is, downriver toward Vauxhall Bridge), if the necessary property rights and contributions were secured within four years. The altered plans were approved by Parliament in September 1853, despite opposition from a major landowner, Lord Cadogan.[44]

Although Works solicitors were still negotiating with riverfront proprietors, Page received permission to begin building the frontage for the northside Chelsea Bridge approaches on 3 September 1851. The next day he sent plans and sections to contractor Thomas Earle, who had obviously been briefed ahead of time, and on the fifth was able to accept Earle's bid. Earle finished the first fifty feet of river wall by the end of the year, scavenging bricks from the demolished wall around Chelsea Hospital orchard.[45] He had built only 150 feet more by November 1852, when a monster tide devastated the whole Chelsea shoreline. The work was then postponed, with the expectation that the incoming Tory government would support yet another embankment bill, for which Page drew up yet more plans.[46]

As soon as the new bill was passed, in September 1853, Page sent out specifications for the embankment to selected contractors, along with an initial estimate of £15,157.[47] The six bids received ranged from £11,571 to £26,168. The low bidder, John Jay, had already built King's Cross Station and the first part of the underground Metropolitan Railway, and was now principal contractor for the new Houses of Parliament.[48] He was to start at Vauxhall Bridge and work his way west, skipping over Cubitt's embankment, until he reached the Chelsea Bridge site. A second contract was let with Thomas Jackson of Pimlico to take over the Chelsea Bridge frontage begun by Earle and extend it to the Ranelagh sewer at the west end of Chelsea Hospital.[49]

Page was encouraged by these contracts to build himself a new office in Middle Scotland Yard near the Office of Works, and to appoint an assistant engineer, Daniel Creegan, for the substantial sum of three guineas per week.[50] He directed John Jay to begin construction near Vauxhall Bridge, in January 1854, even though the Office of Works had not finished writing up his contract. However, when Jay sent notices of occupation to the proprietors, one of them, Smiths' Distillery, threatened legal action if its access to the river was interrupted.[51] Page immediately wrote to the new Office of Works solicitor, John Gardiner, proposing that Jay move to the western side of Cubitt's embankment until the legal problem was resolved. Gardiner replied that Jay's contract did not provide for such a change, and, in any case, the property claims on the western side were not all settled. Page ignored this message. As he later explained, contractors like Jay and Jackson who do good work at reasonable prices expect that, if they go to the trouble to locate plant and equipment at the site of a large project, they will be moved from one assignment to another to avoid work stoppages.

At the new site, Jay had to lay the embankment foundation below low water, anywhere from 50 to 170 feet out from the old shoreline. This required the use of sheet piling to hold back the river while the concrete cured, and brick arches had to be added to dock entrances for extra support (see figure 3). Whereas Cubitt had built his embankment around an old river wall that had stabilized the shoreline, Jay and Jackson encountered quicksand and shifting mud shoals and had to accommodate dock and canal traffic during construction. As a result, Jay's costs soared to £17,000, almost 50 percent above his original bid. When Gardiner finally produced a contract in May, he discovered to his horror that Jay had already completed half the work and that Page had approved the extra expenses. The chief engineer was sent a severe reprimand for these irregularities. A new contract was drawn up and signed in late July: Gardiner reduced Jay's portion of embankment to what was already done and awarded the rest (from the Chelsea waterworks to the Grosvenor Canal) to Jackson, whose prices were lower. Jay finished his contract in 1855, but Jackson took another three years, completing his work little by little as outstanding property claims were settled.[52] The western end of the embankment was left unfinished, in case Parliament decided to allocate funds for an extension. The usual petitions and plans were drawn up for this purpose, but to no avail.[53]

Page now had a new project, however. Parliament had appropriated funds, in 1846, to create a public park on 320 acres of low-lying land on the opposite side of the Thames around Battersea Common. The river there was bordered mostly by a string of timber wharves. Page designed a simple embankment for it, a brick wall reaching down about twenty feet to the riverbed clay, supported by wooden piling and filled in with dirt. The result would be a level walkway running 3,800 feet between Chelsea Bridge and Prince Albert Road. The appropriation quickly proved inadequate, however, as property negotiations dragged on. The project limped along on Exchequer loans until 1853, when a special Park Commission was formed to award annual parliamentary grants. Construction actually began in 1854, when the Office of Works was offered a million cubic yards of fill dirt excavated from the Royal Victoria Docks. The Chelsea Bridge approaches were formed first, to permit better access to the rest of the site, and fill was dumped along the shoreline to support an embankment. The £25,000 promised for that part of the project, however, had disappeared over the years, while the estimate for its cost had risen another £10,000. Works architect James Pennethorne put a good face on it: he wrote to Page, in March 1857, that the higher shoreline had improved navigation along the river, and the lack of a brick wall made it easier for barges to unload fill dirt for the rest of the park. For that reason, he advised waiting another two years to put the wall in![54]

By that time, Page was in very hot water with the First Commissioner of Works. His overoptimistic estimates and freewheeling style of contracting had already drawn criticism from the commissioner's staff, but it was compounded when he added yet another project, the rebuilding of Westminster Bridge, to his list of accounts. Because this led to his downfall and rejection from the Thames Embankment, it is worth a short digression.

The first bridge at Westminster, built about 1760, was a ponderous stone affair with a high, arched roadway and tall parapets that blocked the view from either side. Like other metropolitan bridges, it suffered erosion around its piers as a result of the replacement of London Bridge. It was stripped of 30,000 tons of stone, in 1846, but continued to deteriorate. The Metropolitan Improvement Commission of that year finally accepted the need for a new bridge, and Page offered the elegant arched design one sees today. After the usual delays for further study, negotiation, and appropriation, Page was able to start building in May 1854. He devised a new type of iron cofferdam that allowed river barge traffic to continue during construction and replaced the old bridge in stages so that cross-river roadway traffic never ceased, an achievement applauded by the press.[55]

When Page started work on Westminster Bridge, he had already been engaged on the Chelsea embankment and bridge projects for about two years. The First Commissioner in 1854, Sir William Molesworth, consolidated Page's schedule of payments for all three projects, promising him 3 percent of the total cost for his consulting services, .625 percent for all the drawings he made, and 1.375 percent for expenses of assistant engineers and clerks of works. The total was £19,611, which was to be spread out in quarterly payments of £711 over five years, the estimated life of the combined projects.[56]

Given the delays and legal entanglements of the Chelsea Embankment and the uncertainties of bridge building, it was probably rash of Molesworth to set out a fixed schedule of payments. If Page had been a contractor, the agreement would have designated 10–15 percent of the total sum for contingencies. It would have been assumed that unexpected complications or additions to the work specified in the contract would be charged as "extras." No such arrangement was made in this case. Page himself forged ahead without bothering to keep the Office of Works informed of his dealings. The result was a series of charges, denials, recriminations, and demands that escalated over the next four years, until it was no longer possible to sort out who owed what to whom.

In 1856, the original contractor for Westminster Bridge went bankrupt, leaving the foundations unfinished, and Page elected to take over the contracting himself. This was an unusual arrangement for a mid-Victorian consulting engineer, since contracting was a lower-status enterprise and involved possible conflicts of interest. The Office of Works was understandably reluctant to proceed on this basis, especially since they had had no correspondence from the engineer for two years. After some delay, however, Sir Benjamin Hall agreed, and £500 was sent as an initial payment. Hall also asked Page for a full report on progress at Chelsea. At first, Page promised completion of both bridge and embankment by the end of the year. After Hall had reported this to Parliament, the engineer told him the bridge would probably not open until the summer of 1857, at a higher cost than previously estimated. Hall was furious at being publicly embarrassed. In fact, the bridge was not finished until March 1858. Although it was acclaimed as "one of the handsomest bridges which now span the silent highway of the British metropolis," Page's reputation at the Office of Works had become as odoriferous as the Thames itself.[57]

From 1858 on, the engineer and the Office of Works fought a kind of guerilla war. Page kept recalculating his fees retroactively, while the office staff demanded appropriate accounts and receipts. Not even arbitration by the eminent Robert Stephenson could resolve the struggle. In April 1860, William Cowper, already a veteran of the office in one capacity or another, sent a sharp note to Page threatening "extreme measures" if the accounts were not sent in as demanded by "numerous official letters" over the years. Page replied that he was "hurt" by Cowper's attitude, so contrary to his reputation for kindness and justice: "a slight reminder would have been more in accordance with your own position ... but it is too much the habit of the Board to treat professional men as their dependents." After recalling his seventeen years' experience with the Office of Works, and hinting that "certain influences" among the staff were against him, he assured Cowper that the necessary accounts were "in the mail." However, the Works secretary found them quite insufficient to make any determination of expenses or fees. When Page applied again for an advance of £2,000, he was refused pointblank. Facing financial difficulties, he began to make "intemperate" demands and complaints, which the Works staff tried to overlook.[58] When his accounts for Westminster Bridge were finally submitted in January 1864, they were deemed "overstated and inadmissable." Page retaliated by sending in, almost randomly, bills for old services such as drawing maps and writing reports in the 1850s, and charges for interest on the unpaid claims.[59] The Office of Works countered with a lawsuit. Page, in default on bank loans, facing bankruptcy and the seizure of his property, agreed to another arbitration.[60] In 1868, he was finally awarded £15,166 in fees, mostly for his work on Westminster Bridge.[61]

It would be misleading to say that Page's dispute with the Office of Works undermined his career as an engineer. Despite his singular methods and missing accounts, he was employed by Works from 1842 until 1864 and paid over £30,000 all told, including £4,250 for the Chelsea Embankment. He was working steadily at other projects during the 1860s and consulted with several towns on proposed municipal improvement schemes. Writing to Cowper in 1862, he remarked that the respect and cooperation accorded by his other clients "would rather surprise you."[62] Outside the corps of London engineers, contractors, and government employees who probably knew of his financial troubles, he enjoyed a favorable reputation. As he himself wrote, in a line that could stand as a motto for all engineers, "when the remembrance of my omissions in correspondence shall have passed away, the Engineering Works constructed by me under the Department will be regarded with satisfaction."[63]

Nevertheless, Page's reputation at the Office of Works was at its lowest just when the question of who would build the new Thames Embankment was settled. The first Commons select committee on the project met, in 1860, just after Cowper severely reprimanded Page for his failure to provide accounts and answer letters. The Chelsea Embankment, a relatively simple project, was then seven years old and barely finished, with cost overruns and controversial alterations in the contract. In 1861, when the Royal Commission considered designs for the Victoria Embankment, Page had taken over contracting on Westminster Bridge and was becoming "intemperate" in his demands for payments.

It is significant that Page, unlike most leading engineers in London, did not enter into partnerships with his colleagues. His employment at Works was always as an individual consultant, and, so far as I know, none of his other contracts involved a partner. At one time, he had an office in Great George Street, where all the leading civil engineers congregated, but he soon moved next to the Office of Works in Scotland Yard. Even though he was a member of the ICE, he does not seem to have been active in discussions regarding the River Thames in the late 1850s and early 1860s. Thus he appears not to have been intimate with that community of leading engineers and contractors who collectively, both in their testimony and behind the scenes, influenced the select committees and Royal Commissions on the Thames Embankment. They knew Page's local projects, of course, and undoubtedly knew something of his troubles with the Office of Works. But they were unlikely to engage him in the daily, informal conversations that often constitute an invisible dimension of the arena where commerce and politics mix.

Had Page been more astute in his professional relations, and had William Cowper been an ambitious politician of modest origins like Sir Benjamin Hall or John Thwaites, the two might have used the Embankment project as a means of securing greater power and furthering their careers. But Cowper belonged to a class that prospered through kinship and patronage more than through professional achievements, and he regarded government service as a duty or a sinecure rather than as a "career." By the end of 1860, he must have trembled at the thought of another project involving Thomas Page, or indeed anything involving embankments. Besides, Works had no engineering department of its own to expand for the occasion, with or without Page, whereas the Metropolitan Board of Works did. So when John Thwaites approached him in July 1861, to negotiate the assignment for the MBW, Cowper readily agreed.

Parliament's investigation of the Embankment question gave Page one last opportunity to present his own designs, virtually unchanged since 1844. Unfortunately, the select committee appointed in 1860 was a creation of Cowper, already working with Thwaites. Its members included Sir William Tite of Chelsea, who was familiar with Page's career, and Lord John Manners, the former First Commissioner of Works. Minutes of the committee proceedings show that Page, as usual, overstated his case. Leading engineers and knowledgeable members of Parliament paid respect to his previous plans but did not take his current ideas seriously. J.W. Bazalgette agreed that his own design was essentially similar to Page's (figure 4), but the latter's work was discussed in the past tense and then dropped, while Bazalgette's plan was published in the committee's final report. During his testimony, Page also declared that John Thwaites's idea for replacing the unsightly wharves with a solid embankment was absolutely wrongheaded. This, too, doomed his chances.[64]

One could argue that Page was a victim of bad management by his superiors and of the contingencies inherent in any attempt to modify the environment through technology. But when his style of engineering is compared with that of Bazalgette at the Metropolitan Board of Works, it becomes apparent that he had not kept up with advances in project management. It was not, after all, a grand vision of architectural splendor that mattered most in engineering the Embankment. Rather, it was the ability to organize, delegate, and supervise hundreds of separate but interlocking tasks over a long period of time, while convincing the political authorities that mistakes, cost overruns, and delays were all part of a flexible and professionally competent plan. Leaving behind the freewheeling, hands-on engineering practiced by Page, Bazalgette created a new institutional arrangement—a municipal department of engineering, with a large, trained staff, operating under the general direction of a politically oriented Metropolitan Board of Works.

The 1858 amendments to the Metropolis Management Act gave the MBW most if not all the power it needed to secure a construction site and adequate finances, and it specifically mandated construction of the Main Drainage system. MBW engineers had relatively clear authority over project management, despite the political machinations and fiscal pettiness of the board. Most board members were businessmen, tradesmen, solicitors, or local magistrates. Many were retired, and very few had technical building experience or vision beyond the parish level. David Owen and Gloria Clifton conclude that the board members' general mediocrity kept them from interfering in projects and allowed the increasingly specialized engineering staff to get on with first-rate professional work.[65] On the other hand, under the chairmanship of John Thwaites the board developed a sense of public responsibility for London's infrastructure and architecture that complemented the outlook and abilities of its engineering staff. Compared to the Office of Works, with its rotating First Commissioners, its confusing methods of project management, and its vague constituency, the MBW exhibited a remarkable consistency of purpose. Even though Thwaites was sometimes accused of despotism, the engineers gained a good deal of influence over day-to-day business and even promoted project ideas for the MBW agenda.[66]

Since the general community of engineers had not been disposed to work in large, bureaucratic public offices (in fact, there were none for them to work in until the MBW appeared), how then did the MBW's engineering department develop the outlook and organization needed to complete the Main Drainage, the Embankment, and all the streets, bridges, buildings, and other large-scale projects it was assigned? In part, it was due to their years of working together at the MCS and, in part, to the leadership of Chief Engineer Bazalgette (figure 7).

Joseph William Bazalgette (1819–1889) was sixteen years younger than Thomas Page—in terms of the development of Victorian engineering, almost a generation younger.[67] The son of a royal naval commander who distinguished himself in the Napoleonic Wars, he was educated privately and apprenticed, in 1836, to Sir John McNeil, a prominent Irish road and railway engineer. Bazalgette supervised land drainage works in Ireland during his pupilage and, in 1842, at the age of twenty-three, established himself in Great George Street as a consulting engineer. He was elected a member of the Institution of Civil Engineers in 1846. At the time he was engaged primarily in laying out railway lines. The great railway "mania" drove him to overwork, his health broke down, and he had to leave London, in 1847, for almost a year of convalescence.[68]

He returned just in time, in August 1849, to join the staff of the new Metropolitan Commissioners of Sewers (MCS) as an assistant surveyor under Chief Engineer Frank Forster. The MCS used a number of consultants like Robert Stephenson, I. K. Brunel, and Sir William Cubitt, but it also hired one engineer for each of the five sewer districts under its supervision. By 1851, this small in-house staff had worked out the essentials of London's Main Drainage scheme. Bazalgette was soon promoted to assistant engineer, along with three other men—Thomas Lovick, Edmund Cooper, and John Grant—who were to remain together for the duration of their careers. This continuity of association was unprecedented among Victorian engineers and does much to explain the success of their undertakings.[69]

Forster died in 1852, worn out from the anxieties and criticism incurred in his job. Bazalgette was promoted in his place. He worked closely with William Haywood (1821–1894), engineer to the City Sewers Commission, and the two men developed a mutual respect that lasted until they both retired half a century later. When the Metropolitan Board of Works was created in January 1856, Bazalgette was given temporary charge of the sewers and other works inherited from the MCS. It was assumed by some newspapers that this appointment was a move by the City of London to gain an edge for the old-fashioned anticentralizing party, thus forestalling innovations.[70] If so, it was a major miscalculation.

Bazalgette won the subsequent election for permanent chief engineer over nine other candidates and was given a salary of £1,000 a year. He promptly named Lovick and Cooper his assistants for the northern sewer districts and Grant for the area south of the Thames, each for £350 a year with £50 traveling expenses.[71] And he began to build a department of engineering that could cope effectively with the many duties laid upon the board by the Metropolitan Management Act of 1855. A decade later, at the peak of construction on both the Main Drainage and the Thames Embankment, the engineering department included twenty draughtsmen, five clerks, and an accountant at the main office in Spring Gardens, plus two surveyors, twenty-two sewer technicians, and fifty-nine clerks of works.[72] Larger than any commercial engineering firm, this hierarchy rivaled anything developed by contemporary railway companies or building contractors. It provided a fairly reliable if gradual route toward respectable careers for trained young men. Gloria Clifton has calculated that about 20 percent of staff were drawn from the working classes, with the rest from all levels of the middle class. Just under half kept at least one servant in their household (commonly perceived as a mark of middle-class status), which would have taken an income of at least £150 per annum (though one or two bachelors did it on less). The two senior engineering clerks, who had held their posts since 1844, earned about £250, while the clerks of works earned between £75 and £200.[73]

As the scope of its activities broadened, the MBW classified and graded the various types of jobs, fixed salary scales and increments, introduced competitive examinations for office clerks, and adopted policies on pensions, discipline, and lunch breaks. It was, in this respect, well on the way to becoming a municipal civil service. However, it would be a mistake to think that employment in a nascent bureaucracy made the engineering staff immune to their previous professional culture. They were still influenced by kinship and patronage networks operating in the London engineering community (see chapter 7). Over half the permanent staff of the MBW was drawn from the ranks of temporary employees, and these, in turn, were selected for the most part by the senior officers. The board encouraged Bazalgette to nominate candidates and, in some cases, authorized hiring a certain number directly.[74]

Moreover, although Bazalgette was employed full-time by the MCS and MBW for forty years, he was not exactly a modern career civil servant. While building the Main Drainage and Embankment, he shared offices in Great George Street with Thomas Telford and Robert Stephenson and consulted in partnership with James Walker and John Fowler. Some of his collaborations led to an inquiry by the MBW before he was confirmed as chief engineer in 1856, and they came up again during a dispute over contracting the first section of the Victoria Embankment. Like Thomas Page and other consulting engineers of his time, Bazalgette recommended fellow engineers and contractors for commissions with local authorities, private firms, and foreign governments. When he first appeared before a Commons Select Committee on the Thames Embankment in 1860, he presented plans in association with G. W. Hemans, who led a joint-stock company.[75] The committee did not realize at first that Bazalgette was testifying in a private capacity. When they did, they asked, "Doesn't your appointment as Engineer to the Metropolitan Board of Works prevent independent projects?" Bazalgette replied, "Yes, generally it does, but it does not prevent my appearing before a Committee of the House of Commons, and preparing plans to express my views."[76] In fact, Bazalgette had it backward: after an investigation for possible conflict of interest, he continued to accept outside commissions, but he ceased representing associates in private or parliamentary business.[77]

By 1860, when the question of the Thames Embankment came before Parliament, and when Thomas Page was receiving harsh reprimands from the Office of Works, Bazalgette was well known in London and Westminster as a hard-working, sometimes irascible chief engineer who combined a rare genius for synthesizing and rendering the ideas of others into workable engineering specifications with a talent for explaining the results to parliamentary committees and board members.[78] He had, among other things, supervised the construction of over twelve hundred miles of sewer lines. Denis Smith describes him as a quiet man, slight of build and of uncertain health, who was later considered to have a "cold and phlegmatic temperament." Yet his "prominent acquiline nose," muttonchop whiskers,

Figure 3. Changes in the scale and complexity of Embankment construction. *Top:* The first Chelsea Embankment by Page and Jackson, 1857. *Bottom:* The Victoria Embankment from Westminster to Waterloo Bridge, by Bazalgette and Furness, 1865. (*Illustrated London News,* 24 January 1857 and 22 April 1865.)

strong eyebrows, and "keen grey eyes" gave him an air of resolution and power (figure 7). Thoroughly engrossed in his work, he yet fathered ten children and maintained a large family establishment at Wimbledon Park. Once, overcome by his lifelong asthma, he had himself carried on a litter to inspect a work site near the Embankment. Even in old age, he challenged what he called "his sedentary life" by riding two or three hours a day.[79] These qualities can still be discerned in the bronze relief portrait mounted on the Embankment wall at the foot of Northumberland Avenue. They are also readily apparent in the transcripts of his monthly reports to the board and his testimony before Parliament and Royal Commissions. Bazalgette not only prepared plans and reference books for all the bills to be sent to Parliament (a task that had to be completed by 30 November each year for the ensuing session), but also acted as expert witness during the committee stage. He testified, too, on the efficacy of all private enterprises likely to affect the public interest.[80] The result was a truly remarkable outpouring of professional reports that extended over forty years and covered every conceivable aspect of metropolitan life.

When Bazalgette's training, temperament, institutional connections, professional associations, and organizational skills are compared with those of Thomas Page, the question that opens this chapter is not hard to answer. Although resident in London, Page was not part of the Great George Street circle, and his style of project management was not suited to the scale of planning and organization needed for the Main Drainage and Thames Embankment.[81] Page never outgrew his experience with the early Chelsea Embankment, whereas the team of Bazalgette, Cooper, Lovick, and Grant built upon their experience at the MCS and continued to expand their vision and management of public engineering after 1855. That growth, in turn, led to new ideas about how an embankment of the Thames could transform the urban environment.

CHAPTER V

The Genealogy of the Thames Embankment

It should be obvious by now that the Thames Embankment recommended by the Royal Commission of 1861 was neither an original idea nor a simple extension of the London Main Drainage. The concept of embanking the river had been part of a public conversation among London-area engineers, government functionaries, newspaper editors, civic improvers, wharfingers, contractors, and entrepreneurs of all sorts since before 1800. As this conversation developed, the traditional meaning of "embankment"—a utilitarian, local, small-scale wall or dike—began to take on the qualities of a collective noun and to exude overtones of grandeur and crisis, depending on a person's feelings of promise or threat. From a recommendation and an idea, it evolved into a concept and then into a *project*. This designation, as noted previously, encodes the goal of technological control over the environment within a social narrative which acknowledges both purpose and contingency.

As a project, "embankments" became "The Thames Embankment," capitalized in writing before being capitalized in a budget. And, as a public project, its design experienced two kinds of metamorphosis. First, as one would expect, a variety of interested people beyond the inner circle of professional engineers and politicians came forward with ideas about its architecture, operations, financing, methods of construction, management, dimensions, location, and use. Second, the emerging project, although based originally on the need to improve river navigation, and subsequently on a convenient route for sewer lines, attracted all sorts of other purposes. Once it was perceived as likely to happen, it acted as a magnet for sewer proponents, railway interests, gas and water utilities, advocates of public health, transportation experts, imperialist politicians, steamboat proprietors, fertilizer purveyors, and more. It crystallized the aspirations of a whole metropolis as if they had been suspended in some conceptual solution, a happy analogue to the "fuliginous matter" in London's atmosphere which crystallized on windowpanes during foggy nights.

The public discourse which framed the Embankment in the period from about 1800 to 1862 reveals that the definition of a cultural artifact, even one so massive and concrete as the Embankment, is not inherent in its technology, but derives from a kind of negotiation among relevant social groups, which may or may not be resolved. The meaning or meanings attached to an artifact at its first appearance often change or evaporate; meanings attributed to it by advocates when it is finally constructed may not be shared by those whose perceptions or aspirations were denied. This is especially true of public works projects, which take a fairly long time to complete by individuals' daily reckoning, and which gradually encroach upon the adjacent social and physical environment in unanticipated ways. Like many such projects, the Embankment was admired in its design phase, criticized during the initial stages of construction, held up by the vagaries of weather, financial markets, and collateral projects, and widely praised upon completion. The Embankment which emerged by 1874, like a butterfly emerging from its chrysalis, tended to subsume and conceal all previous interpretations. To understand its metamorphosis, we need to reconstruct those interpretations, especially the ones that failed of realization.

Local embankments had been commonly built, maintained, and regulated along the Thames from medieval times. The parish and borough Commissioners of Sewers, authorized in 1531, had embankments as part of their oversight. In 1608, for example, a jury called by the Surrey and Kent commissioners ordered William Hall of Rederhithe [Rotherhithe?] to "pile and boorde one pole of the bancke lying againste his house and againste the River of Thames" at a cost of 13s.4d. Gilbarte Este and John Barlow of Bankside were ordered to repair two and a half poles of the Thames wall on their premises, and similar orders were obtained against other proprietors.[1] Periodically, each borough commission prepared plans of the extant river walls in its jurisdiction. In 1832, when James Walker was contemplating a line for regulating embankments on the north side of the river, the Surrey and Kent commissioners prepared longitudinal elevation plans of the south bank all the way from Westminster down to Greenwich.[2] The legal authority of the local Commissioners of Sewers eventually passed to the Metropolitan Board of Works, and, in the 1870s and 1880s, when flooding prompted the need for additional embankments on the south side of the Thames, the MBW claimed this ancient prerogative to assess local property owners the costs of planning and construction.

On the north side of the Thames, embankments were commonly regulated by the City of London Navigation Committee, acting in their capacity as Thames Conservators. Their authority was first granted by James II and then, following the Great Fire, by the Rebuilding Act of 1670, which first envisaged an extension of the royal customs quay upriver from London Bridge. This plan had ever since been called Christopher Wren's "forty-foot way." Wren's plan failed of realization because of repeated encroachments and arguments among wharfingers over right of way. By 1762, little of the original line remained, and the act was eventually repealed.[3] Robert Mylne's Blackfriars Bridge of 1767 included an embankment extending about 1,000 feet upriver to Temple Gardens and 100 feet downriver from the bridge abutment, effectively moving the shoreline about two hundred feet out into the river. One reason given for the extension was that the irregular line of wharves in the area caused the river's current to shift toward the south bank, leaving mud flats and shoals on the north side to be covered with refuse from the local sewers.[4]

All sorts of private embankments, wharf walls, and terraces had been licensed over the centuries. Before 1660, a line of aristocratic residences to the west of the old Somerset House—Exeter House, Worcester House, Suffolk House, and Whitehall—displayed a fairly straight sequence of individual river walls and stairs, but from Somerset House down to Blackfriars there was a hodgepodge of irregular house walls, warehouses, steps, and wharves. When Somerset House was rebuilt, in 1774–84, by Sir William Chambers, it was equipped with a broad, fifty-foot terrace raised on high arches over an embanked quay.[5] The line of the terrace extended that of the Adelphi Terrace, built just to the west by the Adam brothers in 1768–72. In 1814, it was linked to the east by the embankment for Waterloo Bridge. Up toward Westminster in early modern times, Whitehall Gardens was edged by a low river wall that ran past the Bowling Green and the Privy Garden, interrupted by the Whitehall Palace landing stairs. The old Houses of Parliament projected into the river on a stone wall that ran up to landing stairs next to the first Westminster Bridge. However, the new Houses of Parliament, built between 1837 and 1852, had a 1,200-foot embankment; half the ground for the "New Palace" was reclaimed from the river.[6] Finally, embankments fronted Millbank Penitentiary and the Grosvenor Estates west of Parliament (1813–30).[7]

The south banks of the Thames were much less developed. The land was low and liable to flooding, and there were few mansions or public buildings. Only Lambeth Palace and Doulton's pottery works stood out across from Westminster, and contemporary prints show their foreshore mostly unembanked. Wharf pilings and piecemeal flood walls made up the bulk of construction.

Despite the constant addition and modification of embankments along the north side of the Thames, and the efforts of the City of London Navigation Committee to bring them into line, the general effect remained uneven. Walls were generally built close to the high-water mark, following the undulations of the shoreline, and they were broken by landing stairs, temporary piers, wharves, and dock entrances. The mills and warehouses upstream from Parliament and downstream from Somerset House stored excess coal, stone, or other goods on barges anchored nearby, while lumber merchants allowed large flotillas of logs to build up alongside their premises. The jerry-built floating docks and causeways of the steam ferry companies added a finishing touch to the agglomeration.

The initial purpose of embankments was either to prevent flooding or to improve navigation on the river by straightening its shoreline and reducing the accumulation of mud and debris. It was obvious to engineers and government officials alike that, if the Thames could be narrowed and its banks made more regular, its current would flow more swiftly, and its channel would deepen and stabilize. The wharfage line proposed by James Walker, in 1840 (map 2), was approved both by the City of London Navigation Committee and by the Admiralty with that goal in mind. Improved navigation continued to be a prime purpose for the Embankment, even after other purposes had been grafted onto it. But the likelihood of obtaining an effective line of embankments through piecemeal licensing of individual proprietors clearly had been disproved in the case of Wren's forty-foot way, and it was disproved again in the case of Walker's line. The alternative was to replace the existing waterfront with an entirely new structure, or line of structures, designed from the ground up. Walker himself proposed to embank the length of the river from Vauxhall Bridge to London Bridge on both sides with nothing more than a line of pilings, which would be filled in on the land side and faced with a uniformly sloping bank on the river side.[8] Barges could lay up on the slope at low tide and float easily off at high tide, as they do today. Other men were more visionary. They saw that a level embankment could support new wharves, office blocks, warehouses, even residences, and could be designed to turn the riverside from a sewer into the urban equivalent of a landscaped park.

One of the earliest of these visions was published, in 1766, by John Gwynn. It is outlined by Felix Barker and Ralph Hyde in their interesting collection of maps, designs, and descriptions called *London as it Might Have Been*, which devotes a whole chapter to "the Thames quay."[9] Gwynn planned a line of quays, topped by houses and commercial buildings, on both sides of the river from Westminster to London Bridge. He was concerned more with the overall visual effect than with the feasibility of its parts and made no provision for traffic or sewage. The design was probably linked to the construction of Blackfriars Bridge (1767) with its plans for embanking part of the river; but nothing came of it. A similar plan was proposed at the same time by John Lacy, who had developed the Ranelagh Gardens near Chelsea. He imagined a raised roadway with warehouses underneath and elegant houses above, extending from Hungerford Bridge down to Blackfriars, with Mylne's new embankment included.[10]

A more ambitious and certainly more celebrated attempt to embank the river was made by Lieutenant Colonel Sir Frederick Trench (1775–

1859).[11] Sir Frederick was an Irish Protestant who represented Tory Cambridge in Parliament. He had already published designs for a giant pyramid to cover Trafalgar Square and a new royal palace for Hyde Park when, in 1825, he announced plans to embank the Thames. The timing of this venture was evidently inspired by the recently announced plan to rebuild London Bridge. Sir John Rennie, engineer for the bridge, advised Trench on his project and prepared general plans for him. In his *Autobiography*, Rennie also describes a "Committee of Taste" set up, in the early 1820s, by Sir John Soane, Sir Robert Smirke, John Nash, and others who thought London looked aesthetically wretched compared to Paris or St. Petersburg. This "Committee" had planned Trafalgar Square and helped improve the Strand, Oxford Street, and Holborn with new architecture.[12] John Nash had also designed an arched embankment for the north side of the Thames but had abandoned it for the Regent Street development. Given this background, it is hardly surprising that Trench's proposed embankment, stretching from what is now Northumberland Avenue down to Blackfriars Bridge, would have a unified, classical facade of cream-colored stone supported by a continuous archway. It would clearly have linked the Thames architecturally to the rest of Regency London.[13]

Trench's prospectus claimed that the new embankment would counteract the navigation problems to be expected from the removal of London Bridge. It would also advance commerce and support roadways for free and rapid land transport between the City and Westminster. It also envisaged "an airy, healthful, and beautiful Promenade, for all Classes of the Community."[14] Finally, despite the promise of free transport, it claimed that the company would make money from rents and tolls. Whether this last claim was realistic remains a question. Unlike other Embankment visionaries, however, Trench formed a joint-stock company to promote his scheme and persuaded the Duchess of Rutland, a patroness of art and of new town planning, to join the board of directors. Other aristocrats joined as well, including the Duchess's lover, the Duke of York. Trench then applied to Government for a subsidy to get the project started, confident in his ability to recoup expenses through leases.

Although the City of London Navigation Committee approved his plans, and Lord Palmerston was one of his leading supporters, Government refused to commit tax revenues to the venture. It was also denounced by aristocrats who lived along the river and by Strand shopkeepers who feared a diversion of business. When the Duchess of Rutland died, and then the Duke of York, the project fell through.[15]

The recruitment of peers to sponsor commercial and industrial ventures was not new with Colonel Trench, and it continued throughout the Victorian period. In the 1850s, the practice started to be criticized as a kind of fraud revealing the public's "feeble worship of aristocracy." The *Economist* snorted that "the aristocratic puff is tried when no other puff would have much chance of success."[16] As private partnerships and joint-stock companies gave way to the modern incorporated firm, and the great agricultural depression of the 1870s forced the landed classes to look for other sources of income, company directorships began to take the place of such sponsorships.

Barker and Hyde present Trench's embankment as one of a group of failed schemes to remake London, making it seem eccentric and irresponsible. But, in the contexts of Nash's Regent Street and Rennie's London Bridge, it made quite good sense. James Walker later testified that Trench had taken him down the river to indicate the project's features, and he found no problem with its general design. All the right people were in on it, and it reflected the best thinking of the day. Its central problems were two: it was a private venture depending for its success on questionable projections of rents and tolls, as well as a government subsidy; and it aroused the opposition of powerful interests along the Thames shoreline.

Many of the plans put forward in the 1820s and 1830s aimed at architectural grandeur, a kind of facade for the river's edge that would hide its workaday ugliness under layers of marble or decorative ironwork. Most of the schemes involved a mix of office or residential blocks combined with renovated wharves and warehouses. Few dealt realistically with the problems of barge and lighter traffic or the consequences of channelling the flow of river currents. Most assumed that the costs of construction would be borne by government subsidy or joint-stock issue, recovered through property rents.

Because the City of London was exempted from the Municipal Corporations Act of 1835, and later from the Public Health Act of 1848, there was no metropolitan authority capable of studying or authorizing improvements outside the square mile of the Corporation itself. Improvement projects, therefore, became the concern of a series of Parliamentary Select Committees and Royal Commissions on Metropolitan Improvements, as described in chapter 4. These temporary bodies studied problems of street building and maintenance, housing and sanitation, gas and water supply, open spaces, bridges and railway terminals, as well as sewer drainage and embankments. As they pursued their charges, both members and witnesses perceived functional relationships among the several types of improvement. A new street, for instance, could also carry a new sewer line; bridges would affect traffic and might require new approach routes; both sanitation and passenger ferry facilities were linked to riverside restructuring. In this way, the Embankment, which started as a simple plan to improve navigation, or as an architectural display, became a magnet for other schemes of metropolitan improvement.

One difficulty of explaining the agglutination of other schemes is that they did not attach themselves to the Embankment in chronological order. Some appeared very early, like Trench's architectural design, only to disappear before the Embankment was seriously considered. Others, such as a provision for gas and water lines, were added almost as an afterthought. The exact combination of features was in doubt for a very long time. This may be seen by comparing two memoranda. The first, privately circulated to members of the Select Committee on Metropolitan Improvements, on 6 July 1843, asked several questions: Is the proposed Thames Embankment meant only to improve navigation, or for other objects? Should the river be embanked on both sides for the whole distance from Vauxhall to London Bridge, given the experimental nature of the project? Should it accommodate existing wharf traffic? Could the top surface be used to satisfy the demand for new thoroughfares? The writer suggested priorities: navigation first, then accommodations for trade, and, finally, the provision of an east-west thoroughfare.[17]

The second memorandum was also a set of questions, returned to the Metropolitan Board of Works, in 1856, by Chief Engineer J. W. Bazalgette, in response to an invitation to expand on his suggestion for an embankment. Did the board want a simple report on the possible inclusion of the low-level sewer line, or a "broad and comprehensive view of the whole subject?" Should he include the south side of the Thames? What proportion of the Embankment should be devoted to metropolitan traffic? And "Should the primary object of such design be public convenience, or should it rather be so constructed as to render it, to some extent, remunerative?"[18] Each possibility had social, political, and financial implications and involved a different conglomeration of interests. As they were articulated in two decades of meetings at the MBW, the City of London Common Council, the First Commissioner of Works, and various parliamentary committees, they began to overlap and influence each other. We can sort them out in rough chronological order, mindful that they came and went on the stage of public discussion like actors in a play.

The architectural splendor of early plans like Trench's assumed that London was a place of seasonal residence for the aristocracy and gentry, and a permanent home for the middle and working classes. But already suburban housing estates and the availability of transport were moving people who could afford it out of the central city, even as its overall population grew. When plans for embankments reemerged in the 1840s and 1850s, a different purpose for erecting buildings along the river became evident. Lionel Gisborne, an engineer interested in riverside improvement, explained the change to the First Commissioner of Works in a letter of 18 August 1853. Many have complained about plans to erect buildings along the Thames, wrote Gisborne; the people want a public esplanade as in Paris or St. Petersburg. But the omnibus and the railway have made it possible for merchants to live outside of London, turning the city into a giant warehouse and business office. Thus there is no need to reserve space for residents, or embellish the Embankment for urban beauty or recreation. Improvements should recognize the trend toward utility in urban planning.[19]

It is evident from an inspection of the plans later submitted for the Thames Embankment that many architects and engineers agreed with Gisborne. Government ministers such as William Gladstone also conceived of the Embankment largely in terms of ground rents to be charged for new buildings. But the people who still lived in London had other ideas. The translation of the garden-park concept from the countryside to the city had hardly begun, but already residents thought the land to be reclaimed by an embankment should be developed as open space. Based on their belief in the miasmatic theory of disease, they regarded the Thames as a kind of ventilation shaft bringing fresh air into London from the sea.[20] Their voices were barely heard, however, until the Embankment was actually under construction, and uses for the reclaimed land became an urgent issue.

In the meantime, commercial buildings had to compete with other possibilities. In the 1840s, as another railway boom began, it seemed likely that an embankment would support one or more rail lines. Prince Albert had ridden a train in 1839, and Queen Victoria rode from Windsor to Paddington in 1842. L. T. C. Rolt claims that the Queen's journey marked the start of royal and public acceptance of railways as against roads and canals.[21] But by that time lines had proliferated throughout Britain, bringing railway terms into everyday language. There was not yet a comprehensive rail network. The short-lived empire of George Hudson crashed in 1848, and Gladstone's railway regulation bill of 1844 was emasculated by the railway interests. But many of the new lines terminated in London. Although the ordinary Londoner did not take to commuting by train until later (only about six thousand people rode the local and suburban trains to work by 1854), and though the "boom" in railway financing and construction slackened in the 1850s, prospects for expansion were good and expectations were high.[22] Suburban growth to the northwest and southwest of London made it appear commercially feasible to bring a line into London by using the Thames foreshore as a right of way. Already, in 1841, the ever-hopeful Colonel Trench, assisted by engineer George Parker Bidder, had submitted a bill to Parliament for authorization to construct an elevated railway along the river. His bill was rejected by Peel not because of its limitations, but because of its wildly inflated estimates of profit.[23]

In 1853, about the time that Lionel Gisborne was arguing that the railway had made residence in London obsolete, a group called the Thames Railway Company, headed by engineer Hamilton H. Bird, submitted a bill to authorize a rail line along an embanked north shore of the river, bringing his trains right down to Blackfriars Station. Other projectors followed his lead. By 1859, there were dozens of private railway bills before Parliament, so many, in fact, that Gladstone at the Board of Trade took to combining them into public bills for expediency. Even after the Embankment was approved and construction begun, new railway bills were deposited. Stanford's map of London for 1863 shows lines projected along the Embankment site by the Metropolitan Grand Union, the Metropolitan District, the London Union (to be built along the older Chelsea Embankment and Cheyne Walk), the "London Low Level" railway, and even a Thames Viaduct Railway to be built on arches.[24] Of these, only the Metropolitan District Railway, the shortest of the projected lines, was ever built (figure 6). The obstacles to an Embankment railway were formidable. The traditionally obstreperous wharfingers, warehouse owners, and ferry companies were, in this case, joined by the barristers who guarded Crown or City rights over the river. As G. P. Bidder complained, recounting his collaboration with Colonel Trench, as well as his own projects:

> The complaints against Government interference were but too well founded, as was unfortunately felt, not only by the City, but by all who had Bills before the House, for any works on the banks of the Thames. The Bill might go on very smoothly, until it was nearly granted, when a sleek gentleman made his appearance with an application "merely for the insertion of a little clause for the protection of the rights of the Crown," which meant, that all plans should be submitted to, and be approved by various heads of "Circumlocution" Offices; the inevitable result being, that the unlucky wight seeking a Bill, wasted his time, and then had to pay a good round sum for some imaginary right supposed to be conceded to him, for doing something which could be of no earthly use to him, and for proceeding without let or hindrance, to some place, whither he had no intention of, or interest in going.[25]

Hamilton Bird agreed with Bidder's complaint when he testified before Parliament in 1861. He had resubmitted his bill for the Thames Railway Company every year since 1854, but he had been stymied by the sheer number of conflicting interests involved. Projectors did not have enough information about these interests beforehand and were thus incapable of satisfying government reservations and winning the required legislative backing.[26]

Although Londoners recognized sewage and river pollution as problems by the 1830s, they did not connect them with the concept of the Embankment until later. Even then, the prevailing assumption was that, if the river were embanked, sewage would more readily be carried downstream instead of building up along the edge. Trench's scheme called for existing sewers to empty into tidal locks or basins, which would flush their contents into the river at ebb tide.[27] John Martin's three-level embankment included intercepting sewers, but they, too, discharged between Westminster and Blackfriars.[28] The first plan to carry the sewage out of London altogether was that created at the MCS by Edwin Chadwick, which was then modified and adopted by Bazalgette and the other professional engineers. The first contracts on the London Main Drainage were let in February 1859. By the end of 1861, when the Royal Commission on the Thames Embankment finished its deliberations, the high-level lines were finished and the mid-level lines well under construction.[29] The low-level sewer, as we have seen, could have been laid along several routes, but the Embankment was the preferred one.

Traffic congestion made an embankment between Blackfriars and Westminster even more appealing. A traffic survey, in 1855, was said to have counted 200,000 people travelling from Charing Cross to the City in cabs and omnibuses in one day, and most of those were confined to Fleet Street and the Strand.[30] There was no other east-west thoroughfare of comparable volume (the Holborn Viaduct to the north was not built until 1864–69). The first Commons Committee on the Thames Embankment (1860) was, therefore, charged primarily "to consider the best means of providing for the Increased Traffic of the Metropolis." It joined the ranks of previous commissions and committees on metropolitan improvements and inherited many plans for new streets projected over the years. Street improvements for Belgravia and Westminster were already under construction, along with Lambeth Bridge. The Grosvenor Road fronting Pimlico and Millbank was in place, and the problems associated with surveying new thoroughfares and fairly compensating property owners were understood.

As one can see today, roadways took the place of railways as the transportation mode of choice for the Embankment. Few direct comparisons were made by contemporaries in public debate, but reasons for preferring roadways can be easily imagined. Roads were cheaper to build and easier for a public Board of Works to control (the railways were all private ventures). They drew on more immediate populations in suburbs and City. They would be linked easily to the Strand and other established streets. A roadway could give access to warehouses situated on an embankment despite cross traffic, whereas railway tracks would create impediments. Finally, as Donald Olsen reminds us, "to people of taste and discernment, roads were good, railways bad.[31]" Railways endangered health, ruined architectural vistas, and interrupted all other forms of traffic. The forty-five-acre clearance for King's Cross Station, in 1852, and the new town created around the Great Western terminus at Paddington disrupted London settlement patterns, boosted property values, and reconfigured the urban landscape with extravagant, monumental buildings.[32] As the Embankment became more and more a public project rather than a private venture, the likelihood that it would be used for a railway diminished. Only the Metropolitan District line, discreetly hidden underground, gained admission.

Many London commuters favored the ubiquitous steam-driven passenger ferries that had plied the river since the late 1820s. The traffic survey cited above noted that the "penny packets" carried thirty thousand people a day east to west through the metropolis and across the Thames. During the year 1861, over three million passengers landed and embarked from one of the main piers on the boats of just one company.[33] The ferry companies were notorious for deceptive, cutthroat competition. Their rickety piers, reached across the muddy shores by way of labyrinthine plank walkways, were crowded and confusing. Rival steamer crews routinely hustled passengers onto their craft by calling out false destinations.[34] Due to the tides and the lack of adequate landing sites, some of the loading platforms were located alongside bridge piers (as at Waterloo Bridge) to which they were connected by stairs (figure 8).[35] The Embankment was expected to provide new, easily accessible facilities for safe, orderly passenger loading.

All of the above functions were considered by the three main agencies for London improvements: H. M. Office of Works, which had jurisdiction of all Crown territory, including the bed of the river; the Thames Navigation Committee of the City Corporation of London, which claimed authority over the foreshore of the river and matters concerning navigation; and the new Metropolitan Board of Works. The Office of Works, as we saw in the previous chapter, had been building a section of embankment along the Thames near Chelsea, but had faced difficulties financing it through local property assessments. Works had conducted a long running feud with the City Navigation Committee over rights to develop the banks of the river. The MBW, for its part, wanted desperately to solve the problem of the low-level intercepting sewer line, but it lacked authority and funds to proceed on its own.

After considering several multifaceted schemes submitted to it in the late 1850s, the MBW developed a list of objectives for a Thames Embankment: (1) to improve the Thames as a navigable river, with due regard to the safety of the existing bridges; (2) to increase wharfage accommodation and shipping; (3) to improve the river's sanitary condition; (4) to improve the appearance of the "unsightly" river banks; (5) to open a new east-west thoroughfare; and (6) to facilitate the construction of the low-level sewer. Under the leadership of Chairman John Thwaites, the MBW also argued that commercial development of reclaimed land, a feature of most private schemes, should be avoided. "A work of such magnitude and of so peculiar a character," wrote Thwaites, ". . . ought not to be left to commercial enterprise, but should be carried out by a public body and by means of public funds, and conducted solely with a view to the public advantage."[36] David Owen comments that the MBW may have been concerned to keep control of the Embankment project in connection with its work on the Main Drainage, but its belief in public oversight was ultimately correct.[37]

The rival agencies, and a host of other interested parties, finally realized that the development of the river front was far too complicated to be sorted out privately among themselves. The Commons select committee headed by Joseph Paxton, in 1860, also decided, after taking evidence from over two dozen witnesses with contracts and plans in hand, that the Embankment was a problem exceeding its charge. There were too many schemes to be considered in one parliamentary session, and the legal, technical, and financial difficulties were enormous. Paxton recommended the appointment of a Royal Commission. However, the committee did support the idea of embanking the north side of the river from Westminster Bridge down to Southwark Bridge, with a roadway on top to relieve traffic congestion, and a sewer within. It accepted Thomas Page's idea, now part of other plans, that the Embankment should extend parallel to the shore, with openings to allow barge traffic access to the wharves inside. The improvement of the wharfage and warehousing facilities would yield higher rents and help repay the cost of construction.[38]

The Royal Commission appointed in 1861 issued a general call for proposals in November, listing in the official *Gazette* the main requirements as indicated by Paxton's select committee. By this time, the Embankment was known throughout the engineering community, and more than fifty designs were submitted. Dozens of witnesses appeared to defend their plans or to offer miscellaneous suggestions. The commission might have benefitted from an independent critique of all these proposals, as was done on other public projects, but, as Hamilton Bird commented, it was hard to find any of his colleagues who had not already submitted one.[39]

Because the published criteria accurately reflected years of discussion, the plans presented in 1861 were roughly similar to one another. Most featured a roadway built on arches above an enclosed sewer line, and many had tidal gates fitted into the archways to accommodate barges. Some proposals tried to encompass all manner of public and commercial objectives in large, complex projects. Others, usually drawn up by amateurs or men trained in peripheral fields, emphasized a single device or an architectural design. It is worth looking at a selection of these proposals to see contemporary visions of beauty and utility and to appreciate what the commission, like any other public body holding open hearings, had to deal with. For example, Thomas Weller, who owned an iron works in Wales, proposed a double-decked viaduct with roadway on top and railway below, all made from cast iron. Like many other engineers of his time, he believed ironwork was aesthetically more pleasing than nature itself. "In front of Temple Gardens," he enthused, "there would be nothing but a beautiful viaduct."[40] A long-time Thameside resident named Edward Walmsley, "born on a wharf and nursed and bred on a wharf," presented plans for a "Crystal Arcade," an iron-and-glass pavilion spanning a solid quay set out from the shoreline, filled with shops and galleries like the Crystal Palace itself.[41] A rope-pulled railway operating with a stationary engine was featured by John Fordham Stanford, whose plans were originally drawn for a company formed in 1841.[42] Benjamin Bush proposed to install sewage filtration beds within the embanked area to receive the low-level sewage and pump "pure water" back into the Thames. The extracted solid waste would be processed and sold as fertilizer, as other firms had already proposed. Ernest E. Gohns also had a plan for sewage disposal built into his embankment design, but "It is at this moment my secret."[43]

Members of the Institution of Civil Engineers who served on the Royal Commission could be hostile to surveyors or architects claiming the status of professional engineer. "Outsiders" and obvious amateurs were often treated with polite condescension. Architect Charles Henman introduced his design for a double-level roadway with lengthy digressions into finance and urban development. These evoked no response from the commission, but Chairman Cubitt graciously complimented him on his "beautiful designs." "They are all very nice drawings indeed," he repeated, "showing a great deal of talent and inventive genius."[44] Cubitt greeted another witness wryly, "Well, Mr. Bardwell, you have been about your plan a long time?" William Bardwell, it seemed, had carried his plans around to any official who would listen to him, including the Lord Mayor. Such visits were common in an age where patronage was still competing with more professional and bureaucratic processes. Other witnesses were less ingenuous. When the Mayor pointed out that Harry Newton's scheme for a double embankment supporting government buildings went far beyond the commission's instructions, Newton refused to answer any more questions and stalked out.

Of the many proposals entertained by the Royal Commission of 1861, two were given serious consideration. Each was presented by a prominent London engineering partnership, and each was linked, in curious ways, to a companion plan. The first proposal, as described by Hamilton H. Bird of the firm of Loder, Jackson, and Bird, would create a solid embanked roadway with an underground railway from Westminster to Southwark Bridge, with provision for new wharfage. It did not have the required sewer line, because Bird assumed the MBW would opt for the Strand-Fleet Street route. Bird's plans had been before various public offices since 1854, usually paired with a similar design by Lionel Gisborne, who had suffered such frustration at the hands of "circumlocution offices." Gisborne's version would add a roadway to the south side of the Thames from Vauxhall to Blackfriars Bridge, and on the north side a glass-covered walkway along a row of shops, again reflecting Paxton's Crystal Palace. Gisborne promised that the reclaimed land would yield at least a 7 percent return after being "developed," so the Embankment would be built at no cost to the public.[45]

After the MBW organized its operations in 1856, Gisborne and Bird submitted their plans to its new Committee of Works. Gisborne had now added a series of cellars and warehouses in the arches under his elaborate promenade, which, if added to wharfage fees and railway passenger fares, was calculated to generate £100,000 per year in income. The MBW reject-ed this plan as too commercial.[46] Gisborne and Bird incorporated their ideas in Parliamentary bills introduced annually from 1854 through 1860, and Bird appeared again with his plans before the Royal Commission of 1861. By that time, however, John Thwaites had convinced the commission that the Embankment should feature a broad, open roadway and recre-ational gardens, and so the commercial plans were finally abandoned.

The second prominent Embankment plan was submitted by the firm of Fowler, Fulton, and Hemans. John (later Sir John) Fowler and Hamilton H. Fulton were prominent railway engineers, and George Willoughby He-mans had worked with them on many projects. Their Embankment would support both an eighty-foot roadway and a subterranean railway, with openings for tidal docking only at Hungerford and east of Somerset House. The roadway would rise on a viaduct up to the Blackfriars Bridge approach road, and the enclosed sewer would connect at that point to an unspecified new line in the Main Drainage system.[47] The Fowler, Fulton and Hemans plan is interesting because it anticipated the roadway, sewer line, and viaduct of the later Embankment and reduced wharfage facilities to a minimum. It indicates that, by 1860, a consensus was forming among leading engineers and officials about essential Embankment features. Al-though Fowler's railway would not be part of the Embankment design au-thorized in 1862, it would reappear two years later, approved by a private (non-government sponsored) Act of Parliament.

Hemans's alliance with Fowler and Fulton in 1861 is curious because, a year earlier, he had submitted a similar design in partnership with MBW Chief Engineer Bazalgette. Most of the London engineers and engineering contractors formed such ephemeral partnerships among themselves to bid on specific projects. It was possible for a single engineer to be involved with a dozen or more at one time, some of them competing for the same contracts. John Fowler, for example, served as consultant for about ten different railway companies planning extensions into London.[48]

Bazalgette, as noted earlier, was questioned about serving as Hemans's partner while engaged as the chief engineer of the MBW, and he did not participate in the competition before the Royal Commission in 1861. His plans, however, are of more than passing interest. Like Fowler and Fulton, Bazalgette and Hemans would have enclosed the low-level sewer line within a concrete quay set at the level of water at low tide. Enclosing the sewer within a wall was preferable to simply sinking it into the Thames riverbed, because the wall channelled the river and allowed dredging for mud and gravel shoals, thus aiding navigation. A roadway would be raised on cast-iron arches above the quay, parallel to the shore from Westminster to Blackfriars. Barges could float underneath the arches and across the quay at midtide, to access the wharves and warehouses lining the tidal "pools" between quay and shore.[49] Cranes mounted on the edge of the upper roadway could unload barges directly from the river. The roadway was intended for both commercial and passenger traffic.

A roadway raised on arches had been a feature of other Embankment plans, but Hemans pointed out to the select committee of 1860 that his, like Thomas Weller's, was constructed of open ironwork and devoid of interior buildings, so that light and air could pass through and riverside tenants could still view the river. He defended the aesthetic quality of the plan, reporting that the new library at Temple Gardens had been built higher than originally planned in anticipation of a raised embankment. A drawing of the proposed structure, decorated with barges, cranes and coaches, was included in the Royal Commission report (figure 4). Some Victorians, who considered wrought iron bridges almost more beautiful than the rivers they crossed, might have agreed with Hemans, but, compared to the clean, monumental lines of the solid embankment actually built two years later, his design was an aesthetic disaster.[50]

As mentioned before, many of the plans submitted to the Royal Commission of 1861 provided for substantial commercial or public office facilities, all of which were intended to generate rents and recover the cost of the Embankment. Even plans that reflected the Commission's preference

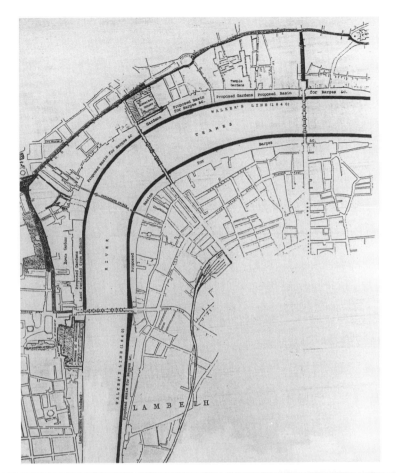

Figure 4. Original designs for the Thames Embankment. *Top:* The "open" quay proposed by Page, Bazalgette, and others, 1844–1860. The wall would parallel the shoreline, with barge access to wharves at high tide. Gardens would be formed on reclaimed land at Whitehall. *Bottom:* Cast-iron arched roadway proposed by Bazalgette and Hemans, 1860. Barges would be unloaded by cranes and goods carried across the roadway to riverside wharves. (House of Commons *Sessional Papers* 1857–58 (442) 11 and 1860 (494) 20. Redrawn by Christine Porter and Dale Porter.)

for public rather than commercial development assumed that the existing wharves and warehouses would be renovated and served (for fees) via new docking and handling facilities. One impetus for this assumption, of course, was the uncertain state of financing for new public projects. Under Gladstone, first at the Board of Trade and then at the Exchequer, tariffs, fees, and other burdens on commerce and industry had been severely curtailed. Eighteen-sixty was the year of the Cobden-Chevalier treaty liberating trade with France. Government subsidies for public projects were correspondingly frowned upon. Improvements should be made to pay their own way. Most of the engineers and contractors who proposed Embankment schemes followed Government's lead. They were used to commercial projects and designed this one the same way.

Public or private, Embankment plans faced a stiff charge for compensation at the outset. The wharfingers and warehouse proprietors lining the Thames from Westminster to Blackfriars had vociferously and litigiously protected their property rights during every previous attempt to improve the shoreline. Aside from residential properties at Whitehall Gardens and Adelphi Terrace, the north shore of the Thames was home to nineteen wharves (two of them large complexes), a mechanical engineering firm, a draw dock, and the City of London Gas Works.[51] Most of the wharves supplied timber for building and coal for heating London houses. They were attended by some 150 colliers and lighters moving 350–400 barges on an average day. Proprietors demanded open river frontage with easy access, although they were in the habit of storing unused barges and large rafts of timbers in the Thames, to the point of impeding river navigation.[52]

One has only to recall the fate of Walker's line of 1840, drawn in the hope of voluntary compliance, to see how reluctant public officials were to confront this phalanx of interests. Any plan that envisaged eliminating the wharves would not only give up the promise of tax revenues but would also incur unknown costs for compensation. All of the private bills for embanking the Thames in this area which had been submitted to Parliament in the 1850s were abandoned due to the opposition of the wharfingers, and the engineers involved all agreed that compensation was a legal nightmare.[53] The Admiralty's engineer, one of many officials who reviewed Embankment plans, drew up a contingency schedule to prevent encroachments on the river by proprietor-built piecemeal embankments, but he could do nothing more.[54] Wharfingers interviewed by the Royal Commission of 1861 gravely announced that their livelihoods would be utterly ruined and valuable trade lost if the Embankment were built without considering their needs.

However, between 1840 and 1860, two significant changes had occurred in the infrastructure around London. First, new docking facilities downriver from London Bridge were claiming more and more of the coastal trade. The West India Docks (1802) had started the movement, but they and the East India, London, St. Katherine's, and Surrey Docks (all built by 1828) were still basically extensions of the original Port of London. The Royal Victoria Dock, built in 1855, however, initiated a substantial shift.

Second, the railways had arrived. We have already seen how many proposals for embankment involved railway lines into the city. By 1850, there were six thousand miles of track in use around Britain, with more under construction. The Great Exhibition of 1851, with its cut-rate excursion trains, made railway travel immensely popular. Financial difficulties in the 1850s slowed the pace, forcing consolidations, but, by 1860, another construction boom was gathering steam. It was initiated more by independent contractors like Sir John Fowler than by established railway companies, and the aim was to fill out existing routes, especially around London, rather than to build new main lines. The volume of freight traffic now equalled or surpassed that of passengers, and goods could be booked through different lines of the same gauge. London was connected with Birmingham, Manchester, and Liverpool by the London and North Western from Euston; to the north through Peterborough by the Great Northern; to the west by the broad-gauge Great Western and the standard-gauge London and South Western; and to the south by the London, Brighton and South Coast, as well as the London, Chatham and Dover lines, furiously (and ultimately ruinously) competing for business. Michael Robbins believes these lines were integrated and supervised by Parliament to a degree sufficient to call them a system.[55]

The first load of coal was delivered to London by rail in 1845. Thereafter, as the railways crept closer to the east end of the metropolis, they began to bring coal, hay, and other bulk goods from the new downriver docks. Less was carried up the Thames to Whitefriars, Hungerford, and Millbank by colliers and lighters. Coal companies began to consolidate, requiring larger premises, and to market more to gas companies and large wholesale merchants, who chartered their own ships rather than buy from independent colliers as before.[56] An act of 1854, restricting the more noxious manufacturing processes within the metropolis, led many riverside proprietors to move eastward towards the River Lea.[57]

The south side of the Thames developed differently. Railway lines there were built to cross the river rather than parallel it, and the extensive wharves were interspersed with industrial and commercial buildings of much greater size and density than on the north. For example, after the Victoria Embankment was authorized and under construction, in 1865, George Locket Jr.'s coal firm at Northumberland Wharf in Hungerford Reach amalgamated with two other companies and moved to the south side of the river on the Jamaica Wharf near Blackfriars Bridge.[58] Paradoxically, then, as railway promoters were drawn to the concept of the Thames Embankment during the 1850s, seeking to push their lines ever closer to the heart of London, they also made it feasible to remove the existing wharf and warehouse trade along the proposed Embankment site from Westminster to Blackfriars.

Replacing the wharves with a solid embankment had two technical arguments in its favor. One was simply that a continuous wall built along the shoreline without consideration for existing facilities would be easier and faster to construct. More immediately important was the fact that wharf pilings, even protected by an outer embankment wall, encouraged the buildup of mud and gravel that impeded navigation. This was the reason for proposing an embankment in the first place. Constructing a solid quay up to low-tide level, as a cover for the sewer line, only solved part of the problem. Bazalgette's scheme of 1860 and the Fowler-Fulton-Hemans scheme of 1861 still envisaged openings over the quay, under the arched roadway, to accommodate barge and lighter traffic in a series of tidal pools along the shore. During the Royal Commission hearings of 1861, Captain Galton, a naval officer from the Thames Conservancy who was in charge of dredging the river channel, asked all of the important project promoters how they would deal with mud and silt around these openings and along the wharves. He pointed out several times that the goal of improving navigation would be met much better by a solid embankment.[59]

Neither the problem of silt nor the increased use of railways would, in themselves, have generated the concept of a solid, public embankment. That concept had its origins in a quite different region of Victorian London life, the region of civic pride, of the culture of "Improvement" as described by Asa Briggs. The rapid growth of the metropolis, its transformation as a financial and political center of a world empire, and its central role in overseas trade made its inhabitants conscious of its world status. It also made them aware of the shabby, provincial appearance of London's streets, public buildings, and waterfront. Unflattering comparisons with St. Petersburg, with Venice and Rotterdam, and especially with Paris, appeared in the newspapers and journals. In 1838, a correspondent asked rhetorically, "Is there, indeed, a capital in Europe with so noble a river flowing through it, whose citizens would not turn it to high account, both to ornament and utility?" Twenty years later, the river was no longer noble, yet the capital had grown even greater. Just as the Great Stink of 1858 finally drove officials to authorize a sewer system, so it also made them aware of the disparity between the metropolis and its wretched riverfront. The late eighteenth century image of the capital as a gigantic "wen" or parasite upon the country was overlaid, in part, by an image of an imperial city whose appearance, especially to foreigners, had economic and cultural importance for the whole kingdom.[60] Many critics, it is true, thought the solution was to build more commodious, modernized wharves and docks all along the Thames through London. Others argued the need for public access to open spaces by the river. They saw an embankment as a symbol of London's imperial grandeur, a promenade upon which to greet the world as it approached the capital.

John Thwaites (figure 5) was a leader of this latter group. Although labelled "a natural product of London matter-of-factism" by a contemporary newspaper, the MBW chairman had a vision of London as a *public* arena, rather than as a congeries of semiprivate corporations and vestries fighting for local advantage and cheap rates. Testifying before the Commons select committee of 1860, Thwaites argued, in line with contemporary opinion, that the proposed Embankment would improve the existing

Figure 5. Fathers of the Embankment: *(clockwise from top left):* Sir Edwin Chadwick, Sir Benjamin Hall, Sir John Thwaites, the Rt. Hon. William Cowper. *(Illustrated London News,* 13 March 1848; 30 April 1859; 24 July 1858; 25 February 1860.)

wharfage between Westminster and Blackfriars. But even then he had claimed a different role for the land to be reclaimed in Hungerford Reach. To give that up to commercial exploitation, he said, would "spoil the Embankment as an Embankment."[61] By the time he was appointed to the Royal Commission of 1861, he had come to the conclusion that the wharves were unnecessary and could be bought out for a manageable sum of money. His questions and comments were aimed at convincing witnesses and fellow commissioners that the Embankment should be a public rather than commercial enterprise, that the "interest and beauty" of the design should be considered as much as its utility, and that London deserved a waterfront fully as noble as those of other great European capitals. Thwaites was supported outside the commission by the *Builder,* whose editorials on the "new London" of the future advocated the building of large, important, conspicuous structures situated on wide thoroughfares on the Parisian model.[62] He also had a key witness under his control: Chief Engineer Bazalgette was an experienced and formidable figure in Parliamentary committees.

It is always curious how a set of assumptions about a public project accumulated over a rather long time can be suddenly overthrown, so that the conception changes quite fundamentally. The new assumptions may have been voiced before, as these were, but on grounds considered weak or irrelevant. Then other grounds appear, having developed in separate circumstances for different purposes; the alternative assumptions are revived and suddenly flourish. In this case, the moment of change can be pinpointed to the 13th of May 1861. Bazalgette, testifying in the afternoon before the Royal Commission, was asked directly by Thwaites whether it was possible to eliminate the wharves between Westminster and Blackfriars and construct a solid embankment. Bazalgette replied that "a recent survey" of the wharves showed that, indeed, such a plan was now feasible. In light of that fact, he had added a new line, in red ink, to his previously submitted plans to indicate the probable dimensions of such an Embankment.[63] Bazalgette's red line (still visible in the original plans) essentially followed the line laid down by Captain Walker in 1840. It is entirely probable that Thwaites had ordered Bazalgette to calculate and draw it after discussion at the MBW the day before. At any rate, Bazalgette's design for a cast-iron arched roadway fronting new wharves served by tidal pools, which had been discussed by the Royal Commission that very morning, was never pursued again. The commission report did not speak in terms of grand architecture and left open the question of building on the re-claimed land. But it did advocate a solid embankment of "interest and beauty" for the specific reason that the coal trade, having moved to larger docks and railway terminals downstream, could be eliminated from the construction site.[64]

From 13 May 1861, the essential features of the Thames Embankment were agreed upon by the Royal Commission after negotiations with Her Majesty's Commissioners of Works, the Metropolitan Board of Works, and a host of civil engineers, architects, contractors, wharfingers, Admiralty officers, and other interested parties. It is worth noting that many ques-tions about the basic functions of the Embankment (as a railway or road-way carrier, as a commercial development, as a wharfage improvement, as a sewer line enclosure, and as civic architecture) were treated more as so-cial and political issues than as technical problems. The questions of clean-ing up the banks of the river and preventing the buildup of mud and silt were dealt with more technically, but they still turned on negotiations among several interested groups and agencies. Laying the sewer line along the Strand and Fleet Street was negated as much by the traffic chaos it promised to cause as by the discovery of quicksand under the route. And the decision to eliminate the wharves in favor of a solid embankment was based on a judgment about the efficacy of the railway system rela-tive to the perceived cost of moving and compensating the wharf propri-etors.

Once these social, political, and economic judgments had been made, the technical design of the Embankment became relatively straightfor-ward. It would be a solid wall of brick and concrete faced with smooth stone to channel the river's flow, with the area behind the wall filled in up to the high-water mark (figure 6). The reclaimed land within this "fill" area would enclose the low-level sewer line from Westminster to Blackfri-ars Bridge, from which it would angle up a new street to the City. The sewer line would intercept all of the drains and sewers snaking down to the Thames from the heart of the metropolis. A new roadway and pedestrian

Figure 6. Cross section of the Victoria Embankment at Hungerford/Charing Cross Railway Bridge, 1867. The underground Metropolitan District Railway line is on the left. The low-level intercepting sewer for the Main Drainage lies within the river wall, with utility pipe vaults on top. The failed Whitehall-Waterloo Pneumatic Railway tunnel crosses under the Thames riverbed at bottom. *(Illustrated London News,* 22 June 1867.)

walkway would be built on top and connected to the appropriate streets at either end.

Although these features required a good deal of expertise to design and construct (as Bazalgette was to remind critics later on), they were not great specimens of technological innovation. The Embankment as a whole could have been built to a similar design at almost any time in the previous half century; except for one or two developments during construction, it involved engineering methods already tried elsewhere and well known in the profession. As a public works project, it is more important for its social and environmental aspects than for its technical problems. Even after its general features had been established by the Royal Commission of 1861, the meaning of those features to the various institutions and groups involved in the Embankment's origins remained open. How smooth and solid would the wall actually be? To what other uses could the reclaimed area be put? Where would the new streets go? What type of pedestrian activity should be accommodated? Just what is a "public" project? As we will see in the following chapters, the crystallization and "closure" of the arguments regarding the Embankment's design did not bring an end to the public discussion of its significance and purpose.

CHAPTER VI

The Financial and
Institutional Environment

When the Royal Commission on the Thames Embankment decided, in
May 1861, to recommend the construction of a solid embankment, it left
unsettled some very important questions. Who would direct the project
and how it would be paid for? What would be the terms of compensation
for the wharves and warehouses to be eliminated between Westminster
and Blackfriars Bridge? If there was to be a new roadway atop the Embankment, how would it be connected to the Strand and other established
streets? And what would be the legal status of the land reclaimed from the
riverbed, whose potential for development excited the public's imagination?

Technical features of the Embankment cannot be separated from these
institutional, legal, financial, and policy questions. They are as much a
part of the human ecology of the project as the changing attitude toward
sewage and the traditional passion for property rights. For example, if the
Embankment had been designed to accommodate a railway, it would have
been built by a private company, because the government did not finance
such commercial ventures. As soon as it was decided to create a public
thoroughfare along the riverside, the Metropolitan Board of Works became an obvious candidate to supervise construction.

Private institutions and groups played active roles in the definition of the Embankment. The MBW was linked closely to the London engineering community and to its professional "club," the Institution of Civil Engineers. The engineers, in turn, worked hand in hand with a select group of major metropolitan contractors who had the resources necessary to undertake such a large project . As we have seen, riverfront proprietors organized to resist the improvement and hired attorneys to represent their interests in Parliament. Other Thames-side residents did likewise. Finally, the "public interest" was represented by certain journals and newspapers and by members of Parliament for London boroughs.

Linking technology to its social contexts means examining each of these institutions and organized groups and sorting out the legal issues they raised, even if they were not initially or obviously crucial to the Embankment design. In the case of a public project like the Embankment, such agents are likely to negotiate through legal and political channels and to leave evidence of their views among the records of Parliament, City, and the MBW. But we must not overlook smaller, more informal groups such as hackney drivers, ferryboat operators, lightermen, and even members of the criminal underworld who stood to gain or lose their livelihoods from the improvement or elimination of Thames-side commerce. These groups also imagined and interpreted the Embankment as it emerged from a vague possibility towards actuality.

It is important to remember that, at the time the Embankment was conceived and built, the variety of official functions associated with the Greater London Council (before Mrs. Thatcher dismantled it) had not yet emerged. The Thames, England's chief commercial river as well as one of its naval stations, was subject to a myriad of local, regional, and national offices, often with overlapping jurisdictions. Almost none had a budget sufficient to underwrite extensive construction projects. During the period from 1840 to 1860, when the first section of the Thames Embankment was being designed and constructed, the lack of a central authority and ready financing created nightmares for promoters. Lionel Gisborne's frustrated attempts to find some agency willing and able to consider his plans (described in the previous chapter) was unusual only in the duration of his quest and the degree of his tenacity.[1] Gisborne must have thought he had been fictionalized in Dickens' *Bleak House,* published in 1853, in which interminable litigation devours the lives of the characters. Death, in late 1860, was his only release. His last effort had been a letter to the MBW claiming assurance, by previous commissioners and members of Parliament, of his right to share in the design and construction of any embankment approved and built by a public body.[2]

Gisborne was not the only victim of confused jurisdictions. All of the early attempts at embankment were throttled by the legal battle between City and Crown that dragged on for some sixteen years. At issue were rights over the bed and banks of the Thames. Both the City Corporation and Her Majesty's Office of Woods and Forests held competing claims to the tidal portion of the river.[3] The Lord Mayor of London was the official Conservator of the Thames according to royal orders going back (he claimed) through William the Conqueror to Edward the Confessor and the Romans. His deputy chaired the City's "Committee for Improving Navigation of the River Thames," which supervised the dredging of the river channel, the removal of mud and gravel shoals and other impediments, the anchoring of timber and barges near the shore, and construction of docks, steps, wharves, and buildings that might impede navigation by encroaching on the river. The Crown, in traditional law, had property rights over the bed and soil (including the foreshore) of all rivers in England and Wales. Such vaguely overlapping jurisdictions are, of course, common in English law, and they may remain benignly muddled unless and until fees, fines, rents, and other charges come into play. Then lines are drawn, claims made, and barristers engaged. In 1770, the Crown tabled a bill in Parliament to grant the brothers Adam the right to embank a portion of the Thames fronting their proposed Adelphi Terrace. The Lord Mayor's counsel appeared to argue for restrictions on possible encroachments on the river. To justify the Mayor's authority, he produced a grant from Henry VII of all the soil and bed of the river from Staines Bridge down into Kent, near the Medway. He also proved that the City had been collecting rent from a riverside lease near Vauxhall for the past sixty-six years. Counsel for the government displayed a charter from Charles II confirming the City's authority over navigation but reserving the bed of the river for the Crown. He contended that the City, by accepting this later charter, had abandoned its former rights to the bed. The House of Lords ruled with the Crown, and the bill passed through Parliament.[4] Despite losing its claim, the City Corporation was reconfirmed in its role as conservator of Thames navigation by an act of the following year.[5]

In December 1830, the City Navigation Committee considered the recommendation of Sir John Rennie and Robert Mylne that the replacement of London Bridge should be followed, after a period of study, by an embankment of the river. Another report, from City clerks Benjamin Scott and James Firth, detailed encroachments by various wharves and charges for repairs in the vicinity of Blackfriars Bridge, built in 1767.[6] It was evident to the Navigation Committee that, if no other agency was prepared to embank the river, they should do it, using revenues from the ancient coal and wine import duties that had long been assigned to the maintenance and repair of London Bridge. At first, they planned only to extend the legal quays along the Pool (the official Port of London) up to Blackfriars Bridge, on both sides of the river. For the district from Blackfriars to Westminster, they hired James Walker to survey a line of possible future embankments and tried to encourage the wharfingers to stay behind it (figure 4).

By 1838, the firm of Walker and Burgess had surveyed the changes wrought by the new London Bridge and had drawn up tentative plans for embanking both sides of the river all the way up to Vauxhall. After negotiations with H. M. Commissioner of Woods and Forests and hearings by several Metropolitan Improvement Commissions, the City petitioned Parliament, in March 1840, for permission to carry out the project. It pledged some £300,000 of its own funds, with the proviso that it retain rights over the soil and bed of the embanked portion of the river, so that it could recover its costs through ground rents and improvement fees.[7] Referred to a Commons select committee, the project was studied until the end of the session without result. The City renewed its petition, in 1841 and 1842, only to have Crown attorneys challenge its claim to the river bed. The argument ended up in the court of chancery, where the issue languished until 1857.[8]

At first glance, the Crown's objection to the City plans seems outrageous. Here was a project deemed necessary and desirable by all concerned, and a local authority ready and willing to carry it out with its own finances. Given the reluctance of both Liberals and Conservatives to expand the influence and expenditures of the central government, they were unlikely to mount a project of such dimensions by themselves. Private schemes were not a clear alternative in these years: Sir Frederick Trench was still hopeful, but his plans were outdated, and the railway speculators had not yet extended their visions to the banks of the Thames. So the City's offer was the only feasible one. Why not embrace it with enthusiasm?

The problem, of course, lay with the revenue that would be generated along the banks of the river by any improvements, especially if additional ground were also reclaimed through embankment. Both the City and the Crown at this time thought of the Westminster-Blackfriars stretch as a commercial district where proprietors could be charged for new facilities, and the recovery of costs in this way was accepted practice. But the Thames Embankment promised more than that: the twenty-seven acres of land to be reclaimed from the muddy shores of Hungerford Reach, the Adelphi, and Somerset House could constitute some of the most prestigious real estate in the metropolis (see map 2). It would not only return huge dividends through leases and right-of-way fees but would also control future development along the Thames. City officials saw it as a way of extending their jurisdiction over the warren of local vestries, paving boards, and navigation interests in the area. Parliamentary ministers were more interested in revenue as a way of reducing the cost of government.

There was an additional possibility, a constitutional wrinkle that perhaps only a legal mind could have envisioned. Charles Pearson, solicitor for the City during this period, argued that what was happening along the Thames was also happening elsewhere in Britain. If the Crown were to retain its traditional monopoly over the bed and foreshore of every river in the realm, while development surged through the exploding cities, the Crown could conceivably accrue enough new revenue to make it independent of Parliament, as it had tried to do before the Civil War. Control over the Thames Embankment, he said, was the "thin edge of the wedge" reopening historic constitutional issues.[9]

Whether or not other participants shared Pearson's alarm, the chancery lawsuit lay unsettled from 1844 until 1857, frustrating the efforts of both public and private agents who tried to promote embankment schemes.[10]

The only construction during this period was Thomas Page's short embankment associated with Chelsea Bridge. The Westminster-Blackfriars project remained at an impasse.

Although the City and the Crown were the two main contestants in this battle, other agencies were also involved. Trinity House, with offices at Tower Hill and Deptford Strand, was associated with an ancient order of mariners. It had the power to regulate moorings, pilings, and various nuisances, but not to dredge out mud and gravel shoals. Its main concern was piloting seagoing vessels through the port. It was the source of the "Trinity Standard" high-water mark, fixed by law in 1800, that guided the placement of wharves and the design of proposed embankments.

The Admiralty office at Greenwich also concerned itself with the navigational quality of the Thames. It routinely reviewed plans for embankments, wharves, and other riverside facilities that were attached to private Parliamentary bills. These were forwarded by the First Commissioner of Works, who had primary jurisdiction over them. The Admiralty did not interfere with public construction projects and had only the right to make suggestions about private plans. But it obviously infringed on the City's claims to jurisdiction.[11]

The lawsuit was finally settled out of court, in 1856, after protracted negotiations. Both sides wanted an end to the costs of litigation, and the City was anxious to improve its riverside facilities against competition from railway terminals to the east. The City Navigation Committee withdrew its claim to ownership of the bed and soil of the river and paid a £5,000 settlement to the Commissioners of Works, Woods and Forests. In return, the Crown conveyed its rights to the tidal Thames to a new independent authority, the Thames Conservancy. Like the Metropolitan Board of Works established two years earlier, the Conservancy was a council of twelve delegates from its constituent boards: the Mayor and six members from the City, two delegates from Trinity House, two from the Admiralty, and one from the Board of Trade. In 1864, six delegates from private interests—ship owners, steamer proprietors, lightermen, and wharfingers—were added. The Metropolitan Board of Works, however, was not represented—a deliberate oversight which led directly to bitter recriminations.[12] The Conservancy (later folded into the Port of London Authority) inherited the duties of its constituents with regard to river navigation and encroachment, along with authority to maintain and improve steamboat piers and landings. Even the steamboat facilities built later by the MBW along the Thames Embankment were transferred to the Conservancy.[13]

The creation of the Thames Conservancy broke the legal logjam to allow construction of a new embankment, but it did not completely solve the problems of jurisdiction along the river. The Conservancy had plenty of responsibilities, but its budget was limited and its operations unsettled by the mutual jealousy of its constituents. Members of the City dominated its meetings and kept the MBW from obtaining a seat, despite the obvious advantage of such representation. In 1860, when the MBW asked Conservancy help in funding the proposed Thames Embankment, the latter asserted it lacked the resources for such a project. All it could do was convey to the MBW the revenue it might receive from land to be reclaimed along the shore. Yet later it harassed the MBW over a clause in the Embankment contract that forced contractors to obtain fill material from the Conservator's river-dredging operation.[14] The other partners to the Conservancy also retained, in practice, the right to review projects, even though their authority was supposedly subsumed in the new agency. Thus, a private contractor wishing to embank a portion of the Thames would apply first to the Conservators, who would first assess a fee against the estimated amount of land to be reclaimed and then split this with the MBW or the First Commissioner of Works. The contractor's plans would then go to the Admiralty and to the Office of Works for approval by their engineers. Finally, the contractor or his client would endure the complexities and expense of a private member's bill in the House of Commons.[15] Even J. W. Bazalgette, the MBW's chief engineer, had to go through this process, in 1863, with plans for the Victoria section of the Thames Embankment.

Given the rivalry between the Thames Conservancy and the Metropolitan Board of Works, and the uncertain jurisdiction of either the City or H. M. Commissioners of Works, it is no wonder that, when the Royal Commission of 1861 decided on a public, solid Embankment, it immediately opened up the questions of who would build it, how it would be financed, how it would connect to existing public facilities, and who would control the land reclaimed from the polluted mud banks. These questions had to be settled in Parliament. To understand *how* they were settled, we need to review some aspects of parliamentary government in mid-Victorian England.

In the 1850s, the wounds of the Corn Law reform, Peel's subsequent death, and the wilderness years of his followers were not yet healed. Although voters tended to follow party lines more consistently than was once thought,[16] party organization was still rudimentary. Not until the Reform Act of 1867 did a wider electorate and several three-seat constituencies (such as Birmingham) lead to concerted constituency organization on a national scale. The range of patronage available to the aristocracy (and to the ministry of the day) had been severely reduced since the eighteenth century, but that class had survived both Reform and Repeal in good order. In the Parliament of 1865, according to a contemporary analyst, half the members of the House of Lords were related to one another in a complex "cousinhood," and no fewer than 225 members of Commons were of the peerage and baronetage, with another 100 connected to them by descent or marriage.[17] These circumstances made midcentury Britain "the Golden Age of the private member." Men thought a seat in Parliament the pinnacle of a public career, but they were loath to change the existing order in any comprehensive way. Governments were expected to administer the state economically, without introducing too much legislation. Both Gladstone and Disraeli were in the House of Commons, which had already become a grand public arena in which the policies of the realm were debated with wit, classical references, and an impressive level of practical knowledge. Parliament, aided by expanding press coverage, educated its constituents through these debates; but, in a period of relative prosperity, few leaders wanted radical change. Nor, as Walter Bagehot observed, were they in a position to do much: "While the individual members of the House were loath to vote in such a way that they would be involved in the discomfort and expense of frequent dissolutions, the weak governments of this period, unsure of stable majorities, were anxious to avoid a mass of contentious legislation which might split the cabinet and threaten their precarious tenure in office."[18] Although four different reform bills were introduced between 1851 and 1861, they were, according to Asa Briggs, "less triumphant vindications of principle than useful manoeuvres to attract small, marginal groups of members, and they had the special advantage that they seemed highly unlikely to succeed."[19]

During the 1850s, Disraeli, trying to reconstitute the Tories as a vigorous Conservative Party, was frustrated by Lord Derby's indifferent leadership; the Whigs fought among themselves and negotiated with the Peelites and the Irish. After Derby's first ministry fell in 1852, a government was formed by the Earl of Aberdeen with an equal number of Peelites and Whigs. A notable addition, the first deliberate recognition of the radicals in Whitehall, was Sir William Molesworth in the "lowly" position of First Commissioner of Works.[20] Palmerston and Lord John Russell ("those two dreadful old men," said Queen Victoria) had been active in politics since the Napoleonic Wars and were well past their prime. Lord John, small and delicate, in uncertain health, overwrought and sometimes confused, held onto power through sheer longevity and connections. He was deemed largely responsible for the debacle of the Crimean War, which brought down the Aberdeen ministry in January 1855. Even then, he continued at the Foreign Office under Palmerston until the latter died, in October 1865, and Russell took over as prime minister the following year. Palmerston himself had been in the House of Commons since 1809. When he formed his first ministry, in 1855, after both Derby and Russell failed to gain support, he was already in his seventies. Failing eyesight, gout, and a habit of falling asleep during Cabinet meetings alternated with periods of extraordinary health and energy that put other ministers to shame.[21] Primarily interested in foreign affairs, Palmerston cared little for domestic legislative issues. R. J. Evans concludes that "he had no understanding whatever of the task of constructing a civilised industrial society with a suitable government machine. . . ."[22] Nevertheless, Palmerston had proved a competent Home Secretary under Aberdeen, and, in March 1858, he sat on the select committee of Commons that investigated the "Great Stink" caused by Thames river pollution. When asked by a group of Church of Scotland ministers to approve a fast day for prayer against the cholera epidemic expected that summer, he reportedly advised them to improve the sanitation of Edinburgh instead.[23]

In 1859, Palmerston and Russell patched up their quarrel and put together a deliberate coalition of Peelites, radicals, Whigs, and Irish. The ministry was thoroughly aristocratic: it contained three dukes, a duke's brother, four other peers, two sons of peers, and three baronets of aristocratic lineage. But it also included William Gladstone, leader of the Peelites, who severed his last ties to his old Tory associates and accepted the Exchequer. He and Palmerston disagreed about many policies. Palmerston undercut Gladstone's bill to encourage free speech and free trade by removing duties on paper, and pursued preparations for increased defense against a possible French war, despite Gladstone's desire to cut military expenditures. Palmerston wrote in scathing terms to the queen about his colleague, while Gladstone's contempt for the sinful old aristocrat was common knowledge.[24] During the construction of the Thames Embankment, the two argued over the right of aristocratic residents in Whitehall to divert the Embankment roadway and over Gladstone's assumption that the reclaimed Embankment land should pay for itself through leases on new office buildings.

There is no need to detail Gladstone's passion for Free Trade and reduced budgets, which had marked his career ever since he became president of the Board of Trade in the early 1840s. His famous budget of 1860, together with the Cobden-Chevalier treaty, not only raised economy to a high moral virtue but doubled the volume of exports. It promoted a view of England as the center of world commerce, with London as its main entrepôt. After 1861, Gladstone produced increasing budget surpluses, despite lower taxes and tariffs. An era of unparalleled prosperity seemed to arrive. There were dramatic increases in speculative building projects and overseas investments, both of which affected the progress of the Thames Embankment.

The latter years of Palmerston's ministry also saw a stream of humdrum, noncontroversial, but essential reforms in the Poor Law, factory regulation, local government, public health, and criminal and commercial law (including the Companies Act of 1862, which encouraged engineering contractors to form joint-stock companies). These reforms tended to be reactions to developments at a technical and local level rather than initiatives by the House of Commons, and, in fact, many important changes occurred in commerce, industry, agriculture, and public utilities without the sanction or understanding of parliamentary leaders. Many of the legal, financial, and engineering details of the Thames Embankment fell into this category. They were worked out in negotiations among the relevant commissioners, solicitors, and engineers. Even at this level, technical knowledge was at a premium. The First Commissioner of Works, as mentioned previously, usually delegated construction details to his staff, which was still small and appointed mostly through patronage. Civil Service examinations, instituted in 1855, were still a mere formality.[25] By contrast, the legal, banking, and engineering worlds were growing increasingly professionalized. Successful engineering firms, with experience of property compensation, construction finance, and large-scale urban development, could easily surpass the resources of a government department.

The emergence of the Metropolitan Board of Works and its designation as the authority to build the London Main Drainage system and the Thames Embankment represents both the influence of the new professionalism and the suspicion with which traditional government viewed it. In the first three years of its existence, the MBW's efforts were hampered by a lack of sufficient funds and by its dependence on the favor of the First Commissioner of Works. Funding was the more difficult problem. The Metropolis Local Management Act of 1855 gave the board the right to levy rates through the vestries, but this was a troublesome and uncertain process. The MCS had bequeathed a debt of almost £250,000 in uncollected rates, which had to be settled before funds for the new drainage could be gathered, and the vestries sometimes resisted the board's precepts for rate collection in their districts.[26] The board's political dependency was also irksome. Some of its members sat in the House of Commons, but it had no formal representation there except through the First Commissioner of Works.

Yet MBW Chairman John Thwaites had not been shy in seeking responsibility for new projects. He solicited memorials from vestry boards in favor of building the Embankment, negotiated with both the Office of Works and the City Navigation Committee, and published declarations of the MBW's intent in the principal newspapers. Meanwhile, Chief Engineer Bazalgette built mile after mile of sewer lines. Although the sewers disappeared from view, everyone noticed the construction sites and soon realized that the system was working. By the time the Royal Commission on the Thames Embankment met in the spring of 1861, therefore, the MBW had established a stronger power base and had won widespread praise for its engineering efforts.

Thwaites's success in amending the Royal Commission report to give the MBW full control of the Embankment was, nevertheless, guaranteed to raise the hackles of competing agencies. William Cubitt, the Lord Mayor of London who chaired the commission and wrote its original report, discovered that a revised version authorized the MBW to build a new street connecting the Embankment at Blackfriars Bridge with the Mansion House in the City (i.e., Queen Victoria Street). Cubitt professed to be amazed that anyone could conceive of such a thing, since the whole length of the street lay within the City limits. It was an unprecedented breach of ancient prerogatives. Most Londoners, however, saw Queen Victoria Street as a simple accommodation for the new low-level sewer and dismissed Cubitt's apprehensions.

Sir John Shelley, MP for Westminster, was more successful at changing the Embankment plans. He sat on the select committee for the express purpose of defending the wharfingers in his constituency. Sir John harped on the economic value of the wharf and warehouse trade along the Thames and finally persuaded his fellow members to cut off the Embankment at Blackfriars, eliminating a section from Blackfriars to Queenhithe (halfway to London Bridge) that had been part of the original plans.[27]

A third group who sought modification of the Embankment design was made up of aristocratic residents leasing parts of the government-owned Whitehall Estates near Scotland Yard. The group was represented in the House of Commons by the Liberal MP Edward Horsman and allied with the Duke of Buccleuch, who was just completing a £5 million rebuilding of his ancestral mansion, Montague House, southwest of Whitehall near the York Stairs. Hitherto fairly discreet in their opposition to new construction, the lessees now appealed to their landlord, the First Commissioner of Woods and Forests, for protection. They argued that a roadway crossing between their residences and the river would ruin their view and lower the value of their property. Charles Gore, the First Commissioner in 1862, readily took up their cause against William Cowper and the Office of Works, who defended the original plan. Parliament was faced with a kind of civil war between two government agencies which had until recently been combined under a single head. Gore proposed that the roadway be diverted up Whitehall Place to Parliament Street (the westward extension of Whitehall), instead of running directly to Westminster Bridge. He had his consulting architect, Sir James Pennethorne, draw up plans to that effect. Other members of the committee tried to compromise by providing a footway along the river, rather than a roadway. The *Times* and the *Builder* gleefully jumped into the fray, holding up Gore and the Crown lessees as absurd exemplars of aristocratic privilege, quite prepared to sabotage a great public improvement for the sake of their private gardens. For their fractious opposition, "you could run the road through Montague House itself, and through Mr. Horsman's premises also," wrote George Godwin, editor of the *Builder*. The debate was finally ended by Lord Palmerston, who cut through the cloud of irrelevant diversions to state firmly that as Londoners were going to pay for the Embankment, Londoners should have the use of it. The roadway stayed.[28]

The Duke of Buccleuch had prudently stayed away from the select committee hearings to defuse the public outcry, but he was not done protesting the Embankment. When the MBW came to consider compensation for riverside proprietors affected by construction, the Duke submitted a huge claim for injury to his new mansion and grounds. A claims umpire awarded him over £8,000 damages, but the board refused to pay. The Duke sued and fought the case through every stage until he won in the House of Lords in 1872.[29] Perhaps it was not a coincidence that the MBW began construction of the Embankment by driving the first pile for the retaining wall directly in front of Montague House.[30]

In these Parliamentary contests, the concept of the Embankment as a public improvement was further refined. It was important enough to override the objections of commercial proprietors from Westminster to Blackfriars (although these may have been raised to ensure generous compensation rather than to prevent development), but not important enough to extend to Queenhithe. Wren's vision of a continuous forty-foot way from the Tower to Whitehall gave way to the more utilitarian goal of providing a route for sewer lines and streets. The public interest was to override inconveniences suffered by individuals or groups along the main avenue of construction, but more significantly, the "public interest" was now defined in terms of the general population of London, rather than the Crown, the English people, or the government of the day. The *Times* and other newspapers, alert to the possibilities of mass circulation, championed this perception.[31] While political leaders like John Thwaites continued to see the Embankment as an architectural monument to London's imperial stature, or as a bold solution to problems of sewage, transport, and navigation, ordinary subjects began to focus on its emerging identity as an agent of popular access to privileged space—space formerly commanded by vulgar commercial enterprises and self-serving aristocrats. This function was to become even more important as land was gradually reclaimed from the river.

The question of how to finance construction was easy, compared to the struggle for control of the project. An act of 1862 permitted the MBW to draw upon a Metropolis Improvement Fund created expressly for the Embankment by Parliament the previous year out of ancient duties on imported coal and wine. "This decision did not delight all hands," wrote David Owen, "because the tax was inherently objectionable, but the act was probably as fair as was politically possible in the 1860s."[32] In fact, the tax was a key element in the designation of the Embankment as a public architectural space rather than as an arena for commercial development. The coal and wine duties were anachronistic, and the fact that they had became available for this purpose in 1861 was something of an accident; but having become available, they were far too expedient for even a Liberal government to ignore. Some background on British public finance will show why these levies remained so useful.

During the nineteenth century, the Bank of England only gradually built up its institutional control over the London money market. Until the 1840s, it competed with other private banks, both in the capital and in the provinces, and sometimes speculated as irresponsibly as any other investor. But the bank had the only sizable gold reserves and came to monopolize government stock issues and short-term loans. Along with the East India Company and a few large insurance firms, it administered the financing of the national debt, subscribing its share through competitive bidding and then reselling it in small amounts to ordinary investors at a premium. The smaller City banks, private and independent, acted as agents for provincial investors, rural estate mortgage funds with private industrial and business capital. After the Joint-Stock Act of 1833, the latter type of banking proliferated and led to speculative surges and crises. In the troubled economic climate of the late 1830s, Robert Peel and other business-oriented political leaders saw the need for tighter credit control. The Bank Act of 1844 retained the decentralized system of private provincial banks but allowed the Bank of England to compete with them in the speculative loan market, on the theory that it would stabilize rates of interest. Unfortunately, the bank's competition had exactly the opposite effect. Coupled with the railway mania of the 1840s, it led to a stock market panic and financial crash in 1847. Thereafter, the bank withdrew from the speculative market, fixing instead a minimum discount rate that, along with a statutory level of gold reserves for currency issues, controlled to some extent the expansion and contraction of credit.[33]

The average investor in mid-Victorian England had fewer choices than his twentieth-century counterpart. If he wanted low risk and moderate returns, he could tell his solicitor to put his money into national debt shares or into real estate rents and mortgages. The debt was made up of a variety of bond and stock issues supporting specific government projects, together with the 3 percent consolidated bank annuities ("consols") available since 1751. Rents and mortgages in the rapidly expanding London building estates were discounted by small and medium-size builders to obtain capital for further speculative construction.[34] Bills of exchange, first used as negotiable paper and then as credit instruments, were riskier. They were sold along with railway and other construction stock issues at a highly variable discount rate, usually for short terms of less than three months. Returns on domestic railway investments varied enormously but averaged just under 8 percent in the 1840s, better than most other domestic opportunities.[35] Overseas railways built by British contractors also matched high dividends with high risk, with the exception of the Indian railway system, which was guaranteed by British authorities. Finally, one could speculate on loans to foreign governments, on imported commodities such as tea or cotton, or on gold and other precious metals.

Until 1856, most business investments involved the assumption of un-

limited liability by everyone in a company, a condition that tended to restrict the supply of venture capital. A company could incorporate, after
1844, simply by registering at the Board of Trade; in doing so, however, its
prospectus was made accountable to government supervision and its operations opened to public scrutiny. Most commercial, industrial, and construction enterprises were still operated as partnerships or sole proprietor-
ships, and, in the case of the Thames Embankment, such contractors often
had trouble raising capital for large-scale projects. Nevertheless, the railway
boom got the public used to investing and reassured individuals that the
failure of a company did not usually mean the loss of one's entire assets.[36]
With the advent of Free Trade and the expansion of the Empire, British investors mounted a tremendous outward flow of capital. Gold discoveries in
California and Australia at midcentury led to another round of speculative
trading, exacerbated by legislation, in 1850, allowing limited liability for the
first time. The British, as Checkland put it, simply ran out of capital to in-
vest overseas and began speculating on credit. Wage increases and inflation
followed. When the government began borrowing to support its involvement in the Crimean War, the financial markets collapsed once again,
bringing down over 125 important business firms by the end of 1857.[37]

Since the crisis of 1857 was circumstantial rather than structural, the financial markets recovered fairly quickly. Brokers gradually persuaded their
clients to leave the traditional Consols for more sophisticated investments.
When the U.S. Civil War broke out, Indian cotton growers quickly exploited the opportunity, and capital once again poured into the construction of
Indian railways and port facilities. During the next decade, India absorbed
over £61 million in British capital.[38] The Limited Liability Act of 1862 consolidated previous advances in company financing and made investment
at home and abroad even more attractive. There was another surge of railway building around London, connecting the capital with the great regional systems. This boom, in turn, came to a halt with the end of the Civil
War and the failure of many small joint-stock companies. In May 1866, the
leading brokerage firm of Overend, Gurney collapsed. Capital again disappeared, contractors failed by the dozens, and unemployment soared. The
market was restored by July 1867, but overseas demands again drained
London of capital. The agitations of the Reform League and the gyrations
of Parliament over the Second Reform Bill did not help matters.[39]

The ragged pattern of national finance and capital investment in the mid-Victorian period points up the uncertainties faced by government officials, engineering firms, and contractors involved in the planning and construction of large public projects like the Thames Embankment. Government stock and municipal bond issues had to compete with a variety of risky but attractive investment alternatives in the London money market, and tax revenues were always uncertain. The prevailing public philosophy favored reduced government expenditures, and, with Gladstone at the Treasury after 1850, parsimony was the rule. But private enterprises, especially in the building trades, were equally strapped for funds to build large-scale works. The financial instability of the mid-1840s was one reason why Thomas Page's plans for embanking the Thames failed. The speculative overseas investments of the 1850s, and the crisis of 1857, made capital difficult to come by for projectors like Lionel Gisborne, who had to face the complications of government bureaucracy at the same time. Even in the early 1860s, when stability seemed to have returned, the Metropolitan Board of Works had trouble with its sewer and Embankment contractors because of their inability to secure adequate capital. In 1866, when the Victoria section of the Embankment was nearing completion, the crash of Overend, Gurney brought construction to a sudden halt. For the Albert Embankment on the south side of the river, the MBW relied on a single major contractor, William Webster (figure 7), who somehow managed to remain strongly capitalized through the whole period. Webster was also awarded the contract for completing the Chelsea Embankment, 1871–74, after three previous firms, having won the contract, pulled out for financial reasons.

It may be recalled that when the MBW first considered embanking the Thames, they approached the new office of the Thames Conservancy for help but were turned down on a plea of poverty. The MBW was hardly better off. It was empowered to collect a portion (2d. on the pound) of the rates in its member vestries and districts, but much of that revenue was earmarked to pay off the debts of its predecessor, the Metropolitan Commissioners of Sewers, or to underwrite construction of the London Main Drainage. Moreover, Chairman Thwaites advised a Commons select committee, in 1860, that, although the MBW had the legal authority to raise the rates for new construction, he doubted that they could actually collect the money. The memorials from vestries in the 1850s made clear that they expected the Embankment to be funded at the national level.[40]

The MBW's uncertain revenue sources made it difficult even to underwrite Embankment legislation. Since it had no representation in Parliament, the Board was required to seek authority for projects through private bills, which required the services of counsel and expenses for a parliamentary agent, witnesses, advertising, document printing, and (in the case of most MBW projects) lithography for plans. Public bills involved similar costs: the Thames Embankment Act of 1862 cost H. M. Office of Works over £10,700. William Cowper, however, inserted a clause in the act charging the whole amount to the government's Metropolis Improvement Fund. The MBW did not obtain permission to do this until 1868, when the Victoria and Albert Embankments were nearly finished.[41]

Parliament had four traditional ways of funding large urban improvement projects like the Embankment. One way was to charge costs to the rates in the metropolitan area affected by the improvement. As already stated, this was unlikely to yield sufficient revenues even if the vestries agreed to pay, and the MBW could collect. Second, the owners of property along the improved route could be assessed prorated amounts based on the assumption of higher property values, just as local sewer and street projects are often financed today. The local sewer commissioners had traditionally ordered repairs to drains and river walls at the expense of individual proprietors. In the 1870s, the MBW was to use this precedent to assess wharfingers along the south bank of the Thames for flood wall construction. However, in the absence of clear legal authority, assessments required a tedious process of negotiation with individual property owners (or their estates, heirs, guardians, et al.). Early attempts to embank the Thames along Chelsea Reach, in the 1840s, had come to grief because government could not reach agreement with the many residents who would be assessed for part of the costs. It was one thing to charge rents for new land created by embankment—that was a simple prerogative of the owner of the new land—but quite another to persuade residents to pay in advance for often intangible degrees of improvement to come.

A third way to fund the Embankment was to charge it against general Crown revenues as a project of benefit to the whole realm, to its commercial or naval strength. Given the penny-pinching tenor of mid-Victorian governments, and Gladstone's budget reductions in particular, there was little hope or expectation for a direct grant of money. Members of Parliament from other regions and cities would not stand for a special subvention for London. The most Government could do was to loan funds, in the form of Treasury bills, to a legally incorporated local Improvement Commission. But such loans had to be secured by some form of anticipated local income, such as bridge tolls or higher rents on Crown property.

Directing specific taxes, fines, fees, or tariffs into a special fund assigned to one or more projects was the fourth and most common method of funding improvements. This has always been a staple of government finance, blending simple logic with opportunities for plunder and patronage. The early Stuarts were notorious for exploiting special fees and fines in lieu of Parliamentary grants. In the eighteenth century, annuities for friends and clients of government ministers were skimmed off import tariffs on colonial produce like West Indian sugar. All the bridges over the Thames were originally toll bridges, and, in the vestries, paving and lighting costs had always been assessed against specific properties. Before the Royal Commission of 1861 decided to eliminate commercial establishments from the Embankment, it was assumed that assessments and improved rents would pay for it. Now some other source of revenue was needed. The choices were diminishing, since Gladstone's budget for 1861 proposed to leave only forty-eight articles subject to impositions.[42]

Two of the remaining tariffs, possibly spared because their revenues were already assigned, were a .5d. duty on each cask of wine and a 13d. per ton duty on coal brought into the London metropolitan area (defined, in 1854, as the area within a twenty-mile radius of the general Post Office). These duties, traditionally combined as a revenue source, were of ancient origin, probably descended from medieval charges of tunnage and poundage. The coal duty was, in fact, a combination of at least three previous duties: 8d. imposed for the upkeep of London Bridge and its approaches; 4d. collected by the City Corporation as part of its ancient estate; and 1d. imposed, in 1807, as a war measure but extended, by various acts, in subsequent years. At first, only seaborne coals were taxed, but, after 1845, inland (railway) coal was also covered. The coal and wine revenues were formally collected by the City Corporation, but portions were assigned to specific projects for stipulated periods by Parliament, and the accounts were kept in the Bank of England. The funds were apparently used in the seventeenth century for supporting orphans, and then for rebuilding St. Paul's and other edifices after the Great Fire. In the eighteenth century, they were applied to the construction of the first Blackfriars Bridge (1762) and its attached embankment. Sir John Rennie tried to get the coal duties increased in 1825, partly to fund the new London Bridge and partly to support the embankment schemes of Col. Sir Frederick Trench, but the government would not agree.[43] However, in 1830–31, Parliament assigned 8d. of the coal duties and .5d. of the wine duty to the building and maintenance of the approaches to the new London Bridge (the rest was held by the City for its own use). Technically, the annual income only guaranteed a public loan floated by the City Corporation, which was to be repaid by 1858. However, the coal and wine fund built up much more rapidly than expected. The interest it earned was applied to the principle, and the London Bridge venture was paid off in 1852. Still named the "London Bridge Approaches Fund," the coal and wine duties account then became a flexible source for financing metropolitan improvements. It was tapped by the Commissioners of Woods and Forests and the Westminster Improvement Commissioners in 1851 for various small projects. In 1853, when the Office of Works was separated from Woods and Forests, Parliament granted the former the sum of £122,000 per year from the fund to pay off debts inherited from the latter. At that time, the fund had a principle of £665,000 and previously accrued interest of £352,000. When the debt was paid off, the First Commissioner of Works continued to use the annual interest to fund a series of Metropolitan Improvement Commissions, appointed all through the 1840s and 1850s, which basically provided an infrastructure for the rapidly expanding population of London.[44]

The improvement grants authorized by the act of 1853 were due to expire in 1862 and, in fact, were estimated to be paid out of the London Bridge Approaches Fund by the end of 1860.[45] There was no reason, except sheer expediency, to renew the coal and wine duties at that time. Everyone who paid, collected, or assigned the duties regarded them as a peculiar anachronism, quite at odds with prevailing economic theory. The 1d., 8d., and 4d. charges on coal and the .5d. charge on wine had no intrinsic relationship to their respective prices, especially after Parliament extended the original charge on seaborne coals to those brought in by the newer (and cheaper) canal and railway routes. London newspapers and consumer organizations periodically protested against the injustice of the duties, arguing that they were imposed on poor people for the necessities of life, and on suburban dwellers who had no use for the improvements they were meant to subsidize.[46] On the other hand, the duties were apparently easy to collect. They added little to the retail price of coal and wine because they were spread over a large population. The fund was large enough in 1861 (some one million pounds, now yielding about £160,000 per annum) to underwrite sizable metropolitan improvement projects, and there was ample precedent for the practice.

John Thwaites, naturally enough, spotted the duties very early on and strenuously lobbied William Cowper, the Commons select committee of 1860, and his fellow Royal Commissioners of 1861 to get them assigned to the Embankment. In his annual report for the MBW, in 1859, Thwaites devoted the better part of three pages of footnotes to a detailed examination of revenues and a review of each of the coal and wine duties. He argued, contrary to what many members of Parliament were saying, that coal duties were not anachronisms unique to London but similar to those levied in "a large number" of other British seaports.[47] Cowper was easily persuaded, since his office had already used the duties for many years. His colleagues, no doubt, also saw the opportunity to renew the duties for a large and popular metropolitan development as a fortuitous coincidence.

The London Coal and Wine Duties Continuance Act (24 & 25 Vict. c. 42) was signed in July 1861. It assigned .5d. of the wine duties and 9d. of the coal duties to a new "Thames Embankment and Metropolis Improvement Fund" until 1872, and transferred to it the balances from the previous Bridge Approaches Fund and the Metropolis Improvement Fund.[48] Thwaites quickly called a special meeting of the MBW, which voted to proceed immediately with plans for construction.

It was quite fortuitous that the Embankment project was articulated in a specific parliamentary bill just when the coal and wine duties, committed elsewhere for thirty years, came up for reconsideration. Of course, once the Main Drainage system was started, the Embankment became a high possibility, and once it was chosen as the preferred route for the low-level sewer line, the construction schedule required that it be commenced by 1862 or 1863. But some type of embankment could have been constructed any time before 1861. And the First Commissioner of Works, whose metropolitan projects had been subsidized for a dozen years out of the fund, might have raised serious objections about diverting the money to another agency. It is a measure of Cowper's confidence in Thwaites and Bazalgette that the MBW won its case.

Given its new authority, the MBW felt able to raise the rates in its member vestries and districts from 2d. on the pound up to 7d. by 1867. Rate revenue, added to the fund interest, guaranteed a strong, steady income for years to come. But to ensure the large sums necessary to begin construction and meet outlays on schedule, the MBW asked for and received permission to borrow against future rate and fund collections. This approach had been tried successfully with the Main Drainage system and was, according to Gloria Clifton, "an innovation in local government finance."[49] The Thames Embankment (North) Act of 1862 authorized the MBW to borrow up to £1,000,000, an amount doubled the following year. Its companion act for the South (Albert) Embankment, in 1863, provided another £480,000.

Having delivered these plums to the capital, Parliament then provided similar assistance to provincial towns struggling with their own sewage and pollution problems. The Public Works (Manufacturing Districts) Act of 1863 made low-interest government loans available through a Public Works Loan Commission specifically for sanitation improvements. These loans expedited reforms in many cities where fear of high taxes, lost business profits, or litigation had prevented construction of water treatment and sewer systems.[50]

In 1864, Thwaites asked Gladstone for a Treasury guarantee for all MBW loans, as had been done previously with the Main Drainage project. He calculated that a guarantee would lower the interest rate from 4.5 percent to 3.75 percent, saving about £18,600 per year. He reminded the prime minister that the MBW had existed for only a short time and lacked the sources of revenue other large city administrations enjoyed. He also argued that the MBW deserved Treasury backing by virtue of London's unique status as commercial and royal capital: "They [the MBW] have neither real nor funded property, nor tolls, nor dues, as have other corporations; and though a large portion of the commerce of the empire passes through their streets, it bears no share in the formation or maintenance of those thoroughfares. Neither does the mass of Government property, royal palaces, public offices, dockyards or arsenal, or any of the national establishments contribute one farthing to the Board's revenue, a state of facts which I believe exists in no other capital."[51] Gladstone was convinced of the practicality of the suggestion. He approved the Treasury guarantees and then convinced the trustees of his new monument to thrift, the Post Office Savings Bank, to loan their excess funds at a favorable rate of interest. In return, he demanded that the MBW raise its local rates to cover any shortfall from the coal and wine duties.[52] Under this agreement the Bank of England advanced £1,000,000 and the Post Office Trustees £1,100,000 at 3.75 percent for eighteen years. With the loan guarantee, the MBW purchased consols and Treasury bills, which were tradable on the Exchange, and sold them when large sums were needed for construction.[53]

Despite this innovative strategy, the MBW continued to have difficulty generating the sums required for its ever expanding operations. By 1864, the Main Drainage was nearly finished and had already proved so successful that the Board was beginning to attract other types of improvement projects. Despite its inauspicious start and its poor political reputation, its engineering department filled a very desperate need for a technically staffed public works office. It took on proposals for new streets, lighting, sewers, fire engine houses, and paving projects, and in the process became, willy-nilly, temporary owner or landlord to hundreds of residents and commercial lessees. To meet these unexpected assignments the board had to borrow money outside the Thames Embankment funds, usually from insurance companies, at higher rates of interest. Together with the Victoria, Albert, and Chelsea Embankments and the new street to Mansion House, which were estimated to cost £3 million altogether, the new projects demanded periodic infusions of capital that could not always be met by draws upon their specific funds. One irregular drain on the budget was the MBW's practice of immediately compensating property owners for land and buildings taken over by construction, and then leasing or selling whatever remained afterward to recover costs. Not until 1882 did Bazalgette recommend giving notice of occupation first, and paying compensation later. Also, overhead costs for general administrative staff and the engineering department were growing apace. By 1869, the MBW had accumulated a debt of about £8 million.[54]

The board was able to meet some needs by shifting money between projects. For example, in January 1866, £50,000 was borrowed from the Albert Embankment account for the Victoria Embankment. One month later, it was transferred again to the Mansion House Street account. The board also assigned a portion of its administrative overhead (called "establishment charges") to each of the Embankment accounts twice each year.[55] Creative banking was also tried: on the advice of one of its members, stockbroker and East India agent G. P. Richardson, the board placed its funds in the Bank of London, a joint-stock bank that paid interest on accounts (the Bank of England did not). The Bank of London collapsed during the financial crisis of 1866, but fortunately its accounts were insured. The MBW then transferred to the London and Westminster Bank, which offered £1 million in government stock as security.[56] All these expedients were necessary because the board was trying to underwrite an increasing variety of long-term metropolitan improvements with short-term capital loans.

In 1869, Parliament finally recognized the pivotal role of the MBW in the metropolis and dealt with its financial problems. The new First Commissioner of Works, A. S. Ayrton, proposed consolidating all the board's existing loan obligations, rather like the national debt; consolidating the various rates imposed by the board on metropolitan property, and the various public works accounts; and issuing "metropolitan consolidated stock" redeemable in sixty years. The Metropolitan Loans Act of 1869 provided the long-term borrowing power the board needed to carry out its work. MBW stock, despite earning a modest 3.5 percent per annum, proved very popular with investors, and the board soon began financing improvement projects for its constituent vestries and other London institutions. Within a decade, it became a substantial banking establishment in its own right.[57]

The financing of the Thames Embankment and related projects spread beyond the central Treasury to a variety of public and private banks, insurance companies, and incorporated funding agents such as the Postal Savings Bank. These dealt not just with the MBW but with the consulting engineers and contractors hired to complete each section of the work. They all had to compete for capital in a rapidly shifting marketplace sensitive to national and imperial as well as local changes. The progress of the Embankment was thus intimately affected by the currents of British overseas investment, the London stock market, and government policies both foreign and domestic.

Before leaving the subject of finance, I should mention two schemes that were *not* adopted to fund the Embankment. The 1836 Select Committee of Commons on Metropolitan Improvements suggested, as an alternative to the coal and wine duties, the establishment of an annual lottery. It noted that lotteries had previously raised £1–2 million annually for various projects, before "abuses" killed the practice.[58] From the perspective of the United States in 1996, where many states have instituted lotteries for public income (often dedicated to school funding, a nice moral touch), this suggestion does not seem farfetched. In 1836, it died quickly.

The other scheme was a tontine. It was proposed by an engineer, Charles Henman, in 1862. A tontine is basically an agreement among a group of people that the last one alive (or those still alive on a stipulated date) will receive the whole amount of a given fund, which earns interest in the interim. The method had been used for capital investments, as well as for life insurance and settling estates and trusts (thus creating an ingenious plot device for Victorian detective novels). A relic may still be seen at Ironbridge, on the Severn in Shropshire, where the Tontine Hotel, built by a group of investors in the late eighteenth century, has offered bed and board for two hundred years. Henman proposed that the government issue about thirty thousand shares at £100 per share, and use the coal and wine duties to guarantee a 3 percent return to investors for life. The shares would be variably priced by age bracket, and the surviving investor in each bracket would receive a permanent share of the rents generated by commercial exploitation of the Embankment's reclaimed land. Henman was something of a windbag, and, though the commissioners complimented his "talent and inventive genius," there is no evidence they took him seriously.[59] The Embankment needed more substantial backing than such a dubious plan could produce.

CHAPTER VII

Relevant Interest Groups

The Thames Embankment, like other projects linking technology to the environment, had certain objective features and an internal logic of operation or development, but it was also affected by the perceptions and behavior of legal and financial institutions. While engineers tried to shape the Thames Embankment into a manageable construction project, public officials tended to see it as an object of administrative and territorial hegemony, while Treasury officials, Metropolitan Board of Works accountants, and London bankers and investors developed innovative ways to provide long-term funds for its multiplying functions. Institutional leaders tend to conceptualize technology and the environment in specialized, abstract ways particular to their concerns. Social groups, who are united more by shared experience than by formal precept, think in more restricted and concrete ways, but connect them more directly to their lives. Thus the MBW Chairman John Thwaites, using a standard of public interest, a broad sense of economic development, and a landscape aesthetic of imperial grandeur, could arrive at the conclusion that commercial proprietors along the Thames from Westminster to Blackfriars had been rendered superfluous by the new railways. But a man like Edward Butten Walmsley, "born on a wharf and nursed and bred on a wharf," believed that the new

Embankment should aim primarily to improve his and his neighbor's commercial facilities. After testifying in person before the Royal Commission of 1861, he joined with his fellow wharfingers to hire legal counsel and supported the demands of his MP in the battle to keep the commercial character of the district or else win adequate compensation. Men like Walmsley had a workingman's relationship to the river: it was an immediate, practical part of his experience, and the circumstances of his work carried over into the rest of his life. Water, polluted or not, meant transport. The criteria of people like Thwaites meant little to him.[1]

Just as Victorians of different locations, occupations, and social status held different perceptions of the environment, so those involved with the Embankment had different experiences of the construction project and the completed artifact. One category of people includes the wharfingers and other proprietors along the riverbank who were dislodged or interfered with, the steamboat operators, lightermen, and bargemen whose trade was disrupted, and the lower strata of thieves who worked the commercial part of the river. These people left few records of their daily activity, and I can only suggest the ways in which they were affected. A second category, about which I will have much more to say, includes the London community of professional engineers and contractors, the skilled workers, and the laboring men who together experienced the Embankment as a new opportunity rather than a threat.

The Thames-side proprietors located in the Westminster-Blackfriars stretch of the river included operators of draw docks, wharfingers, mechanical engineers, stonecutters, coal and lumber merchants, millers, and a large gasworks. Most of these were served by sailing barges called "lighters," transferring goods from the ships anchored downriver in the Pool. Loading and unloading at the wharves depended on the tides: barges were ordinarily brought up on the incoming tide, but they might remain through the ebb, resting on the mud banks, like the barges one sees today above Chelsea. The rebuilding of London Bridge had already changed the tidal regimen; the Embankment altered it even more. The smoothly curving river wall on the Victoria (north) side eliminated many mud banks, as it was supposed to, but tended to throw the current back against the southern shore, creating new patterns of erosion and flooding. The wall also contracted the river's width, especially at places like Hungerford Reach, increasing current velocities at ebb tide and forcing the flood tides farther upstream.

J. W. Bazalgette's original plans included embanking both sides of the river. However, the commercial district on the south side, home to the Lion Brewery and the giant Hay's Wharf among other enterprises, was deemed more valuable than that around Hungerford and Whitefriars, so the Royal Commission of 1861 excluded it from development. (Perhaps it was just coincidence that the commission chairman, William Cubitt, had built much of Hay's Wharf and married his two daughters to sons of the wharf's owner, Alderman John Humphery).[2] The stretch between Vauxhall and Battersea, where another gasworks operated, was also excluded. These alterations allowed many of the proprietors dislodged by the Victoria Embankment to relocate on the other side of the Thames or downstream below the Pool. The seed merchants Beck, Henderson and Child moved from Adelphi Wharf to Upper Thames Street near Southwark Bridge; Gwynne and Company, mechanical and hydraulic engineers, moved to Battersea. No doubt, the lightermen continued to service them at their new locations.

Relocation may actually have been an improvement for some of the proprietors, but none of them admitted this when presenting the MBW with claims for compensation. The law regarding compensation for business lost to public works was notoriously murky, and the claims were inflated with every possible injury. David Owen has described in some detail the tortuous negotiations carried on by the board's architect, solicitor, and property appraiser. For the forty-four claims settled between November 1863 and December 1865, the board paid £200,977 out of the £297,765 originally demanded.[3] Some claims were arbitrated, and others went to court.

Along the site of the Albert Embankment across the river from Millbank and Westminster, relocation was much easier. Most of the grounds were occupied by Doulton's pottery and china works, which was pressed for space and willing to sell. At the same time, the trustees of St. Thomas's Hospital were searching for new quarters, as their London Bridge location was being encroached upon by new railway lines. The MBW was thus able to use the area for construction without interrupting commercial operations, yet still recover its property costs by selling the finished site to the hospital (figure 9).[4]

One firm that resisted moving was the City of London Gas Light and Coke Company, which had a large works near Blackfriars Bridge, and another just northeast of Vauxhall Bridge, where the Albert Embankment was to go. Serving London householders with coal gas, produced with primitive technology, the gas companies took in huge bargeloads of waterborne coal on a daily basis. Although small in scale compared with later industrial organizations, they enjoyed considerable influence at Westminster and among City bankers. The MBW negotiated fruitlessly with the "City" Gas Company throughout the 1860s as the Victoria Embankment grew down the riverbank from Westminster, and the Albert Embankment stretched upriver toward Vauxhall. The company would not relinquish its access to the Thames. Finally, Bazalgette had to alter his original plans for the Victoria section and carry the Embankment roadway, as it approached Blackfriars Bridge, up a viaduct over the barge entrance. He planned a similar viaduct to carry the Albert Embankment under Vauxhall Bridge and along the foreshore to Nine Elms, but the gas company there finally decided to relocate and the other wharf owners made so many objections that the extension was abandoned.

A more cooperative group were the Thames steam ferry operators, who required piers at strategic locations. As mentioned previously, the "penny packets" carried thousands of commuters daily but were notorious for sharp practices. To load and discharge passengers, they anchored old barges beyond the low-water line, covered them with planks to fashion rudimentary landing platforms, and built winding, rickety walkways and stairs to reach them from the shore. These were, in the words of one historian, "an eyesore, a public nuisance, and an obstruction." Despite such drawbacks, the steam ferries were so popular that they were expected to compete successfully with the new railway lines; more than one Embankment design submitted to the Royal Commission of 1861 omitted a railway for this very reason.[5]

The steam packet owners could reasonably expect the new Embankment to incorporate landing platforms and thus improve and increase passenger traffic. New piers were indeed built at Westminster, Charing

Cross, Lambeth, and Battersea Bridge, and the floating platforms with hinged gangplanks located there today are essentially those designed by Bazalgette in the 1860s. In its review of the decade of construction ending in 1874, printed as a souvenir of the new Chelsea Embankment, the MBW confidently promised that passengers would appreciate the new facilities.[6]

The steam packets did not, however, frighten proprietors of the new Metropolitan District Railway. They used the filled-in space created by the Embankment to extend their new line underground from Westminster down to Blackfriars Station. Together with the new railways crossing the river from the south and west into Victoria, Charing Cross, Blackfriars, and Cannon Street, the Metropolitan District garnered the bulk of passenger traffic within a decade. Commuter ferries became redundant and then virtually disappeared. The new piers were used by tour boats (which had been around for a long time), as they are at present. Only in the 1970s, with the growing congestion of the underground and the riverside roadways, was it feasible to think of the river as a thoroughfare again. Suburban commuters began riding the tour boats from Greenwich, Hampton Court, and Kew, and, in the 1980s, a modern airfoil craft began carrying serious businessmen down the river to Hay's Wharf pier, all renovated glass and steel, where they could catch the Hydrofoil to France or walk to their new offices in the gentrified Surrey Docks development.

Another group of Thames denizens who were dislocated by the building of the Embankment, but not by any means deterred from their business, included dredgemen, lumpers, mudlarks, and other petty thieves. They had been around for as long as the port of London. When Parliament considered the state of Thames navigation in 1800, Thomas Telford testified that pilferage was "the leading object" of improvement. True river pirates were rare above London Bridge, but the young lightermen who navigated barges from the Pool upstream along the wharves pocketed items from their cargoes along the way. The bolder ones would steal an empty barge, tie it up next to a laden one when the crew were on shore, and plunder it wholesale. Dredgemen and mudlarks, according to John Binney, who wrote a section on thieves for Henry Mayhew's classic *London Labour and the London Poor*, were boys, girls, and women of all ages who waded into the Thames mud banks at low tide to dig up coals, presumably fallen from barges. Ragged and filthy, they clustered around the coal wharves at Blackfriars Bridge, the City of London Gas Company, and Hungerford Reach. They lifted one another up the sides of barges and knocked coal, iron bars, or anything else they could reach into the shallows, to pick up later and sell to the nearest marine stores shop. They also stole copper funnels and ropes from the sides of ships. Some operated in tandem with lumpers—barge workers ("coalwhippers") who tossed items overboard at high water, letting them sink out of sight until low tide, when the mudlarks could retrieve them.

A related group of thieves were the teenage "sweeping boys," mostly orphans or sons of coalwhippers, who bobbed about the river in old skiffs to locate untended barges. Their strategy was to climb on board with a broom and pretend to be sweeping out the barge, while grabbing anything of value. Expert swimmers, they would, if pursued by the police, "take to the water like a rat, splashing through the mud." Theodore Bonnet's sentimental novel *The Mudlark* (1949) includes some good descriptions of their practices. Although Binney described them as swarming around the riverbanks, he was "surprised at the comparatively small aggregate amount of these felonies"—the police reported only 203 cases with a property value of £712 for 1860.[7] It must be assumed that theft did not die out when the coal wharves were replaced by the impenetrable Embankment wall, but whether the mudlarks moved downriver to the new docklands or turned to other kinds of pilfering is a question requiring a different kind of research. They were not the only children at work in the depths of Victorian London, of course. Lord Shaftsbury's Commission on Employment of Children, detailing the wretched exploitation of match girls and pottery workers, made its first report in 1863. On the other hand, the mudlarks and dredgers may have been associated in the public mind with more violent criminals who habituated the waterfront. In 1862, when there was a wave of robberies in which the victims were strangled, apparently by convicts on parole, even Parliament took up the alarm.[8]

In contrast to the current fashion for history "from the bottom up," my account of groups dislocated by the Embankment has proceeded from the aristocratic classes at Whitehall downward to mud-grubbing waifs at the bottom. The same might be done for those who saw the project as an opportunity for advancement—the engineers and contractors, mechanics and laborers—primarily because they were associated in a rough social and professional hierarchy, wherein the fortunes of the lower groups depended on the enterprise of the higher, although laboring men could (and did) hold projects hostage to their demands for better hours and wages. The Thames Embankment offers a nice perspective on these related occupations because, as a well-known, large-scale project that had been promoted for years and offered potentially large returns, it attracted a wide variety of men with authentic or claimed skills.

In the late 1850s, civil engineering was consolidating its status, gained during the previous three generations, as a recognized profession. It followed the path of other professions in the nineteenth century by establishing a formal association, regulating its membership, tightening apprenticeship and examination standards, seeking political and legal recognition of its avowed status, and adopting, through individual initiatives, the conventions of family, suburban life, and church. The eighteenth-century British engineers were "a motley crew," according to R. A. Buchanan: they were generally classed as skilled artisans along with mechanics, smiths, molders, and millwrights. But some, like Robert Mylne, John Smeaton, and Thomas Telford, rose to a higher position by designing, rather than building, public works and large machines. The major canal, railway, and waterworks projects of the first half of the nineteenth century made engineering's reputation. In the great mid-Victorian period of prosperity following the Crystal Palace Exhibition, engineers enjoyed, as Buchanan has said, a unique self-assurance amounting almost to euphoria.[9] The excesses of the railway mania, which led engineers into all kinds of adventure and misrepresentation, were gradually being replaced by the conservatism of success.

Although many of the earlier practitioners had been indifferent to honors and social pretension, midcentury engineers eagerly sought social gentility, "with all the accoutrements of titles, estates, and the way of life to go with them." Samuel Smiles's *Lives of the Engineers* enshrined them as paragons of self-help, and their collective attitudes soon became stereotypically Victorian.[10] By this time, an occupational hierarchy had emerged, led by independent consulting engineers, with assistant engineers, clerks, and pupils ranged below. The Institution of Civil Engineers (ICE), founded in 1818, served (and still serves) as a research center, a London club, and a professional society for British engineers around the world. Mechanical engineers, who worked in tandem with the "civils" on the great railway systems, were at first included in the ICE, but they formed their own institution in 1847. Their work was oriented more toward the provinces, while civil engineers focused on the London area; but many "mechanicals" belonged to both institutes. Membership in the two groups grew from 220 in 1830, to 900 in 1850, and 3,500 in 1870, with over four hundred engineers registered in London alone. The roll of ICE presidents and vice-presidents, elected for two-year terms until 1880, includes almost every notable engineer of the century. Meetings were held weekly during sessions of Parliament, with members presenting papers for extended and often lively discussion. Their *Transactions* were published periodically in respectable formats, and, from 1837, they also produced annual *Minutes of Proceedings*. The ICE kept dues high enough to exclude the casual amateur but voted honorary membership to aristocratic enthusiasts and retired military engineers. Careful to maintain its image as a gentleman's club, the ICE regarded formal discussion of current projects, wages, and working conditions, or related political issues, as taboo. Even an enthusiastic call for a public resolution in favor of the Thames Embankment, proposed after a spirited debate in January 1856, was ruled out of order by the chair of the session, John Fowler (1817–1898). The ICE established a model for most other engineering groups arising in the second half of the nineteenth century. In fact, its determination to maintain an intimate and cohesive club-like ambiance for its members contributed to a gradual proliferation of societies for other types of engineers, which emulated its rules, style, and publications. Nevertheless, until the 1880s the ICE regarded itself as the leading representative of the profession. Having received a royal charter, in 1828, and maintaining premises in Westminster close to Parliament, it jealously guarded its members' interests.[11]

The ICE rule against the public discussion of current contracts did not prevent members from dissecting each other's work after it was completed. A good analysis of the Main Drainage system and the Thames Embankment emerges from a long evening session at the ICE in March 1865 (after a paper by J. W. Bazalgette) and again in April 1878 (after a paper by Edward Bazalgette, J. W.'s son and protégé). Speakers faced an audience of engineers, architects, and contractors who had good heads for figures and construction details, had often served in planning and oversight committees, and had usually competed for the same contracts. These conditions promoted remarkably specific, wide-ranging, and sometimes acrimonious exchanges.[12]

Although the ICE proceedings display an impressive grasp of engineering principles, the acquisition of professional expertise was still a hit-or-miss affair. Entry into the field was not yet regulated by certification, as in medicine or law, and training was still largely private and practical. The situation reflected engineers' uncertain social status as professionals. The gentry did not want their traditional classical education supplanted with practical knowledge, and even reforming liberals objected to state subsidies for science and technology.[13] Chairs in civil and mechanical engineering were established at the University of London by midcentury, but most engineers did not follow a regimen of academic study. Instead, they were typically articled as apprentices to a successful engineer and assigned to various projects until they were thought fit to consult on their own. Such practical training contrasted with preparation in France and Germany, where government institutes provided a more theoretical approach; but British engineers claimed that their training was far more realistic.[14]

Most historians have seen British engineers' individualism and lack of theoretical training as major sources of weakness, leaving common technical problems to be tackled on project after project without deriving a general approach.[15] "Pointed remarks" were made after the Paris Exhibition of 1855 about the sophistication of French and German engineers in comparison to their British counterparts.[16] But a lack of theory did not necessarily mean a lack of knowledge; there was no doubt that, whatever its curricular deficiencies, the profession was becoming more scientific.[17] The building of the Embankment shows that engineers and contractors commonly used methodical experiment and testing to solve engineering problems.

Victorian engineers liked to picture their early apprenticeship as a demanding but close relationship with a well-known veteran. It is unlikely, however, that this was always the case, any more than it is the case today between undergraduate students and famous university professors. Sir John Rennie (1794–1874) described a more credible situation in his *Autobiography:*

> A youth leaves school about the age of seventeen or eighteen, without any previous training, and his parents, thinking that he has got a mechanical turn, as it is termed, decide at once to make him a civil engineer, whether he likes it or is fit for it or not. They then send him, with a considerable premium, to an engineer of some standing and practice, who, unless special conditions are made (and very few engineers will make them), will not undertake to teach him the profession. The pupil is sent to the office, and placed under the direction of the principal assistant, who directs him to do whatever is required, if he can do it, whether drawing, writing, or calculating, or anything else; and if he wishes to learn anything, he must find it out himself; neither the principal nor assistant explains the principles or reasons of anything that is done. If he prove to be steady, intelligent, and useful, keeps the regular office hours, and evinces a determination to understand thoroughly the why and wherefore of every kind of work that is brought before him, and by this means acquires some practical knowledge, he will soon attract the notice of his employer, and will be gradually transferred from one department to another, until the expiration of his pupilage, which varies from three to four years; then, if he really has acquired a competent knowledge of the profession, and the employer thinks his old pupil can be of further service to him, he is engaged at a moderate salary, to be employed in such capacity as he is fit for. If during his pupilage he has made but little progress, nothing beyond mere routine, he is discharged with a certificate according to his merits, and sent into the world, to find his way forward as best he can.[18]

By midcentury, when Rennie was retired, young men of the lower and middle classes tended to have a rosier view of apprenticeship, because of the triumphant achievements of his generation. Samuel Smiles had made the names of Telford, Stephenson, and Brunel into household icons. British engineers travelled the world to build harbors, railways, canals, monuments, and sewer systems. A domestic, government-sponsored training program seemed superfluous. The master-pupil relationship provided ready-made assistants and practical experience in a growing field, as well as a strong professional network among leading consultants. While many apprentices were undoubtedly exploited, those with "a good head for business" were quickly placed in positions of responsibility and recommended to other engineering firms. For example, Thomas Gardner

(1841–1871) was articled to J. W. Bazalgette in 1859, upon the recommendation of I. K. Brunel. Finding the youth skillful and eager, Bazalgette employed him constantly on the design and execution of the London Main Drainage. After his term of service, Gardner was quickly hired by George Furness and Company, a leading engineering and contracting firm known to Bazalgette, and was put in charge of the northern sewer outfall contract when barely twenty-one. Furness then sent him to Odessa on railway construction projects. In 1868, he returned to London to complete Furness's contract for the Westminster end of the Victoria Embankment.[19] Other memoirs of ICE members show that Gardner's career was not atypical. The life of a young engineer on foreign service seemed adventurous and exerted an irresistible charm on many boys growing up in quiet inland villages.[20] It could also mean exhaustion, early disability, or even death from disease or overwork, as in Gardner's case.

Young engineering aspirants with more skills than money often ended up serving as clerk of the works. A clerk's on-site supervision of the construction process featured long hours, close attention to detail, confrontations with foremen and contractors, considerable responsibility without much authority, and a lack of job security. Clerks usually trained as apprentices in the building trades or surveying, and then worked their way through low-level jobs under a major contractor or in the Metropolitan Board of Works. Their duties, outlined by Bazalgette in an 1868 report, were to account for all the contract specifications, measure the contractor's work, regulate construction lines and levels, and assure the quality of materials used.[21] The official workweek was forty-eight hours (about average for other sorts of clerks), but overtime extensions were common. Clerks of works were usually hired on a temporary basis and shifted from site to site as senior engineers saw fit. Their families claimed that constant exposure to sewer gas and other pollutants led to a high rate of illness and early death.[22]

Although engineers as a whole were winning the battle for respectability, only a minority held the coveted status of professional consultant. Most were employed by railway, mining, and building companies or involved in speculative ventures, and the line between professional and journeyman was still vague. Since the ICE was slow to adopt rigorous standards of academic training and examinations, certification was still a hit-or-miss affair, especially in the provinces. Besides, the term "engineer" could also denote a skilled mechanic. If the long parade of witnesses who appeared before the various Thames Embankment commissions and committees is a good guide, it was common for men of questionable competence and experience to assume the title of "engineer" or "architect" and speak assuredly about the design and construction of public works. Such people got short shrift from the Royal Commission of 1861, because the chairman, Sir William Cubitt, was a former president of the ICE.[23]

A young accredited engineer might spend years in the lower and middle ranks of a nebulous system of advancement, since success depended on social and professional contacts, ambition, and luck. If successful, however, he could look forward to an affluent, upper-middle-class life and the possibility of styling himself a gentleman. At the Metropolitan Board of Works in 1870, Bazalgette earned a salary of £2,200 (including £200 for travel and expenses) and his three assistant engineers £1,000 each. These amounts were slightly above average for the profession at that time, but all of the men had been employed at the MBW since at least 1855.[24] Bazalgette supplemented his salary with private consulting contracts, which helped support a large household, with carriage and servants, at Wimbledon in Surrey.[25] His two most senior engineering clerks, who had held their positions since 1844, earned £300 (up from £220 in 1863), while the several dozen clerks who supervised construction at the project sites earned between £78 and £218.[26] Over half of these were able to keep a servant, one of the distinguishing marks of middle-class life, which, according to Francis Sheppard, required an income between £125 and £150 per year.[27]

Most leading engineers, however, did not work for the government. If they served in any official capacity, they did so on commission, as Thomas Page did at the Office of Works. For most commissions, a consulting civil engineer was paid a fee, based on time and costs, for investigating a client's requirements and writing a report. If that was approved, he produced the design, calculations, and general drawings. He might assist the client in locating a suitable contractor but was not supposed to accept any fee from the contractor for doing so. When the work was contracted, the consultant appointed a resident engineer to handle the details of construction. The consultant would make frequent site visits and exercise a general supervision of the project.[28]

Not all engineers maintained such a lofty professional autonomy. The flair with which new railways, docks, or suburban housing developments were often promoted, and the speculative nature of their financing, led the less successful down the path of expediency. Sir John Rennie acidly described the way a contractor might accept shares of stock for backing an engineer's prospectus, inflate costs and profits, and pack the company's board with his own nominees to "look after his interests." As a result, "the engineer, who ought to be his master, loses all control over him, and in many cases becomes little better than his servant."[29]

Of course, many engineers operated as members of a firm, and some did their own contracting as well. Edward Ladd Betts (1815–1872), partner with Samuel Morton Peto (1809–1890) in a leading railway-building company, was a civil engineer by training, as was Hamilton Henry Fulton (1813–1886), who formed a partnership with railway engineer John Fowler and contractor George Willoughby Hemans (1814–1886) to promote a private Thames Embankment scheme.[30] An engineer could also act as agent for foreign governments to procure other engineers or contractors for specific works. John Rennie Jr. and Bazalgette were agents for Russia when they landed the Odessa contract for George Furness.

Leading engineers earned enough to purchase suburban or country houses, as well as to furnish offices in London. R. A. Buchanan has illustrated their social mobility. Many were knighted for their professional achievements, Telford and Stephenson were buried in Westminster Abbey, and Sir John Fowler acquired a Scots highland manor and entertained royalty there. Practical-minded men, they seldom filled public office beyond the rank of local magistrate. Yet they were often called as consultants and witnesses before parliamentary committees and royal commissions. Reading their testimony, one is struck by their command of information, their confidence, and their directness.

The *Post Office London Trades Directory* for 1866 lists about 500 mechanical engineers and about 280 civil engineers of all types. A few are specifically noted as consulting engineers, and members of the Institution of Civil Engineers are starred. Most of the prominent engineers were established near Westminster. One group clustered near Adelphi Terrace, and a much larger group inhabited Great George Street and spread around the corner into Parliament Street, along with a flock of architects, surveyors, construction-related solicitors, parliamentary agents, and a few notable contractors. The ICE itself was located there (and still is, though across the street from its former premises), and more than one advocate of competing Thames Embankment schemes referred to "the Great George Street clique" as an oligarchic presence. Over half the proposals to the Royal Commission of 1861 came from these engineering firms, and their advocates appeared familiar with each other's designs. The rule against discussing current projects in ICE meetings obviously did not hold in the corridors nearby.[31]

It is interesting to compare the professional aspirations of engineers with those of engineering draughtsmen. There was as yet no efficient process for making multiple copies of drawings. Sometimes a very thick, slow-drying India ink was used for the original, and then blank paper was pressed onto it. That yielded five impressions at most, and the quality was uneven. Therefore, engineering firms, company promoters, architects, and builders of all kinds employed platoons of young draughtsmen, usually on a temporary basis, to turn out drawings required by clients, parliamentary committees, and the like. Draughtsmen also produced watercolor "views" of proposed projects, although such promotional work was often done by commercial illustrators. Few standards attached to the trade, and the level of experience varied even more than among engineers. An exchange of letters in the *Engineer* for December 1859 revealed that "draughtsman" was about as descriptive as "clerk." Some were simple copyists and watercolorists, young men with little mechanical knowledge whose parents bought them a set of pencils or paints and set them up in what was regarded as a respectable line of work. Some were architectural draughtsmen who understood elements of design and shading. The elite few were trained in mechanics and the principles of projection. They actually understood the objects they were drawing and could elaborate a set of working plans from a general sketch.[32] Metropolitan draughtsmen sought social respectability in much the same fashion as the engineers who employed them, and probably with some assistance from the latter. The *Engineer* advocated the formation of a professional society, and, by 1866, "The London Drawing Association" was established among the engineering offices on Duke Street, Adelphi, with Frederic Young, C.E., as its manager. By the end of the 1860s, the superintending draughtsman at the MBW, perhaps the premier post a draughtsman could aspire to, earned £500 annually.[33]

A different pattern of advancement can be seen among the "practical" engineers, the mechanics, smiths, molders, engine drivers, and millwrights who, with carpenters and masons, formed an elite within the skilled labor forces building the London Main Drainage and the Embankment. Although the consulting engineers had sprung from their ranks only a generation or two earlier, the engineer-mechanics of the 1840s and 1850s remained proud "aristocrats of labour" who disdained middle-class pretensions, practiced the self-help gospel of Samuel Smiles, and supported the cause of Reform. They were the most technically advanced of all the trades, and they formed the first real trades union, the Amalgamated Society of Engineers, in 1851. Some twelve thousand members of regional societies and branches of unions were united under the leadership of William Newton (1822–1876), who later became a prominent member of the Metropolitan Board of Works. The society's shilling-a-week dues gave it money and power. It was by far the largest trade organization of its time, with salaried officials and an impressive London headquarters.[34]

Asa Briggs remarked that the engineer-mechanics typified their age by declaring their object to be the securing of a vested interest in society, on the same order as doctors, brewers, or authors, rather than the representation of the workers as a general class.[35] But they were quite willing to cooperate with other trades unions to improve their lot, and though real wages rose about 56 percent between 1850 and 1874, local conditions, periodic financial crises, and legal reverses required strong union responses.[36] An act of 1855 had allowed engineers and other skilled tradesmen to register as friendly societies, thereby bringing their funds under the purview of the courts. This reduced the old problem of embezzlement by union officials and strengthened the workers' determination to press for better working conditions, including a reduction of hours from eleven and a half hours to nine hours a day. The financial crisis of 1857, however, put many building employees out of work. When one of their union spokesmen was fired by

Messrs. Trollope of Pimlico in July 1859, the builders, led by George Potter, went out on strike. The contractors not only refused the nine-hour day but threatened to lock the men out until they signed "the document" renouncing union membership. The long, bitter strike closed some 225 firms employing 24,000 men, including those working on the London Main Drainage. It was supported by all the other unions—the Engineers, Bricklayers, Stonemasons, Ironfounders, and Carpenters—and, though the eleven-hour day remained standard until the end of the decade, the "document" was defeated. This action led to the formation, in 1860, of the London Trades Council, no longer just a local group but the common council for a half-dozen strong national unions, strategically placed to influence Parliamentary legislation and bring their cases to the highest courts.[37]

Relations with contractors were not always so bitter. In the winter of 1860–61, the Main Drainage works had to be closed because of freezing temperatures, and a large body of artisans and laborers were laid off. Perhaps through the influence of the London Trades Council, or the mediation of the MBW, the Main Drainage contractors gave private aid to the men and arranged for them to be employed as navvies, moving construction materials about the various building sites and extending excavations wherever possible.[38]

During the 1860s, response to union activity from the middle and upper classes was mixed. Some appreciated the skilled workers' dedication to self-improvement and the restraint of those unemployed by sudden economic changes. They looked to the trades unions for support in the resurgence of radical Reform. Gladstone was one of these. He was certainly impressed when, in 1865, the day before he announced his conversion to Reform and strode "unmuzzled" before the electors of South Lancashire, a deputation from the Society of Engineers asked him to allow union funds to be deposited in the new Post Office Savings Bank he had created.[39] But other people grew alarmed at union expansion and the continuing violence in some industrial areas like Sheffield. Businessmen in Parliament still quoted David Ricardo's "Iron Law of Wages" against worker combinations and wage increases, in spite of the fact that their own workers disproved Ricardo by negotiating raises during periods of prosperity and suffering cutbacks during depressions.[40] Men who opposed further Reform measures certainly linked them with working class radicalism, since the mass demonstrations in support of Reform during 1866–67 coincided with riots by workers laid off during the financial crisis of those years. As a result of these disturbances, a royal commission on trade unionism was appointed in 1867. The London Trades Council got two middle-class sympathizers appointed to the commission and, over the next two years, presented their case with such skill that the majority report recommended the legalization of all unions and the protection of their union funds. These recommendations were incorporated in the Trade Unions Act of 1871 by Gladstone.[41]

The wages paid to workingmen on the Thames Embankment varied slightly from year to year, or from project to project, but the difference between skilled and unskilled was very clear. For the first section of the Victoria Embankment, the MBW specified that bricklayers, masons, carpenters, and smiths should be paid 6s. 6d. for a ten-hour day, and engineers 7s. 6d. Excavators wearing their own "long water boots" earned 4s. 6d. and common laborers 3s. 9d. The same wages obtained in the 1865 contracts for the Albert Embankment, and in the 1871 contracts for the Chelsea Embankment, except that day laborers earned 3s. 6d. in 1865, and bricklayers got 7s. 6d. in 1871.[42] If an engineer worked a ten-hour day (and not the eleven-and-a-half-hour day common among builders before the 1860 strikes) six days a week, with the usual lapses for illness or weather, he could earn about £110 per year, equalling the wages of a young civil engineering clerk of the white-collar variety. Imagine the trials of a newly minted "clerk of the works" sent out by the MBW to supervise a sewer or Embankment construction site, when he encountered his unionized and experienced "engineer" counterpart, who not only knew more about the practice of construction than he did, but earned higher wages.

The MBW specified wages for laborers in order to undercut the common practice of hiring "butty" gangs. These were small groups of men, often Irish, mustered by a tough foreman who obtained jobs by competitive bidding for the whole gang. Foremen typically took on relatives and friends but often shaved their pay. The butty system resulted in uncertain wages for the men and poor workmanship at the job site.[43] Whether rates were set for skilled workers by collusion with the contractors, by arrangement with the trades unions, or for other reasons is uncertain. Stipulating wages in the contract did have the effect of stabilizing income through the financial uncertainties of long projects, thus providing some security for workers and, at the same time, assuring the MBW and its contractors of a relatively continuous work force. After the strikes of 1859–60, this would have been a prime consideration, but it was also beginning to spread as a general principle of labor organization, especially among the large-unit industries such as shipyards and engineering works.[44]

The number of workmen engaged for the various contracts of the Main Drainage system and the three Embankments is hard to determine. The size of the labor force was not specified by the MBW, for obvious reasons, and was not often reported by the board's engineer. Hiring might be affected by the type of construction underway, the supply of building materials, strikes, floods, or weather. Contractors commonly moved workmen from one job site to another as needed. But the numbers were large. In September 1866, contractors Thomas Brassey and George Furness put five hundred men to work on the first of the three sections of the Victoria Embankment. Three years later, they were employing 1,250. William Webster hired sixteen hundred men to construct the two Main Drainage pumping stations and the southern outfall sewer in 1860–61. Figuring in the rest of the sewer system and the four other main contracts for the Embankments, one can get a rough idea of their impact on London's labor force.[45] The large number of skilled workmen employed over a period of a dozen years on highly visible and much discussed projects helped consolidate the position of the metropolitan union officials and the London Trades Council as leaders of Britain's labor movement.[46]

Contractors like Brassey, Furness, and Webster, who could hire thousands of men for large-scale operations, were a relatively new phenomenon around London in the 1850s and 1860s. The great canals of the eighteenth century were built to link up the headwaters of river systems far north of London. The railways, approaching the metropolis from all directions, were regional rather than local phenomena. Until midcentury, civil engineering projects in London, as elsewhere, tended to be relatively modest in comparison, and contractors operated on a workshop scale as individuals or in ephemeral partnerships. The two areas where large-scale organization and financing developed were in the new docklands and in the residential housing estates of Belgravia, Bayswater, Pimlico, and the expanding suburbs. The latter scene was dominated by the figure of Thomas Cubitt (1788–1855), who had a huge works and warehouse on the waterfront at Pimlico and kept a large labor force under continuous contract. Cubitt's biographer, Hermione Hobhouse, says that the scope of his operations and the number of men he employed remained exceptional in his lifetime; other historians argue that his techniques were widely imitated. The question is obscured by the fact that master builders like Cubitt were not always classified as contractors in the engineering sense.[47] Obviously, however, contractors could not tie up capital in a trained labor force and stocks of materials unless they were in a position to bid for and win a variety of contracts, including a few large, long-term ones. The emergence of the Metropolitan Board of Works and its sewer, street, and Embankment projects helped create these conditions.

The large-scale contractors formed a loose network that complemented the metropolitan engineering community, although the two groups functioned somewhat separately. Contractors had no institution or club to mark their social aspirations, and their memoirs often stressed humble (but hardworking) origins. They tended to have less formal education than engineers and to display greater entrepreneurial vigor. Whereas engineers might be sensitive to environmental issues like pollution, landscape aesthetics, architectural perspective, and public green space, contractors viewed the environment as a source of materials and physical site to reconstruct and build upon. Contractors were perhaps more likely to sit on local vestry and district boards to protect their immediate interests, as contractors do today, but less likely to seek higher office or honors. Those elected as delegates to the MBW did not distinguish themselves politically. Their firms were usually family affairs or working partnerships that seldom survived the death of the founder, and their financial operations were speculative and run on slim margins. They lived in a world of contingency, vicissitude, hyperbole, and sharp dealing. Most of their day-to-day agreements with subcontractors, foremen, and laborers were necessarily oral, and it was not in their interest to keep detailed written records of business dealings. What we know about them is reconstructed from directories, parliamentary committee testimony, the occasional memoir, reports by the MBW engineering department, and by extrapolation from more recent practices.

Major contractors tended to set up offices at their warehouse and manufacturing premises around London, although several located in the Great George Street neighborhood. Among those who bid on the Embankment contracts, Samuel Morton Peto and Edward Ladd Betts, George Furness, Alexander Ogilvie, and Samuel Ridley all listed Great George Street addresses; Henry Jackson and John Bayliss were in nearby Parliament Street. William Moxon moved his office next to the MBW headquarters in Spring Gardens, off Trafalgar Square, and William Webster established his around the corner at St. Martin's Place. Thomas Brassey, a major railway contractor who also built docks, canals, streets, and sewers, was located at St. George's Square, Pimlico, near Thomas Cubitt.[48] Contractors often maintained several warehouses and wharves whose importance varied according to the type and location of current contracts. In 1866, Webster kept a special sewerage office on the east side of London, where he had several works in progress; in 1872, he listed an address at Swan Walk, Chelsea, because he was constructing the Chelsea Embankment. Peter and Alexander Thorn had works both at Cremorne Wharf, Chelsea, where the low-level sewer for the Main Drainage began, and at Blackfriars, where the Embankment section ended. Peto and Betts had a depot at Waterloo Bridge Wharf, and the firm of Mowlem, Freeman and Burt gave four locations, including Grosvenor Wharf, near the Grosvenor Estate housing development at Millbank.[49]

Because of the scattered locations of their premises, it is harder to call the contractors a "community" in the same sense as the engineers. Yet they had a common fund of experience building London's docks, sewers, streets, railways, and bridges. They negotiated and consulted almost daily with the engineers (in fact, some of them *were* engineers). They also entered into countless oral and written contracts with one another for staffing, financial support, materials, and equipment.

Flexible partnerships and subcontracts were necessary because of the uncertainties of contract financing. In the first place, a contractor might incur enormous expenses just winning a contract and was not reimbursed by the client until each stage of construction was completed. As we have seen, large-scale projects often required private acts of Parliament, in which case the contracting firm would have to place notices in the major newspapers, notify affected landowners, deposit earnest money with the appropriate government office, and pay for preliminary engineering plans. They had to prepare estimates for the bid and be ready to assemble workmen, materials, equipment, and temporary buildings at the site. Arthur Helps provides a detailed description of the estimating process in his biography of Thomas Brassey. When Brassey received a proposal, usually accompanied by plans and sections, he would give them to a trusted assistant to investigate. The assistant returned with an oral report and Brassey decided, then and there, whether to go forward. He sent the assistant back to check the site, measure the whole project on the ground, examine the contract drawings and specifications, and check costs of materials and labor. Then the agent "took out the quantities" (drew up detailed cost estimates using the contract data) and reported back. The assistant had cause to be diligent, as he was the employee most likely to supervise the project.[50]

Brassey often made final decisions on contract bids after perusing the figures for less than twenty minutes. He could do this because the figures were frequently prepared by his trusted partner and brother-in-law, Henry Harrison. But he was also trained in a system of rapid mental arithmetic, a useful skill in that precalculator age. George Parker Bidder, an MBW consultant on the Embankment design, was the acknowledged genius and exponent of this system among London engineers. As a boy, he had toured sideshows and music halls exhibiting his amazing powers of computation, and once, in the 1850s, during a parliamentary committee meeting, he recalculated all the figures for the London Main Drainage in about five minutes, discovering a significant error in the total.[51]

When the Royal Commission of 1861 had invited proposals and plans for the Embankment, it published criteria in the official *Gazette*, but these were so general that contractors could only guess at the costs of materials and construction, let alone property compensation.[52] Even after the MBW had been chosen to build the Embankment, and had established its functions and boundaries, the plans Bazalgette submitted to the House of

Commons select committee, in 1862, were little more than preliminary sketches. A full set of thirty-four drawings (elevation, section, and detail views) and fifty-six pages of specifications was not ready until mid-July 1863. As a basis for estimating costs, the MBW also displayed a series of sample cores drilled along the Thames low-water line to show the soil composition down to hard clay in each section. These proved to be inaccurate on several counts. Although they could be compared with the ordnance survey maps completed during the previous decade, they led several contractors to adopt inappropriate techniques for excavating and laying foundations. Technically, the Embankment itself was fairly straightforward, but some of the specific methods and materials were new. The utility of Portland cement, iron sewer pipes, and steam-driven excavators was still being argued. Work sites had to be insulated from the river by extensive pile cofferdams, and maintaining excavations against twice-daily tidal changes complicated work schedules and methods.[53]

The contingencies of large-scale construction, and the natural tendency for both parties to interpret contract specifications to their own advantage, encouraged civil engineers and contractors to rely on personal relationships and prior business experience to supplement their solicitor's inquiries. The ICE provided one informal clearinghouse for such judgments, and the MBW, as one of the largest sources of contracts in the metropolis, became another. But kinship remained an important criterion among both engineers and contractors, as it did for many other segments of Victorian society. Thomas Brassey's use of his brother-in-law as agent was not unusual. The great London firm of Peto and Betts, who helped Brassey build the Victoria Docks in the 1850s, had similar features: Samuel Morton Peto's father had been one of the leading contractors in England, erecting the Houses of Parliament in partnership with his cousin, Thomas Grissell. The younger Peto sat in Parliament from 1847 to 1854, promoting himself and his business; Betts, the practical engineer of the firm, was married to Peto's sister Ann.[54]

George Parker Bidder, the "calculating boy," followed his father into the business, as happened with the Stephensons, Brunels, Rennies, and Mylnes. The MBW records of contractors who bid on the Embankment includes John Aird and Sons, George Baker and Son, Henry Lee and Son, and George Myers and Sons of Lambeth. Alexander Meadows Rendel, who prepared estimates and plans for a proposed embankment in 1860, was the son of an engineer who built the West India Docks. The designer of Victoria Railway Station, Charles Fox, formed a partnership with his sons Charles and Francis in an ill-fated attempt to construct a railway under the Thames in 1863 (see chapter 9). Thomas Pearson bid individually on the Embankment, then built sewers in company with his son. Even Mrs. Pearson (presumably Thomas's widow) landed a sewer contract with the MBW in 1878. The record for father-son continuity, however, belongs to Stephen Leach (d. 1843) and his son Stephen William (1818–1881). As clerk of works for the City of London, the elder Leach served the City's Thames Navigation Committee for over forty years. His son succeeded him as engineer in 1843 and continued with the new Thames Conservancy from 1857 until his death, playing an active role in the design and subsequent maintenance of the Embankment.[55]

Thomas Cubitt was also part of an extended contracting family. His brother William was engineer to the London and Croydon Railway, twice Lord Mayor of London, and (as mentioned above) chairman of the Thames Embankment Commission of 1861. A second brother and a cousin were listed as civil engineers in Great George Street. The Barlows were another such family: William Henry (president of the ICE in 1880) and his brother Peter shared offices in Great George Street, while Thomas worked around the corner in Parliament Street.[56]

Just as engineers recommended their pupils to colleagues, contractors employed each other's relatives and the sons of their favorite workmen. George Pauley (1854–1925), a wealthy contractor who began his youth as a brawling railway navvy, recalled being handed on from one construction job to another by friends of his feckless but gregarious father. In 1869, he was hired as a combination draughtsman and night watchman for the Metropolitan District Railway's new Embankment line, then under construction.[57]

Although kinship and patronage were significant features of the engineering and contracting networks in the London region, they appear to have been exercised with discretion in favor of men and likely youths with "a head for business" and appropriate skills for the job. This did not preclude putting worthless nephews on the payroll from time to time, but the increasing technical and financial complexities of engineering tended to discourage amateurism rather earlier than in government, the army, or the established church. And though contractors might make oral agreements with foremen, suppliers, and subcontractors, they dealt with each other and with consulting engineers almost exclusively through the specifications of legal contracts where the bottom line was money, not influence. Furthermore, the patronage exercised within the world of engineering and construction was no substitute for that other, more traditional kind of patronage, represented by the names of City bankers, members of Parliament, and titled aristocrats on the prospectus of every joint-stock venture offered to the investing public.

It was natural that the close relationships among engineers and contractors should carry over into the operations of the Metropolitan Board of Works. Because it was a new element in London metropolitan politics, which led eventually to the London County Council, it is tempting to see the MBW as a departure from old-style kinship and patronage systems. But the MBW's success was due in large part not to its political delegates but to the achievements of its engineering department. The technical staff, though hired into a rapidly developing bureaucracy, not only brought their previous training and associations with them but dealt regularly with the independent contractors and engineers around them. Gloria Clifton's meticulous study of the MBW shows that the chief officers in each department had considerable influence over the selection of temporary employees, and over half the permanent staff came from this pool. This created a substantial arena for patronage. The board encouraged Bazalgette to nominate candidates for jobs or simply gave him power to hire a certain number. Often only one name was proposed for each position. In his turn, Bazalgette let the assistant engineers recruit clerks of works for their assigned territories. As one result, about 10 percent of the engineering staff hired between 1856 and 1888 were sons of MBW employees, including one, a pupil of Bazalgette's, who was given a job to help support the family after his father's death.[58]

That the style of MBW operations was not more bureaucratic, looking perhaps toward the formation of the Civil Service in 1870, was primarily due to the fact that Bazalgette remained chief engineer for its entire life from 1855 until 1889. Although I have contrasted his organizational talents with the idiosyncratic style of Thomas Page, Bazalgette remained in many ways a product of his early training. Many of his own apprentices found their way into MBW positions, including at least three who were related to incumbent employees.[59] His most conspicuous recruit was his son Edward. By all accounts an unusually capable young man, Edward was articled to his father in the 1860s and assigned jobs on the Main Drainage and Embankment projects during his apprenticeship. He won two prizes from the ICE for his early work, and, though he was not a prominent figure in the construction of the Embankment it was he (with his father looking on) who later entertained the ICE with a lengthy description of that complex achievement.[60] Edward was the only member of the MBW engineering staff hired into a permanent position without competition from other candidates and without previous work in a paid temporary post. Instead, his father had his qualifications verified by the president and a vice-president of the ICE.[61] Although his rapid promotion between 1870 and 1880 occasioned murmurs among the staff, his success was probably due as much to the rigorous and extensive training provided by his father as it was to the latter's position at the MBW. None of his five brothers (nor, of course, his four sisters) was trained as an engineer and none was employed at the MBW. Nevertheless, the connection between father and son was viewed, even then, as somewhat inappropriate for a public agency.

A more traditional type of relationship, though one that ran deeper than most, grew up between two men closely associated with Bazalgette: his long-time assistant engineer, John Grant (1819–1888) and the MBW's favorite contractor, William Webster (1819?–1888). Their work together illustrates what the Embankment meant to two of its related social groups, and how difficult it was to distinguish between patronage, professionalism, and bureaucratic practice in this period.

Glasgow-born John Grant was appointed an assistant surveyor at the Metropolitan Commissioners of Sewers office, in 1849, through the influence of Edwin Chadwick. In 1852, he became assistant engineer assigned to the MCS district on the Surrey side of the Thames. When he transferred to

Figure 7. *(left)* Metropolitan Board of Works Chief Engineer J. W. Bazalgette in 1859. *(right)* William Webster, the principal contractor for Main Drainage and Embankment works, in 1875. (Bazalgette: *Illustrated London News,* 12 March 1859; Webster: from an original photograph, courtesy of the Institution of Civil Engineers Library.)

the MBW in 1856, his territory expanded to cover the whole southern part of the metropolis, and he thus came to supervise construction of the southern sewer lines for the London Main Drainage. In 1863, when Parliament approved construction of the Albert Embankment from Westminster to Vauxhall Bridge, he naturally supervised that as well. At one time, when Bazalgette fell ill, he was made temporary chief engineer.[62] An affable, "exceedingly painstaking" employee, he aimed at a high standard of perfection. His claim to fame, however, was not the sewers or embankments themselves, but the development of the Portland cement concrete that went into them.

Although cement and concrete are ancient compounds, the chemical process through which they are formed has not been clearly understood until very recently. Cement, usually a form of calcium silicate mixed with water, gradually hardens by drying. When ground to a powder, burnt in a kiln, and mixed again with gravel, sand, and water, it undergoes a chemical reaction—hydration—to form concrete. In the early nineteenth century, most concrete was of the "Roman" variety, discovered in 1796 and popularized by Smeaton, Rennie, and Telford. Made from stone found near Sheerness at the mouth of the Thames, it set up quickly even when exposed to water and could, therefore, be used for hydraulic works. Portland cement concrete, patented in 1824 by an English bricklayer, not only set up under water but was twice as strong as Roman cement.[63] Until the 1860s, however, it was quite unreliable. Experienced "mixers" at the manufacturing plant, working like brewmasters with secret recipes, could get consistent test batches, but clerks at the work site could seldom duplicate the results. The size of batches was, therefore, limited, and Portland cement was used mostly as mortar for brickwork. Most of the larger sewer pipes, and the Victoria Embankment walls, were made of brick.

John Grant began experimenting with Portland cement concrete in 1858, when the Main Drainage project got underway. Over the next thirty years, through ten thousand on-site tests, he investigated the properties of a vast array of mixes. Through this trial-and-error empiricism, typical of British engineering then and since, he achieved the degree of reliability needed for large-volume applications and found many new uses. In particular, he demonstrated that the large sewer pipes used in the Main Drainage could be formed from concrete rather than from time-consuming brickwork, and that the Embankment walls, faced with granite slabs, could be backed with poured concrete instead of brick or rubble. His three papers on the subject won a Telford Medal and other prizes from the ICE, of which he became a member in 1861.[64]

Grant's work on cement flourished on the site of the Albert Embankment (figure 9). This wall and roadway project was necessitated by the low-lying, marshy condition of the Surrey side of the Thames, where flooding was common. Bazalgette wanted to embank the whole length of the river down to London Bridge, but the commercial interests there were judged too important, and the First Commissioner of Works restricted the plan to the most vulnerable section at Lambeth, where the river, turning north, cast its current. As David Owen remarks, the chief engineer proved more prescient than the commissioner in this case, because the south side was hit by repeated flooding in the mid-1870s and had to be embanked piecemeal, to no one's satisfaction (see chapter 9).[65]

The contract for the Albert Embankment was awarded to William Webster in 1865. Born in the Lincolnshire town of Wyberton, near Boston, Webster (figure 7) seems to have had little experience with large-scale construction prior to his arrival in London. In fact, his past is quite mysterious. Research by Neil Rhind, secretary to the Blackheath Preservation Trust, indicates that he was born either in 1819 or 1820, the first son of a Wyberton wheelwright; that he apprenticed with Thomas Jackson, joiner and builder of Boston; and that he probably married the daughter of a local publican. By 1857, he was set up as a joiner and publican at the "Pincushion" Inn, Wyberton, having bypassed the usual journeyman stage of his career. During the next few years, he is credited with building or restoring several churches in Lincolnshire and Nottingham, and with building the Cambridgeshire Lunatic Asylum at Fulbourne. In 1861, he claimed residence in Arlesey, where he had built the Three Counties Lunatic Asylum.[66]

None of this accounts for Webster's transformation, in 1860, into a London sewer contractor. In that year (apparently while still working on the Arlesey Asylum), he joined John Aird and Sons in a couple of unsuccessful bids for sewer projects and established a residence in the suburb of Blackheath. Soon he won major contracts for the Main Drainage southern outfall sewer and the regional pumping stations, including the picturesque Abbey Mills. Within a year, he was employing 1,600 men. By 1863, he was also starting construction of the southside low-level sewer (supervised by John Grant) and, before that was completed, won the contract for the Albert Embankment and a section of the northern low-level sewer as well. In 1868, the board appointed him to finish the disputed third section of the Victoria Embankment and pave the entire Embankment roadway. When that was finished, in 1870, he also built the approach roadways from Whitehall Place and the streets around Charing Cross Station.[67]

The professional or financial connection in London that would have enabled Webster to make such a spectacular transformation remains obscure. His early partner, John Aird, was an established contractor, and he later built his house very near the Airds' in Blackheath, but the nature of their relationship is not clear.[68] It is also curious, though more comprehensible, that Webster turned from church and asylum building to large-scale sewers and embankments in such short order. He seems to have been a master at organizing men, work, and money. Bazalgette found him efficient, reliable, and financially sound at all times. The chief engineer's monthly reports to the board repeatedly praise his construction techniques and the high quality of his finished work.[69] He even received a public testimonial from the workers he employed on the southern outfall sewer in 1862. When the original contractors for the Chelsea Embankment withdrew, in 1871, the board simply turned it over to Webster without further ado, and before he died, in 1888, he completed at least fifty other projects in and around London. At one point, according to Neil Rhind, the board just gave him a sum of cash to repair one of the sewer outfalls, without even bothering to write a contract.[70]

Webster apparently never studied engineering in a formal way, and his apprenticeship was as a joiner. As his London business grew, however, he was elected an associate of the ICE in December 1868 and began styling himself "Civil Engineer/Contractor." He also designed a seal depicting a leather belt formed into a circle, the buckle looped below, with his name emblazoned around it and the title "Contractor" in the center.[71] By this time, he had commenced building an imposing Italianate mansion, Wyberton, which still stands in Blackheath. Its facade is detailed with marble (of the type used in the Albert Embankment) and the lintel is decorated with relief carvings of the flowers and vegetables of his native Lincolnshire. The iron railings on the interior spiral staircase were cast in the double-dolphin motif used on the Embankment lightposts.

Webster began working with John Grant on the southern sewer system, but the Albert Embankment really shows the degree of professional collaboration they enjoyed. The two men solved a series of excavation problems caused by the low-lying, marshy terrain between Lambeth and Vauxhall, and Grant used the opportunity to expand the testing of Portland cement. As a result of his work there, whole sections of time-consuming and expensive brickwork originally specified for the main sewers and embankment walls were replaced by poured concrete. The design of the Chelsea

Embankment, authorized in 1868, was revised in consequence, and it was fitting that Grant and Webster should end up building it.[72]

Both men enjoyed a respectable family life. Webster had five children, Grant two. Grant became a widower at an early age, established himself in a modest semidetached villa near Webster's mansion in Blackheath, and became an intimate member of William and Ann's large circle of friends. Both worked steadily into old age and died in harness, in the spring of 1888. Grant struggled through severe illness to complete his last work, a memoir of his "friend and neighbour" whose funeral he had attended the previous week.[73]

Grant and Webster represented two interconnected groups, engineers and contractors, for whom the Embankment was an significant opportunity for professional achievement. Before the Royal Commission of 1861 and Parliament decided on the primary elements of the project, these men had a wide variety of ideas about what the Embankment should look like, how it should be financed, and how it might function in terms of navigation, sanitation, transportation, commerce, and so on. Once the basic design was decided, and the MBW assigned to plan and contract for its construction, the Embankment became an object of employment for thousands of assistant engineers, clerks, draughtsmen, skilled workmen, and laborers. Their commitment brought the first, conceptual phase of the project to a kind of closure. The spate of alternatives proposed along the way was largely forgotten. Future amendments and technical innovations, such as John Grant's use of Portland cement, were not ruled out. But the main aspects of the Embankment were established in such a way that almost all of the relevant social groups—wharfingers, engineers, members of Parliament—changed their former visions. The short shrift given to the Whitehall residents, who wanted to reverse this process, demonstrates that the process of closure was due more to social and political forces than to any technical imperative.

Neither Grant nor Webster had connections to leading political groups, and neither was in a position to influence policy decisions about the major economic or social issues of the day. They were, nonetheless, important figures in the building of Britain's urban-industrial infrastructure. Their work on the Embankment reveals a persistence of tradition in the context of technological and organizational change. They belonged to groups that were achieving a new level of social and professional recognition in the metropolis, yet expressed traditional values of apprenticeship, kinship, and patronage. The Embankment, street, and sewer contracts stimulated and justified the growth of the MBW as a new metropolitan administrative body, which inevitably changed the financing and organization of public engineering projects, but the MBW itself displayed an uneasy mixture of tradition and innovation, patronage and nascent bureaucracy right through the second half of the nineteenth century.

CHAPTER VIII

Contingencies

"Architecturally," wrote David Owen, "the Embankment was relatively uncomplicated." It was a huge undertaking, but the technology for building it had existed for half a century or more. H. H. Bird characterized most of the river wall as "child's play," and even Chief Engineer Bazalgette said afterward, "I get most credit for the Thames Embankment, but it wasn't anything like such a job as the Drainage."[1]

Despite these disclaimers, we know that the Embankment had a social and political side to it that must have continued to interfere with project plans. Add to these the difficulty of clearing an extremely cluttered site, building below water level, the vagaries of weather, irregular tides, interruptions of material supplies, accidents, breakage, and simple miscalculation, and we have a large set of contingencies to deal with. The whole idea of a project is to reduce such contingencies to manageable proportions. For Victorian contractors, "manageable" meant 10 to 15 percent of the total project cost.[2] Even that margin, however, proved insufficient for some parts of the Thames Embankment.

It's true that the engineers who replaced London Bridge predicted the impact of that change upon the river fairly accurately, even though they adopted a "wait and see" policy before recommending specific criteria for embankment. But then unanticipated problems and people working at cross-purposes defeated the plans of the most powerful authorities for many years. Investigating these complications in previous chapters, we have seen the combination of old impulses and new departures that contributed to the final plan. The views and actions of many individuals and groups interested in the Embankment were largely ignored or defeated and forgotten. The meaning and design of the Embankment evolved until, at a specific juncture, one design was chosen, and responsibility for the project was captured by John Thwaites and J. W. Bazalgette rather than by Thomas Page at the Office of Works or one of the competing private firms. It would be inconsistent with that narrative to depict the actual construction of the Embankment as a neat, rational path from design to finished product. Engineers coped with unexpected problems, with their own fallibility, and with new possibilities disclosed in the very process of construction. Incorporating their mistakes and surprises into the account will be truer to experience than glossing over them. This approach may also suggest wider application in the field of environmental history. If engineers, those exemplars of planning, assume a certain level of contingency in their projects, how much more ought historians to consider it?

The Thames Embankment Act of 1862 incorporated the essential features recommended by the Royal Commission the previous year (map 2). The Victoria section, the most complex of the three, was basically to be a concrete and brick wall extending a mile and a quarter from the foot of Westminster Bridge to the west side of Blackfriars Bridge. It followed Walker's line of 1840, curving four hundred feet out from the high-water mark at Scotland Yard, three hundred feet from the Marquess of Salisbury's house between Hungerford and Waterloo Bridge, 130 feet from Somerset House, and two hundred feet from Temple Gardens. The Albert Embankment, authorized in 1863, after further investigation and planning by the Royal Commission, would stretch from Westminster Bridge upstream to Vauxhall Bridge. The Chelsea section, authorized in 1868, simply extended Thomas Page's embankment from the west end of Chelsea Hospital up to Battersea Bridge, enclosing the new low-level sewer line. The walls of all three sections would descend at least twenty feet below ordnance datum* and rise four feet above the high-water mark. They would

---

*The datum was a line four and a half feet above average spring low water and twelve and a half feet below average high water, measured and marked at Trinity House, in 1800, to initiate the topographical or ordnance survey of England.

be faced with granite. On the river side, new steamboat piers and landing stairs were designed for river access. A million cubic yards of riverbed gravel and other fill would be dumped behind the walls to create a level bank. Within the filled ground, next to the river wall, the low-level Main Drainage sewer line, averaging eight feet in diameter, with thirteen-inch brick walls layered with concrete and asphalt, would intercept a myriad of large and small sewers. At every major junction, a large penstock would collect storm water and sewage overflows for discharge at ebb tide, as a means of preventing backups in the drainage system and of periodically flushing the mud banks.[3] The sewer line would be tied to the river wall by eighteen-inch-thick cross walls every six feet. Above ground, Bazalgette planned a tree-lined roadway and pedestrian walkways, surfaced with York paving stone, and decorative gas lightposts for the top of the river wall.[4]

The Royal Commission guidelines were, of course, not sufficient to organize construction, nor even necessarily workable in practice. "You know that in coming to Parliament you produce certain plans and sections," said Bazalgette later, "and if you get your Act of Parliament, you can go into things in detail, and prepare detailed or contractor's drawings and plans."[5] The engineering department had first to measure the sites, estimating the types and quantities of materials needed, and obtain samples for contractors to inspect. They took core samples from the riverbed to determine the depth of foundations and the location of difficult conditions like quicksand. Assistants prepared general drawings and sections for the main wall and sewer line, and detailed drawings for the landing stairs, lampposts, mooring rings, and other architectural features. In smaller projects, it was common to settle for general drawings with a few cross sections and details, leaving the rest to the contractor's experience. For the Embankment, however, Bazalgette prepared a set of thirty-four drawings so complete that they are consulted even today by maintenance crews. Every feature, every type of material, unusual procedures, and even wage rates had then to be specified. The specifications ran to fifty-six pages of dense print.[6] The plans were lithographed and the specifications printed so that the MBW could make copies available to all prospective contractors at the same time. These were sold for a fee of five pounds to discourage frivolous tenders.[7]

Bazalgette made several amendments to the plans as they were being drawn up. On top of the large, egg-shaped sewer pipes, he added brick archways that could carry gas and water lines and, later, electric power lines. This addition was questioned by a member of Commons on grounds of added expense and possible gas explosion, but Bazalgette explained that the archway was merely a series of vaults supporting the river wall, which would be built in any case, and perfectly safe.[8] He had originally thought to face the outside of the river wall with large iron plates, but rejected them in favor of more expensive granite facing after a fellow engineer testified (sensibly enough) that iron plates rusted too quickly in the presence of salt water and had a tendency to become brittle and crack in cold weather.[9] Bazalgette also suggested to the board that parts of the Embankment wall could be fashioned as vaulted storage cellars, like those under Somerset House, and leased commercially as a way to recoup expenses. This idea reflected the design of Somerset House and the Adelphi Terrace and was current practice under metropolitan railway viaducts. The board deemed it inadvisable.[10]

The plain, monumental lines of the river wall attracted several proposals from architects and amateurs for the kind of eclectic ornamentation many Victorians thought proper to public works. One of the many suggestions sent in by enthusiastic citizens was a series of tidal fountains described by W. L. Baker in a letter to the *Builder*. The fountains, shaped as mermaids, dolphins, etc., would be mounted on the formal landing stairs that led down from the wall at intervals. They would be fed from reservoirs within the walls that filled at high tide and emptied by gravity through pipes to the fountains at low tide. Baker declared his scheme inexpensive to build and maintain, with a free water supply.[11]

Bazalgette (and probably Thwaites as well) rejected such embellishments. Gas lampposts were practically the only ornamentation on the Embankment wall. Bazalgette took special pains to design them and exhibited his proposals at the 1863 Royal Academy show. The first pedestal treatment was roundly criticized by the editor of the *Builder* and promptly discarded (although the prototype was later installed near Carlyle's statue on Cheyne Walk, Chelsea, where it remains). The distinctive double-dolphin design seen today (figure 10), perhaps meant to symbolize an environmentally friendly river, came from a revised set of drawings published in 1864.[12] The bronze lion-head mooring rings mounted every twenty feet along the river wall (secured by long bolts to plates on the landward side) also evolved through a series of sketches. The MBW Embankment committee records include prototype drawings of landing stairs, the statue of Boadicea at Westminster Bridge, other ornamental stonework, ironwork sewer access covers, railings, and pedestrian benches, some with the gryphon motif still evident today.

The original plans also indicate that the York Gate, a squat stone archway guarding the river's edge at the end of Buckingham Street, was to be relocated on the Embankment in front of Whitehall Gardens, where the Royal Air Force Memorial stands now. The gate was a relic of a river wall built by the Duke of Buckingham, in 1626, around a palace originally erected for the Archbishop of York. During the eighteenth century, it fronted the York Waterworks, as well as a set of residences where Samuel Pepys and Peter the Great are said to have lived; in 1770, the Adam brothers built the Adelphi Terrace behind it.[13] York Gate marked the farthest reach of Hungerford Bay, some 450 feet from the proposed Embankment wall. Bazalgette specified that it was to be moved, "carefully restored, the stones re-dressed, and made good with new stones. . . ." However, the London County Council, successors to the MBW in 1889, found it overgrown with bushes and vines and "fallen into decay" at the back of Embankment Gardens. It was restored *in situ*.[14] The reason for leaving it behind is not clear; Bazalgette cited "legal and other difficulties." Most probably, it was incorporated into the gardens when Bazalgette's original plans for that site had to be abandoned (see below).

Bazalgette originally divided the Victoria Embankment into two contracts. Contract no. 1 covered the area between Westminster Bridge to Waterloo Bridge; no. 2 ran down to Blackfriars Bridge. In August 1863, however, Parliament authorized the Metropolitan District Railway to extend its underground line through the new Embankment to Blackfriars Station. Bazalgette had to create a third contract, which was not let until 1868, to cover the short but complicated stretch from Temple Gardens to Blackfriars Bridge where the railway would intersect the sewer line. This third contract would also include the paving and maintenance of the whole Victoria

Embankment roadway and sidewalk. To anticipate a bit, the whole of the (southern) Albert Embankment was covered by contract no. 4 and the Chelsea extension by contract no. 5.

A number of subsidiary contracts were, as usual, let for lighting, shrubbery, temporary walkways, etc. Each contract set wages for different grades of workers, as well as the prices expected for every type of material. Contractors had to deposit a substantial surety at the outset, from which a penalty of 1 percent of the contract amount per week would be deducted if they failed to meet deadlines.[15] They had to bring materials on the construction site for stated periods so that the clerk of works could inspect them before use, and they had to provide a substantial house and boats for the MBW's engineers. They were paid in installments as the work progressed, with the last 10 percent held until the contract was completely settled. Obviously, a contractor had to have solid financial backing. We have seen that capital was not easy to come by in the early 1860s, and it was even harder to obtain after construction started. Advances on the contract might be made by the chief engineer, but that was rare. If weather, mistakes, or lack of supplies delayed the project too long, the contractor might run out of money.

Thirteen firms tendered bids on the first contract, fourteen on the second, with only five common to both lists. Tenders for contract #1 ranged from Samuel Ridley's £495,000 to George Wythers's £620,000, and tenders for #2 spread from Anthony Ward Ritson's £229,000 to Thomas Pearson's £264,000.[16] All the tenders were opened at a board meeting and announced publicly by the chairman. The board was not required to accept the lowest bid. A well-known contractor whose bid was only slightly higher might be favored over one whose previous work or financial condition was suspect. Ritson and Company had a solid reputation and were awarded the second contract without discussion. But Ridley's bid on contract #1 was another matter. David Owen has described how the board rescinded its initial award to Ridley, giving the contract instead to George Furness, the fourth highest bidder at £520,000. Ridley's experience had been mostly in Canada and his letters of reference were vague, while Furness had substantial railway and other works to his credit. Unfortunately, Bazalgette was tangentially involved in a consulting venture with Furness. Although he had advised against hiring Furness, preferring to readvertise the contract, the chief engineer's apparent conflict of interest was made the ground of attacks upon the board's integrity for years to come.[17]

In April 1863, Bazalgette assured Parliament that three years would be "ample time" to complete the Victoria Embankment. He was not even half right. Although Furness began the first contract in good financial condition, a combination of delays led him into trouble. He gathered an inadequate labor force, but other problems were not of his own making. First, the board's architect George Vulliamy and its property appraiser George Pownall were unable to secure all the wharves and other property needed for access and storage in a timely manner. Owen's account of their frustrating negotiations with riverfront proprietors recalls Lord Lincoln's failed efforts to sign the Chelsea property owners in the 1850s. It confirmed the warning by engineer-contractor Hamilton Bird that England's jumbled compensation law could doom the Embankment, as it had already doomed previous ventures.[18] Furness was not able to occupy the construction site until 16 February 1864. The board had hoped to lay the official foundation stone in the spring of 1863, but the ceremony had to be postponed until the following October.[19]

Furness's second problem was a contract requirement to maintain access to the steamboat landings at Westminster and Hungerford, two of the most popular on the river, while building the Embankment.[20] This meant constructing temporary facilities and putting up with constant boat traffic next to the building site, where his own barges were operating. The situation was exacerbated by sightseeing boats, which made a nice profit from the thousands who wanted to gawk at the new works.

Third, Furness lost time and money experimenting with a new type of cofferdam advocated by Bazalgette. Cofferdams kept water out of the construction site, a crucial factor in building along the tidal Thames. Traditionally, they consisted of two parallel rows of wooden pilings driven into the hard London clay of the riverbed by a barge-mounted pile driver. The space between them was filled with clay and gravel puddle, which formed "a consistent and tenacious mass." Sluice gates were cut into the dam just above the low-water line to let accumulated water out of the site when the tide was out.[21] When the project was finished, the puddle was drawn out and the piles either pulled up or, if the river wall needed additional support, cut off under water. Despite their widespread use and simple functions, however, cofferdams had not been studied systematically. As with other aspects of mid-Victorian engineering, there were no manuals available, so each contractor had to figure out for himself how best to build them.[22]

To avoid possible damage to the foundations of nearby buildings (such as Montague House) from the shock of driving wooden piles, Bazalgette urged Furness to substitute wrought-iron caissons. This was a technique proposed by Lionel Gisborne, in 1854, employed at the Victoria and West India Docks, and used by Thomas Page for the Westminster Bridge. The caissons were large oval pipes with external slots on either side. They slipped over short wooden piles to form a kind of loose picket fence. Then long iron plates were slid down into the slots between pipes, forming a solid wall.[23] An iron cofferdam cost extra time and money to construct, but it was thought easier to dismantle, and its elements could be reused. In fact, the 195 caissons set in place by Furness were later made into pontoons for the floating steamboat piers along the Embankment. Bazalgette's son Edward later characterized the technique as quite successful, but it was not used for most other Embankment contracts, and Edward's claims were challenged by several engineers who considered timber cheaper, faster, and better. Furness certainly would have agreed.[24]

Getting good gravel to fill in behind the wall was a fourth problem, which Furness shared with Ridley and the other contractors. The Thames Embankment Act required them to obtain gravel from the Thames Conservancy, who dredged it from the riverbed as part of their statutory responsibility to improve navigation. But contractors were allowed to go elsewhere if the quantity or quality of Conservancy fill was inadequate. Very soon after construction began, the contractors told the MBW the fill was expensive, slow in delivery, and adulterated with mud. They began buying gravel from land dredgers at lower prices. The Conservancy complained unsuccessfully to the MBW and then appealed to the Board of Trade. After complicated arbitration and nasty arguments, the Conservancy forced the contractors to buy their fill. Meanwhile, Furness was stuck waiting for deliveries and paying more than he had counted on.[25]

A fifth cause of delay was Bazalgette's insistence on facing the Embankment wall with high-quality granite from Scotland. Supplies were especially hard to obtain, and both Furness and Ritson had to send agents to other parts of the British Isles and even to Sweden for substitutes.[26]

New railways were yet another obstacle that Furness had to work around. In 1863, Sir John Fowler persuaded Parliament to pass a private bill allowing his Metropolitan District Railway to extend its underground line from Westminster down to Blackfriars, using the space between the low-level sewer and the old shoreline.[27] The bill apparently went through over the objections of Bazalgette and the MBW, forcing them to negotiate a right of way with Fowler in return for £200,000 compensation. Work on the railway began in 1864. The plans would have allowed Fowler to use inexpensive cut-and-fill methods rather than to dig tunnels, but at Whitehall Gardens the Commissioner of Woods, Charles Gore, once more invoked the rights of his aristocratic tenants and forced Fowler to shift his line towards the river. It ended up running under the Embankment roadway that Furness was trying to construct at the same time.[28]

A second railway proved equally disrupting for Furness. The "Whitehall and Waterloo Pneumatic Railway" was chartered, in 1865, with plans to extend a line perpendicular to the Embankment from a station in Scotland Yard to Waterloo Railway terminus on the other side of the river. This meant driving it under the District Railway tunnel, the roadway, the low-level sewer, and the Thames riverbed (figure 6). Pneumatic or "atmospheric" railways ran on vacuum power. A steam engine generated a vacuum at either end of the line or section, and acted upon a piston set in a tube running below and between the rails. The railway cars were linked to the piston by a metal arm, which traveled along a slot in the top of the vacuum tube. The slot was covered with leather or some other pliable material that allowed passage but closed again after the arm went through, keeping the vacuum sealed—at least in theory. Pneumatic railways had enjoyed a bubble of popularity in the 1840s. William Cubitt built a section south to the London suburb of Croydon, and I. K. Brunel began a full-scale line to the southwest. Both had failed by 1848. A short stretch was built at the Crystal Palace in 1851. The "Pneumatic Despatch Company," registered in 1859, was building a narrow-gauge line under the heart of the metropolis for the London Post Office (which used it for a century) when they proposed the Whitehall and Waterloo Railway.[29]

Sir Charles Fox and Sons began work on the W and W, in October 1865, after the District Railway was already started. The contractors paid the MBW £3,000 for the right to tunnel underneath the railway and sewer lines. Their excavations interrupted sewer extensions being rebuilt for the Embankment, and work proceeded so slowly that Bazalgette threatened to do it himself at the W and W's expense. Late in 1866, the railway claimed it was nearly finished with the tunnel and related sewer work, but, in fact, it had fallen victim to the banking crisis of that year and soon agreed to Bazalgette's terms. The board contracted the job with Furness, since he was already at the site.[30] Furness himself was headed for bankruptcy court, let down by his financial backers. With the help of the MBW, he found a new partner, the railway contractor Thomas Brassey. Brassey not only brought the capital and experience Furness needed, but he immediately took over the remains of the Whitehall and Waterloo. He completed and stabilized the tunnel underneath the Embankment, then bricked up both ends and closed the venture.[31]

Though considerably behind schedule, Furness and Brassey made good progress on the Embankment in the next two years. The extent of their works can be judged from a report Bazalgette made in October 1868, when asked why six of the board's clerks of works were still occupied there. The contractors were still forming the steamboat piers at Westminster, Waterloo, and Hungerford, while prefabricating the steamboat waiting rooms at their own yards. The Thames Conservancy was dredging the pier sites prior to removing the cofferdam pilings. Nearby sewers had recently been altered (again), requiring changes to the penstock chambers. The Metropolitan District Railway works were unfinished, but the company, still in financial distress, was due to resume work shortly. When that happened, the Embankment roadway could be paved and the approach road from Whitehall laid out. Finally, Bazalgette hoped trees could be planted before winter, and permanent gaslights installed.[32]

The chief engineer's reports, and drawings in the *Illustrated London News*, show that the contractors used mostly traditional methods of construction but with some techniques and equipment reflecting the industrial age (figures 3 and 9). Wooden piles were driven one by one into the Thames clay bed for the cofferdams by small, specially equipped steam vessels. Massive timber scaffolding framed the walls and tunnels. Men excavated with shovels and hauled materials with barrows, backframes, and horse-drawn rail trucks. Steam-powered cranes and water pumps were commonly used alongside manual winches. Furness apparently possessed a machine for mixing cement that Bazalgette termed "novel and effective." Both Furness and Webster made use of traveling cranes to carry the granite facing into place; Webster powered his with a small locomotive.[33]

Breakage and accidental injury or death were common in this type of construction. In February 1866, one of the iron caissons being moved into place near Westminster collapsed, throwing a man into the Thames and chopping him in half. Later that year, another man fell into the cofferdam when pouring cement and was drowned. There were also many claims for illness caused by effluvia from all the old sewers being opened and rebuilt.[34]

The contingency fund included in each contract was used by the board's engineer at his own discretion to compensate the contractor for legitimate overages. Most contracts had a separate schedule setting out a price for each anticipated problem. There was also provision for crediting items omitted from the contract by the board. The contingency funds for most of the Embankment contracts amounted to £25,000, but Furness's was £50,000. Furness actually spent £87,607 over contract, mostly in the form of extra excavations and steamboat pier materials. He was then credited with £36,629 for additional excavations required by the board. These included £1,300 to finish eighteen feet of river wall that the MBW engineering office had mistakenly omitted from the plans, much to their embarrassment. Furness was also paid £20,000 for "ornamental figures" and lamp standards indicated in the original specifications but not included in his tender.[35]

Such credits and debits could be subjects of disagreement between contractor and engineer, as Thomas Page's experience makes clear. Embankment contracts stipulated that the chief engineer had the final decision in such disputes, and Bazalgette was an experienced arbitrator. He would call in the assistant engineer and the clerks of works with their contract drawings and specifications, sit them down with the contractor and his agents, and go through the various points. Upon agreement, he sent the figures to the board's accountants for verification and costing, and countersigned the accountant's certificate.[36]

While Furness was trying to finish his contract, work got underway on the Albert Embankment, whose features had been decided by the (reconvened) Royal Commission in 1862–63 and authorized by Parliament early in 1865 after several false starts by the Liberal ministry. The low-lying land on the Surrey side was subject to recurrent flooding, especially around Lambeth where the Thames curved northward, throwing its current against the shore. Bazalgette as well as several private projectors submitted plans to embank the whole south side from Vauxhall down to London Bridge but, as happened across the river, the protests of commercial wharfingers and bargemen led the Royal Commission, and then the First Commissioner of Works, to truncate the project. They reasoned that there were already roads running parallel to the Thames on the south side, so a new thoroughfare was unnecessary. The new sewer line would obviate the need for flood control there. The commission blithely suggested that, if the wharfingers wanted to construct an embankment on their own, with some uniformity of design, they should be assisted in every way possible so long as no public money was involved.

For the mile from Westminster Bridge up to Vauxhall, the chief engineer again assembled elements from other proposals.[37] He first reverted to his 1860 concept of an ironwork viaduct as a means of maintaining access to the wharves. Then he realized that the wharf property was of little value and that a viaduct would hardly contribute to flood control.[38] So he proposed a solid embankment along the lines of the Victoria section. His design was further modified in negotiations with Lambeth vestry officials, several riverfront proprietors, and board members eager for economy. Openings were left for two draw docks in the river wall. The Ferry Street Dock, just southwest of Lambeth Bridge, was built for Doulton's pottery works (Frederick Doulton was an MBW delegate from Lambeth). This dock later served the Lambeth Pottery and, at some point, was converted to the double culvert visible from the opposite bank today. The White Hart Dock farther upstream was in constant use, but it extended straight back from the river, allowing high tides to flow through it and flood the surrounding district. It was rebuilt about fifty feet to the west, with an L-shaped approach leading down from Black Prince Road to a subway under the river wall. Now owned by Lambeth borough council, it is still used for regatta landings.[39] Another opening was planned for Stiff's Dock next to Westminster Bridge, but it was eliminated when St. Thomas's Hospital bought the property. Bazalgette accepted the fact that mud would continue to accumulate around the dock openings, a problem the Embankment was originally designed to cure. He only required that contractors dredge the dock areas when they were through building.[40]

The special Embankment extension, 1,000 feet long, which had been planned to run on a viaduct under Vauxhall Bridge and up towards Battersea to accommodate the City of London gasworks, was never built. The gas company moved farther up the shore to Nine Elms, in 1865, and, although the Royal Commission recommended keeping the viaduct, it was soon scrapped in the interest of economy. All that was left was a river wall 4,300 feet long, with a paved roadway and a pedestrian walkway. A small garden would be tucked in at Gunhouse Alley (affording a clear view of Millbank Penitentiary across the river), and the river was widened at that point by 125 feet to obtain a uniform line.[41] Several wharves and docks and a gin and vinegar distillery remained situated between the end of the Embankment and Vauxhall Bridge.

The board received tenders for the Albert Embankment in July 1865. Samuel Ridley was again the low bidder, but again the board rejected him. The contract went instead to William Webster, who had already proved himself on the southern outfall sewer and was now halfway through constructing the southern low-level sewer line, supervised by his friend John Grant as the MBW's district engineer. As mentioned previously, the construction site included the extensive Doulton pottery works, which was soon replaced by the new facilities for St. Thomas's Hospital. This reduced the total cost of the Embankment, but Webster had to work around the hospital construction in order to carry out his own.[42]

The Albert Embankment was built on low-lying, marshy land that had to be pumped free of water to allow foundations to be poured. John Grant later remarked that fine sand was likely to be pumped out also, leaving dangerous cavities under the works and under nearby building foundations. His solution was to sink deep wells between the works and nearby buildings to collect the groundwater, then pump it out through a bed of gravel to filter the sand.[43] Since the high water levels would make future repairs on the Embankment difficult and expensive, Grant and Webster decided to exceed the contract specifications for both brickwork and concrete. Grant was then concluding his experiments with Portland cement concrete and began substituting it wholesale for brick. The Albert Embankment, he later boasted, was "built to last."[44]

Grant's economies, Webster's excellent management, and the board's good luck in selling off unused property made the Albert Embankment a bargain at just over £1 million. Furthermore, it was finished on schedule early in 1869. Webster was thus free to take on the third of the Victoria Embankment contracts, comprising the difficult section from Temple Gardens up to Blackfriars Bridge.

If ever a project suffered from cross-purposes, it was the Temple-Blackfriars section. The term "contingency" took on a whole new meaning. Each of the participants would have applied it to their own experience of contrary conditions and behavior by others, but the interruptions had less to do with chance events or erroneous decisions than with the stubborn pursuit of conflicting goals. The solid embankment envisioned by Thwaites and sketched by Bazalgette, in 1862, encountered two adversaries. The first was the City of London Gas Light and Coke Company, one of thirteen that serviced the metropolis, which had its largest depot near Blackfriars Bridge, inside the City boundaries.[45] Unlike the coal wharfingers, who were generally lessees or individual proprietors, the "City" was a large company with many shareholders and influential friends. The gas it produced was piped into the surrounding urban districts, so it could not easily move downriver without disrupting service to its customers. The 1861 Royal Commission on the Thames Embankment and the 1862 Commons committee on the Embankment bill both assumed that it would remain in place after the other riverfront property was cleared. To accommodate the "City's" considerable barge traffic, Bazalgette designed a pair of subways under the Embankment, with the roadway being carried over them on arches before curving to join the Blackfriars Bridge approach road.[46] Neither the MBW nor the City Corporation was very satisfied with this arrangement, since it left the noisome gasworks to "anchor" the beautiful new Embankment, precluding more aesthetic and profitable development.

The second and more formidable adversary for the board was the Metropolitan District Railway. We have already mentioned that Fowler's company complicated Embankment construction near Whitehall in 1864. It was even worse at the Blackfriars end. The MBW had tried unsuccessfully to persuade the Commons select committee on the railway bill to insert a clause requiring the Temple-Blackfriars section of rail to be built under board auspices or at least at the same time as the Embankment. Incredibly, however, the bill allowed the railway to cut through the Embankment arches where they curved inward, extending the rail line along the shore before plunging back underground to Blackfriars Station. The railway would thus block access to the gasworks from the river, rendering useless the arched subways, and its shoreward extension would preclude granite facing along the incline to the Bridge. The District was to pay £200,000 in compensation to the MBW for the use of its right of way, but the board in turn would be liable to the City of London Gas Company for damages to their business. It was an altogether intolerable situation.[47]

The board immediately tried to negotiate some workable alternative with the railway's chief engineer, John Fowler. They soon realized that Fowler wanted to stall until the Embankment was nearly finished and the City Gas Company's claims resolved, leaving him a clear building route with no hidden expenses. He told the board, in February 1865, that the railway would not begin construction until the reclaimed land was nearly filled in, a strange idea for an engineer using cut-and-fill methods. After further queries, however, he agreed to meet with Bazalgette. Over the next year and a half, the two engineers traded proposals for revising their respective plans and appeared to reach agreement in October 1866. The board accepted the results but inserted provisions protecting themselves from legal liabilities, to which Fowler objected. The two parties then tried negotiating with the City of London Gas Company to move its works downriver, but that, too, proved fruitless.[48]

By this time, the financial crisis of May 1866 had affected all of London.

In the previous two years, the rate of interest had averaged 7 percent, the highest of the century. Railway companies, like so many others, had borrowed money for current operations while investing their capital in high-risk joint-stock ventures and foreign loans. When the bubble burst, the giant discount house of Overend, Gurney failed. The "wildest panic since 1825" ensued, the Bank Act was suspended, and the Bank of England's official interest rate rose to 10 percent for the next three months and then only gradually declined. Investors took flight for the next three years.[49] The Metropolitan District Railway, having gambled like all the others, was soon facing bankruptcy. Its directors refused either to sanction changes in the original plans or to proceed with construction, and Fowler found it convenient to go abroad for much of 1867. In January 1868, the railway's counsel wrote letters to the *Times* blaming the board for delay. The board replied with a pamphlet containing the relevant correspondence, then sought to force the issue through arbitration under a prominent engineer, Thomas Hawksley. Despite serious reservations, the board accepted Hawksley's "compromise" plan because they couldn't stand further delay, which was driving up the cost of the other portions of Embankment construction. When the railway directors confessed that they had no money, the board even agreed to waive the interest and the first year's payment on the £200,000 right-of-way compensation owed under the original contract. Fowler and the railway finally resumed construction, helped perhaps by the opening of other portions of their line in late 1868. They were able to open service on the Westminster-Blackfriars line May 30, 1870, just a month before the Prince and Princess of Wales formally opened the Victoria Embankment.[50]

In the meantime, efforts by the Corporation of London, the MBW, and the railway to force the City Gas Company out were having some success. The first two allies sponsored bills in Parliament either to purchase the main gas companies or to supply gas themselves, and the railway, ironically, supported them. The legislation was unsuccessful, although the board was given the right to test the quality of gas from one company.[51] Private enterprise then came to the aid of public authorities. The coal gas industry began an internal consolidation led by Simon Adams Beck, governor of the Chartered Gas Light and Coke Company, whose chief station was at the Horseferry Road, Pimlico. In 1868, Beck won parliamentary approval for the construction of a massive new works downriver at a site appropriately named Beckton. The City of London Gas Company immediately sought consolidation, and Beck took over that company and two others in April 1870. A year later, the new enterprise was supplying two-thirds of metropolitan gas customers. Despite these changes, however, the "City" company officials did not immediately close their Blackfriars works. They merely reduced it to winter peak-load service and kept it operating until 1873. Some years later, the MBW filled in the double archway leading to the abandoned gasworks, completing its original design for a granite-faced solid Embankment wall.[52]

The board's plans for developing the thirty-five acres of land reclaimed from the northern shore of the Thames came up against similar obstacles. In one case, they were defeated; in the other, they prevailed. Both results benefited the public.

The most controversial parcel of reclaimed land was a four-hundred-foot-wide section of prime new real estate lying between Hungerford Bridge and Waterloo Bridge, sloping down from the Strand with the Adelphi Terrace on the west side, the ruins of the old palace of the Savoy on the east side, and the imposing mansion of the Marquis of Salisbury in the middle. Bazalgette's first plan for the site was a grandiose set of garden terraces, fountains, and stairways with classical balustrades leading up from the Embankment landing steps all the way to the Strand. It was essentially a revival of Sir Frederick Trench's grand design from 1825. A row of shops lining the Embankment roadway on either side of the main stairs would form a kind of retaining wall for the gardens, with their roofs supporting part of the promenade. A "handsome crescent" above would be formed by a pair of new streets rising near the foot of either bridge and curving across the top of the terraces to the opposite corners, one from Waterloo Bridge up to the Adelphi and one from Hungerford up to the Strand at Wellington Street.[53] The plan would, said Bazalgette, "afford the most valuable building frontages, and form one of the finest features of London." It would have created a dramatic public space linking the Thames to London's main thoroughfare. George Drummond, owner of the Adelphi, was in favor of it. The Duchy of Lancaster, which owned the Savoy Palace grounds, was agreeable.[54] The only hitch was that it required encroachments on the Marquis of Salisbury's mansion and grounds.

Lady Gwendolyn Cecil has written that, in all the generations between the first and second Cecils (secretaries to Queen Elizabeth and King James I respectively) and her father Robert Cecil, third Marquis of Salisbury, "the general mediocrity of intelligence which the family displayed was only varied by instances of quite exceptional stupidity." Robert Cecil's father, who resided at Salisbury House until his death in April 1868, was a domineering ultra-Tory who "took command of everything, and listened to nobody." Whether the subject was roads, cottages, drainage schemes, or the Poor Law, "he had his own ideas on the subject, and would never willingly yield them in favour of those who could claim a more special knowledge of it." Experts of any kind roused him to fury, and, as a result, "the atmosphere in which he lived was not wholly peaceable."[55] Robert, a younger son who became heir upon the death of an older brother in 1865, shared his father's attitudes. Styled Lord Cranborne, Robert joined the third Derby administration, but he soon fell out with his colleagues over the revised Reform Bill and launched a series of brilliant, satirical, and scathing attacks on Disraeli's "betrayal and surrender" that made him leader of the unreconstructed Tories.[56]

Needless to say, the board's engineer, architect, and solicitor met with contemptuous dismissals when they approached the old Marquis with their plans. The board tried to mediate, but a draft agreement written in 1867 failed miserably. Lord Salisbury wanted his access to the Thames maintained regardless of the Embankment and would brook no encroachments on his grounds. He had more money and solicitors than the board had, plus a son in the Cabinet. Failing to find a resolution, the board had no choice but to change its plans for development. Bazalgette made the best of the space by creating the present Embankment Gardens on the flat ground raised next to the new roadway by the river. After some difficult negotiations with property owners, Villiers Street was brought down from the Strand between Charing Cross Station and the Adelphi buildings to the west of the site, where John Hawkshaw had just finished rebuilding the Hungerford Bridge. A decade later, the board extended a carriageway leading from the Strand past the Savoy Royal Chapel to the Embankment road on the east side, paved it with hardwood,[57] and renamed it Savoy Lane. The ruins of the old Savoy Palace, taken over by the MBW in the process of constructing the Embankment, were eventually (1889) replaced by the new Savoy Hotel.[58] The half-crescent road originally planned to curve up the hill from the East was then built to service the hotel.

Since the Hungerford-Waterloo site had been dedicated to "public amusement and recreation," the new Embankment Gardens were eagerly anticipated by the people of London. The urban parks movement, mentioned in chapter 1, was just underway, and the Commons Preservation Society had been founded. Men and women of all classes who had followed the progress of construction from the decks of special ferries could now see a broad, uncluttered space opening up with a view toward the somewhat cleaner Thames. Stripped of their original terraces and grand pretensions, the Embankment Gardens proved immediately and immensely popular. The flower beds, statues, and curved walks provided a pleasant relief from the surrounding commercial bustle, and, for many people who still believed in the miasmatic theory of disease, the gardens promised healthy fresh air in the heart of the metropolis. They also gave the Embankment a human-scale aesthetic.[59]

These qualities no doubt contributed to the outpouring of public feeling when it was learned, early in 1870, that the Office of Woods and Forests intended to erect income-producing office buildings on two and a half acres of land reclaimed on the west side of Hungerford Bridge, in front of the Whitehall Estates. This property had been ceded to the Crown during the board's dispute with the Whitehall Estates residents, in 1860–62, as a sop to their complaints about invasions of privacy. Nonetheless, it was most immediately accessible from the Embankment roadway, and many people expected it to be turned into public gardens. Disraeli, had he stayed in office, would have readily concurred. But the Liberal triumph, in December 1868, brought a trio of ministers marked by a "chronic, cheese-paring, tax-cutting compulsion," namely Gladstone, Robert Lowe at the Exchequer, and A.S. Ayrton as First Commissioner of Works. They saw the new Whitehall land as an opportunity to raise revenue for cheap government. Charles Gore, the long-time Commissioner of Woods and Forests, who, in 1860, had defended the Whitehall residents' right to an uninterrupted view of the Thames, now agreed that a "handsome building" was quite appropriate.[60]

David Owen has fully chronicled the two-year public campaign waged by W. H. Smith in Parliament and by the newspapers in public to force the Government to relinquish its hold on the property. The campaign undoubtedly succeeded because of Smith's persistent criticism and his zeal in organizing public protests, together with the intransigence of the board. But Smith and the board could take a hard line because the emergence of the Embankment from concept to hard reality had generated a sense of public ownership that had not been seen before. The daily work of constructing a new landscape, visible to everyone, was forging a common identity for London's patchwork of parishes. This new public identity expressed itself not just in economic terms (having paid the rates that fed the board that commissioned the wall that Bazalgette built) but in moral outrage at the proposed betrayal of its trust. Robert Lowe appointed a new select committee and introduced additional bills in the hope of rescuing the government's project, but even he was worn down in the end. The Thames Embankment Land Act of 1873 transferred the reclaimed land to the board in exchange for a small parcel of property on the other side of Whitehall and a negligible sum of cash.[61]

Before the Whitehall Gardens issue was settled, the board provoked an outcry of a different sort by asking for legislation giving it control of the Embankment roadway. Most streets, once completed by the MBW, were turned over to local vestry or district ownership. But the board decided the Embankment was different, a "national" work belonging to the general public, not to some particular locality. Although public opinion was uninformed on this question, and naturally less concerned than for the gardens, all the local authorities, especially the City of London's Sewer Commission, protested any extension of the board's jurisdiction. Nevertheless, the Thames Embankment (North) Act of 1872 gave the board what it wanted.[62] The roadway had suffered through the delays incurred by the Metropolitan District Railway, and the board decided not to open it to regular traffic while the railway was under construction. After 1868, in accord with Embankment contract #3, the road was paved with broken granite, the sidewalks with Yorkshire stone, the river wall adorned with new lights on double-dolphin lampposts, and plane trees (one of the few varieties that could survive in London's polluted air) planted in specially lined pockets along the sidewalks. The pedestrian subway between the Westminster Underground Station and the Embankment was added as an afterthought. William Webster then secured a maintenance contract to remove slop and horse droppings, keep the roadway watered to reduce dust, remove snow, and repair and clean the gulleys and sidewalks.[63]

The broad Embankment boulevard was quite different from the cramped confines of Fleet Street and the congestion of the Strand. To some observers, it evoked memories of the Regency. To others, it suggested Baron Haussmann's Paris. When the board opened the Chelsea Embankment three years later, it called attention to the trees and ornamental gardens "in imitation of the Boulevards of Paris, a feature somewhat novel in this country." Why not then introduce sidewalk cafés, a few restaurants, pavilions in the gardens? London newspapers professed horror at the thought. English men and women, they intoned, preferred home, family, and dinner parties to decadent street life. The Parisian look "was denounced as if there had been suggested another burning of St. Paul's," and the *Architect* concluded, even twenty years later, that "an immense amount of dawdling and frittering away of life must be put down to the broad ways of Continental cities."[64]

While the roadway was still unfinished, the Board decided that the Embankment was complete enough for a formal opening on 13 July 1870. They hoped Queen Victoria herself would attend, but the reclusive widow declined. The Prince of Wales and Princess Louise presided at the ceremonies instead. Viewing stands were erected along the roadway in front of Whitehall Gardens (still in dispute between Government and the public) to hold members of the royal family and household, parliamentary leaders, archbishops, ambassadors, peers of the realm, and delegates from the MBW. The royal carriage traversed the Embankment between regiments of Horse Guards, appropriate speeches were given, and the newspapers heaped praise upon the board and its chief engineer. They had reached, wrote Owen, the zenith of their popular esteem.[65]

But the great project was not yet done. The Albert Embankment on the south side had been finished the previous November, but the Chelsea Embankment extension, from the west end of the Royal Military Hospital up to Battersea Bridge, was not even under contract. The Liberals had introduced a bill to complete it in 1865, and again in 1866, without supplying the necessary funds. It was reauthorized in 1868, but the board had trouble raising its share of the cost.[66] One may reasonably speculate that the MBW's difficulties with the Temple-Blackfriars portion of the Victoria Embankment, together with the developing argument over the disposition of Whitehall Gardens, probably added to the delay. The board finally advertised for tenders, early in 1871, but on three occasions the low bidders, after being awarded the contract, discovered errors in their calculations or problems in the specifications and withdrew. The job was then placed in the hands of William Webster, who completed it in 1874.

A few residents of Chelsea (now grown into a Parliamentary borough) still objected to the loss of their bucolic waterfront. At first, the board agreed to install the low-level sewer extension within the foreshore without building a new river wall. However, this was probably due more to fiscal than to conservationist motives. When Webster substituted Portland cement concrete for the traditional brickwork, as he had done on the Albert Embankment, he reduced the projected cost of construction about 16 percent, so that building a solid Embankment cost no more than putting up temporary cofferdams for the sewer alone. The improvement reclaimed nine acres of land from the river, which the board later leased to residential builders.[67] For its official opening on the ninth of May, 1874, with the Duke and Duchess of Edinburgh doing the honors, the board published a handsome booklet recounting the problems of the Thames and the history of embankments, including the portion built by Thomas Page. It took justified satisfaction in the diversion of sewage from the Thames and the transformation of its banks. For his part, Chief Engineer Bazalgette was made Knight Commander of the Bath, and a bronze relief portrait of him was mounted on the Victoria Embankment wall facing Whitehall Place.[68]

Had the Embankment proceeded on schedule according to the plans and specifications drawn up by Bazalgette in 1863, without all the contingencies described above, it would not have looked much different than it does today. The District Railway line, which caused so much trouble during construction, is hidden underground. The Whitehall and Waterloo Pneumatic Railway lies buried beneath the river wall. Furness's problems with the iron cofferdam, the tides, and his finances were significant aspects of the construction process, but not of the resulting object. Only the gardens were new.

Nevertheless, we cannot ignore the contingencies. The business of historians is not to describe the external features of objects or situations as they are now, but to explore the interactions through which they emerged out of antecedent conditions. And it should be evident by now that uncertainty and chance play a significant part in all events, even carefully planned events such as construction projects.

Contingency is a matter of degree; it lies on a continuum. At one end, we may place conditions known to be required for a successful outcome. We can say, for example, that the building of a solid embankment was contingent upon eliminating the wharves and docks between Westminster and Blackfriars, or that the installation of a crescent terrace garden between Hungerford and Waterloo bridges depended upon an agreement with the Marquis of Salisbury. Such contingencies provide the traditional dynamic for historical narratives. They are typically represented either as steps in a logical procedure ("It was quickly evident that the Embankment could not be financed out of the current rates") or as challenges to an expected sequence ("Why didn't Thomas Page and the First Commissioner of Works build the Thames Embankment?"). In these cases, the identity of the contingency is known or assumed, and the degree of uncertainty is assumed to be low. For example, engineers expected they would have to compensate for changes in the Thames River after the rebuilding of London Bridge. Similarly, London residents knew that water pollution contributed to the incidence of disease, even if they could not say how that happened.

Farther along the continuum appear interruptions, discoveries, and decisions which cannot be specifically predicted but which tend to occur often enough, and within a known range, to enable people to make provision for them and to deal with them when they arise. As we have seen, engineers and contractors knew the general types of conditions (environmental, financial, logistic, human) on which their projects depended, and they were experienced enough with the uncertainty of these elements to make a standard allowance for them in contracts and a procedure for negotiating the actual amounts after the fact. One of the roots of Thomas

Page's problems with H. M. Office of Works was that he presumed the need for such flexibility, whereas his bureaucratic overseers followed the stricter rules of property, accounting, and contract law.

A third degree of contingency covers phenomena which could not be *predicted* by analysis of previous conditions, but which can be *explained* retroactively according to some set of criteria. Such phenomena have been the subject of historians' causal explanations for centuries. Environmental disturbances fall into this category: the appearance of cholera, unusual tides that devastated construction sites, the hard winter of 1860. Human examples include the authorization of the Metropolitan District Railway and the Whitehall and Waterloo Railway (which was untimely from Bazalgette's point of view), the financial "crash" of 1866, the accession of Gladstone and his doctrine of cheap government, problems obtaining granite for the river wall facing, and the relocation of St. Thomas's Hospital. In some cases, at least in retrospect, it is tempting to ask why the engineer could not have anticipated such changes. The engineer would, probably, reply that even had he been aware of the possibilities, and even had he understood how they might impinge upon his own work, the level of uncertainty was too high for him to calculate an appropriate response in advance. A case in point is the consolidation of the coal gas industry around 1870. The MBW tried to accommodate the gasworks at the east end of the Victoria Embankment and the west end of the Albert, while at the same time sponsoring legislation to force them to move. They were on the right track, but they had no control over the process or the timing of consolidation, which depended on Simon Adams Beck's entrepreneurial initiative and success.

It is sometimes argued that contingency arises from our inability to comprehend the sheer complexity of the historical environment. Our causal explanations are too simple, too linear. We plan our projects, and structure our histories, by abstracting narratives and syllogisms from the ocean of experience. It is no wonder, then, that our projects are interrupted by apparently chance events, and that historians, even with hindsight, cannot arrive at a complete explanation of any given event.

One can carry that argument too far, however. It assumes that contingency is only a function of our ignorance, that everything in the world, both in human society and in the natural environment, is really deter-

Figure 8. Before and after the Victoria Embankment. The view from Somerset House terrace toward Blackfriars Bridge and St. Paul's in 1841 *(top)* and in 1872 *(bottom)*. (Before: Courtesy of the Greater London Record Library. After: Courtesy of the Museum of London Picture Library.)

Figure 9. Before and after the Albert Embankment. The view from Vauxhall Bridge toward Westminster Bridge during construction in June 1866 *(top)* and today, showing the original granite wall facing "double-dolphin" lamppost design and lion-head mooring rings *(bottom).* The St. Thomas's Hospital complex replaced Doulton's pottery works at this site when it was taken over for the Embankment works. (Before: *Illustrated London News,* 23 June 1866. After: Photo courtesy of Andrea Harger Donovan.)

Figure 10. Portion of the Thames Embankment near Westminster, showing the pedestrian walkway and the "double-dolphin" lamp design.

mined. If only we were intelligent enough, even omniscient, we could be certain of it. Engineers know better. They do make mistakes, and some are smarter, better informed, or more painstaking than others. But they realize that technology in general, and construction projects in particular, are not isolated endeavors driven by some inherent logic. They occur within a matrix of changing social and environmental conditions where each level of complexity adds another degree of uncertainty. Contingency is therefore, paradoxically, inevitable. The object of engineering, as an interface between society and environment, is not to eliminate contingency but to reduce it to manageable proportions.

CHAPTER IX

The Historical Future

The Thames Embankment originated as an architectural enterprise, but quickly became an engineering response to problems of environmental degradation and urban expansion. Its design and construction engaged the attention of a variety of interests within the metropolitan region. An account of its development, therefore, takes the form of a series of conjunctures at which diverse trends and interests intersect and are resolved in legal, financial, technical, and political ways. As the Embankment neared completion, the technical alternatives tended to narrow, whereas the social implications of the transformed waterfront widened. The finished artifact, however, was no longer merely an object of human aspirations and efforts; it became a powerful agent in the social and environmental contexts that framed its "future." Its subsequent adventures have been characterized by an extension of its design, a fragmentation of its functions, and an uneven impact upon the social ecology of the Thames valley. A review of these adventures will prompt us to ask whether the Embankment enhanced London's metropolitan image at the expense of her working relationship to the Thames.

Technological and environmental history are different from other sorts of history because the objects or physical conditions they focus on can endure for long periods. Machines, electric power stations, and drainage systems typically remain in operation for years or decades. Agricultural field patterns can last for centuries. The purposes for which they were originally designed, and the contexts in which they originally functioned, may change or even disappear, as in the case of the pyramids. As objects, however, they tend to persist more or less in their original form. They not only affect their environment by their first occurrence, like a treaty or a World Series baseball championship, but continue to impact the behavior and perceptions of each new generation of people.

To make the "future history" of the Thames Embankment intelligible, then, we have to reposition it in successive contexts. These are constructed with the same periodization and subject categories used to construct its past: (1) the impact of the Embankment on its structural environment, with respect to sanitation, flooding, and navigation; (2) the consequences of the project for the institutions, concepts, groups, and individuals involved in its construction; and (3) its meaning as an enduring technological artifact during an era of changing perceptions of the city, of technology, and of the environment.

It is not quite clear when the "future" of the Embankment should begin. Public engineering works typically have a projected date of completion, another date for the formal opening, and yet another for actually finishing the work. The Thames Embankment, divided into three sections and five contracts, was marked with multiple completions and two celebrations. The Victoria section was so much larger, more complex, and more significant to the metropolis than the other two that some accounts end with its official opening in 1870, and when people refer to "the Embankment," they mean only that portion of it. Related contracts ran until 1874, of course. The Chelsea section was not finished until that year, and the issues of public access to Whitehall Gardens, and the reconstruction of Northumberland Avenue as a primary route to Charing Cross, were also resolved at that time.

Yet the effect on London's appearance and political image and on the public's concern for the reclaimed land begins from 1870, if not before. The spectacle of its construction alone galvanized the public's imagination. Picture postcards featuring the Victoria Embankment became immediately popular, to judge from the collection in the Greater London Record Office. Tourist guidebooks were quick to incorporate details of construction, along with notes about buildings, wharves, and alleyways that lined the river in bygone years. Since the Albert Embankment was also finished by 1870, the benefits of diverting sewage from the Thames and streamlining its channel were immediately noticeable. The percentage of people dying from infectious diseases, especially the dreaded cholera, declined sharply after 1870 and, by 1900, was little more than half what it had been in 1850.[1] Although the decline can be attributed, in part, to new sources of drinking water, most people credited the Main Drainage and the removal of sewage-strewn mud flats from the river's edge. Newspapers and journals looking back from the 1870s and 1880s often reminded their readers of the noxious effluvia that rose from the river and the foul state of the wharves before the Embankment. The air around the improved river was at least no more smoky than in the rest of London, and the Thames water was perceived to be clean enough that swimming baths were soon installed on the east side of Charing Cross Pier.[2]

When Londoners attributed the decline in disease to the Main Drainage and Embankment, they thought more about the reduction of "pestilential odours" than to water purification. Yet, ailments such as infant diarrhea continued unabated, and the miasmatic hypothesis could no longer explain them. Attention shifted from the visible river and its effluvia to the relatively less visible degradation of the city's water supply. The germ theory of disease finally drew serious attention. Although medical science could still not ascertain the purity or impurity of the water (a problem which "experts" for the private water companies exploited to the full), the public now accepted the notion of sanitary reform and felt confident that technology could improve the environment. Engineers convinced town authorities that no system of household sanitation would work unless the water companies provided a constant supply of fresh, filtered water under minimum pressure. The Public Health Act of 1872 obligated towns to enforce such supplies. When private firms resisted, municipalities took over and operated their facilities.[3]

The relative purification of the river also affected personal hygiene. Regular bathing for cleanliness (rather than swimming) was fairly common among the upper classes by the 1860s and was spreading among the middle classes, but servants and laborers had few facilities. Body odor was as great a class distinction as flush toilets. As we have seen, the poor were believed to be afflicted by moral as well as environmental pollution; they tolerated dirt and stink more readily than their social betters and did not really want to be clean. But the publicity over sewerage and town drainage projects of the 1850s and 1860s made odor noticeable and then intolerable even to the poor. As soon as sanitary authorities provided municipal baths, the poor flocked to them. In London, by 1865, eight boroughs, mostly in the West End, provided baths serving hundreds of thousands of people every year. By the end of the century, towns were competing to build the most impressive monuments to personal sanitation.[4]

Sewage disposal proved more intractable than water supply. The example set by the Metropolitan Board of Works encouraged other large riverside towns to undertake their own drainage projects, and Government supported them with loan guarantees and other inducements. However, sanitary engineering still consisted of moving human and light industrial waste through underground pipes into catchment basins prior to discharge into rivers. In some places, such as London, the sewage was treated; in others, it was filtered for recycling as fertilizer. But a great deal of it was still dumped raw into rivers. By the 1880s, Liverpool, Bristol, Cardiff, Carlisle, Dover, Warwick, Salisbury, Ely, and other sizable towns had all built or initiated water-based sewerage systems. But progress was uneven. Sewer pipes, now cast in clay or iron instead of built from brick, could still break (rats could no longer chew through the sides, but they learned to tunnel under the joints, cracking them open). Sudden storms could cause nauseating backups in the system. The expense of sewerage and household plumbing was often too great for working-class districts. Consequently, a majority of the queen's subjects did not have use of a water closet by the close of her reign.[5]

Moreover, the basic problem of sewage disposal had not been solved. Every town along the Thames found its water supply polluted by discharges from upriver towns and factories and, in turn, dumped its own wastes into the river outside its boundaries. The same was true for northern rivers such as the Calder in Yorkshire and the Mersey in Lancashire. Local authorities were quite aware of what they were doing but could not see any alternative. The consolidated, metropolitan approach used in London proved attractive to northern urban-industrial areas. When the second Royal Commission on Sewage of Towns came to Lancashire, in 1867, officials pleaded for a regional sewerage project funded by a national or county cost-sharing arrangement, arguing that the area was economically vital to the whole kingdom. Unfortunately, the legislative means to organize such a project were not provided until the Local Government Act of 1888, which established county councils.[6]

River pollution was subject to intense investigation by Royal Commissions between 1865 and 1870. Their reports advocated recycling by means of sewage irrigation but provided few alternatives and no substantial methods of funding or enforcement. Their inquests raised public consciousness about sanitation and half-convinced many people to replace fatalism with more constructive attitudes. However, they did not change the fundamental belief in the paramount rights of property and capital, which often conflicted with public regulatory and improvement programs. Advocacy groups finally pressured Parliament (again led by Disraeli) to pass the Rivers Pollution Prevention Act of 1876, but local officials would not enforce it. Although it remained the basic law in its field until 1951, it never achieved its goals.[7]

Londoners, admiring their improved urban environment in the 1860s, sharpened their criticism of towns along the upper Thames that continued to foul the capital's potential drinking water. Yet their own system of disposal turned out to be little better. The MBW took credit for improving the state of the river, citing the return of fish and the reduction of miasmas and diseases. But raw sewage continued to be dumped into the Thames. The penstock reservoirs built into the Embankment wall at the low-level sewer junctions periodically overflowed when flooded by rainwater—a problem faced, even today, by any city that combines storm and sanitary sewers. The riverbank near London Bridge was the site of one such discharge; it had to be spread with chloride of lime, as in pre-Embankment days, to protect steamboat passengers landing nearby. The main outfalls at Barking Creek and Crossness were a disgrace.[8] As with most other cities on British rivers, the principal polluter turned out to be the local authority.

For dumping the contents of the Main Drainage into the river, the MBW was bitterly and publicly attacked by the Thames Conservancy, which had been given powers to police river pollution in 1857. In principle, the Conservancy was formed, like the MBW, as a board of delegates from interested agencies along the Thames. But it had no popular constituencies to satisfy and was not accountable to any other government office. It was a throwback to preindustrial guild corporations. Its members (the City Corporation, the Admiralty, the Board of Trade, large shipping firms, steamboat and wharf representatives) did a good deal of polluting themselves. But the MBW's two great downriver sewage outfalls offered simple, highly visible targets for criticism. The Conservancy obtained legislation, in 1870, requiring Thames agencies and businesses to dredge the river if discharges or construction created mud banks or shoals that impeded navigation. In 1879, it took the MBW to a court of arbitration to enforce its demand that sewage building up near the outfalls be cleared away.

Because the MBW was not represented on the Conservancy Board, John Thwaites felt justified in mounting immediate counterattacks.[9] His agency was taking on a variety of new duties, often without adequate funding, and he was determined to defend his most famous project. The situation was not improved by Thwaites's successor, Colonel James McGarel-Hogg, a courtly upper-class member of Parliament. Under his uninspired leadership, the board lost its former flexibility. The Conservancy claims were either denied or explained away.

The main problem with the outfalls was that they had been designed, in 1855, to carry the sewage of an estimated three million people outside the boundaries of the metropolis. During the next half century, the population doubled and spread out into the surrounding countryside. By 1880, the inner city was starting to lose middle- and upper-income residents from out-migration, and the outer ring of suburbs continued to grow by 50 percent or more every decade. The working classes, however, were not yet using the Underground and commuter railways to any great extent. Their housing in Finsbury, Shoreditch, Bethnal Green, and Stepney was still rated "overcrowded" by authorities, and they began moving into large houses left by the well-to-do as they were subdivided into flats. High density within the metropolitan boundaries, combined with suburban sprawl, rendered the original sewer discharge facilities obsolete. By the end of the century, they would be pouring 150 million gallons of sewage per day into the Thames, almost one-sixth of the entire volume of the river at Barking Creek, and much of it now washed back and forth within residential districts.[10]

Although the Thames Conservators lost their legal case against the MBW, complaints about the deterioration of the river were taken up by others, and the chorus of disapproval increased yearly. Government finally responded by appointing a Royal Commission chaired by an eminent judge, Lord Bramwell. The Bramwell Commission marshalled a great mass of evidence to prove the obvious. Its first report, in 1884, called upon the MBW to deodorize the sewage at Barking Creek and Crossness and then find alternative ways to dispose of it. Despite pressure from the Home Office, the MBW, with an aging Bazalgette still running its engineering department, took its time. Experiments showed that precipitating the solid matter and flushing the deodorized liquid into the river was the best (or the most expedient) course of action. Funding for full-scale precipitation works, however, was not authorized until January 1887 for Barking Creek and May 1888 for Crossness. In the meantime, the sludge, having been refused for manuring purposes by local farmers, was hauled out to sea in barges. The first of these vessels was christened the *Bazalgette*.[11]

Arguments over the degree of improvement in sanitation and river pollution to be attributed to the Embankment obviously differed from group to group. The general public, and most newspapers, perceived a definite change for the better; administrative agencies with more technical or political agendas often thought otherwise. These same divisions are evident when we consider the other functions for which the Thames Embankment was built: navigation, flood control, traffic, and aesthetic considerations.

The men who conceived the Embankment spent many years outlining the improvements to navigation to be expected from streamlining the metropolitan reaches of the Thames. After their experience with the replacement of London Bridge, they were confident that narrowing the riverbed and removing irregularities from the shoreline would deepen the channel, increase the current, and bring slightly higher tides up to Teddington Lock. Mud banks and gravel shoals would thus be scoured away, and the river would become safer and more accessible for the barges and billyboys, the lighters and steamboats that plied this section of the Thames.

For the most part, the engineers were right. The faster current, helped by the Thames Conservancy's dredging program, scoured a deeper, narrower channel, and shallow bays like the one at Hungerford Reach disappeared. However, the shoreward water level was likewise reduced at low tide, exposing previously uncharted mud banks and making dockside access periodically more difficult. Some of the mud banks and gravel shoals located on pre-Embankment maps continued to build up afterwards or merely shifted their location. They still emerge at low tide today, providing just enough purchase for mudlarking tourists.[12] The stronger scour weakened the underpinnings of several Thames bridges, intensifying the damage wrought by the change to London Bridge. Sir John Rennie's Waterloo Bridge (1817) was especially vulnerable. Its foundations were too shallow to start with, and they had been regularly shored up with ballast since it was built. Bazalgette altered the line of the Embankment wall there to encompass and stabilize the north-end support arches, but the bridge was threatening to collapse when the MBW bought it in 1878. The board paid less than half the original cost of the structure, but then had to fund extensive repairs to make it usable.[13]

The Embankment also had an impact on riverside buildings. Some have been mentioned previously, but two are worth remarking here. Somerset House had always faced the Thames over its grand archway and terrace, whose upper balconies afforded a marvelous view. The view remained (see figure 8), but the lower terrace was covered by the Embankment and the new roadway cut straight across the lower arches in a most unsightly fashion. The nearby Adelphi buildings suffered a related injury. Their cross-vaulted basements, a famous haunt for thieves and homeless people, rested on wooden piles set deep into the Thames mud and clay. The river wall deprived the piles of their accustomed moisture. They dried out, began to crumble, and had to be replaced at considerable expense.[14]

Navigation may have improved upon the river itself, but Thames-side proprietors did not always survive to take advantage of it. The wharfingers between Westminster and Blackfriars Bridges shifted their operations downstream or to the south side of the Thames, some willingly and others not. The mudlarks and dredgers undoubtedly followed. Steam ferry operators and their customers delighted in the new landing platforms and stairs during the first years of the Embankment, but a combination of other improvements soon undermined the ferry business. Their first challenge came from the Metropolitan District Railway, which survived its financial crisis of 1866–68. By 1872, the Metropolitan had extended its line under Queen Victoria Street to the Mansion House, enabling passengers to travel relatively quickly from Westminster along the shore.[15] The second challenge was the elimination of tolls from the Thames bridges. London and Blackfriars Bridges, built by the City Corporation with coal and wine duties, had been free and, therefore, received the bulk of the cross-river traffic. Thomas Page's new Westminster Bridge was also free, but his Chelsea Bridge was not. Other bridges were built as private ventures and charged tolls high enough to divert a great deal of passenger traffic to the steam packets. Most of the bridge companies were losing money and wanted to sell out (for the right price), but public authorities lacked funds to purchase them. The problem was studied by Royal Commissions and select committees during the 1840s and 1850s, and almost from the time it began to operate, the MBW tried to negotiate with the City and the Treasury on the matter. Gladstone's government, as usual, was not eager to spend money on such a project, and the City, as usual, wanted to go its own way. But suburban taxpayers began to demand a return for their share of the coal and wine duties, which legally covered the Thames as far as Hampton Court, and urban residents organized a Free Bridges Association to press for reform. When the coal and wine duties were reauthorized, in 1868, provision was made for buying out the bridges at Staines, Hampton Court, Kingston, and Kew. In 1872, the City purchased Southwark Bridge at a bargain price, freed it of tolls, and watched its traffic volume rise tenfold. When the Conservatives took over, in 1874, several toll bridge bills were immediately introduced, and, three years later, the MBW was authorized to buy the rest of them out.[16] The freeing of the Thames bridges, the proliferation of suburban railways, and the opening of the Embankment thoroughfares all undermined the steam-ferry passenger traffic. Most of the packets went out of business in the following decade.

The Embankment boulevard was planned by the parliamentary committees of 1860–62 to facilitate traffic from the City to Westminster by providing an alternative to Fleet Street and the Strand. The committees also designated thoroughfares along the Albert and Chelsea Embankments for access to the western suburbs. None of the new streets met their expectations until long after 1870. The Embankment boulevard (figure 8) was certainly wide and well built, and, after 1872, it could be reached from the City by way of the new Queen Victoria Street. But from that point, lacking the access roads Bazalgette had planned around Embankment Gardens, there was no easy way to leave the thoroughfare until one reached Whitehall Place or Westminster Bridge. Consequently, as the *Building News* observed, the Victorian Embankment was usually deserted except for "a few foreigners or country people, or Hansoms with uncommonly knowing drivers."[17] The situation changed dramatically, in March 1876, when Northumberland Avenue was opened, giving access to Charing Cross. The number of carriages using the Embankment daily, which had averaged eight thousand for the previous two years, jumped to over ten thousand.[18] Perhaps the residents of Whitehall Gardens had been right to demand that the roadway follow that route from the start. Since the advent of automobiles, of course, the Embankment has become as crowded as any other London street.

The other two Embankments fared little better. Traffic coming along the Albert Embankment to and from Westminster Bridge was diverted away from the river along Lambeth Palace Road by the new St. Thomas's Hospital. Also, until 1932, Lambeth Bridge was closed to all but pedestrians. Commuters arriving from the west could cross Vauxhall Bridge (after it was purchased and repaired by the MBW in the 1870s) and proceed along the Grosvenor Road, but there they encountered commuters from Chelsea. The combined traffic would then enter the labyrinth of Abingdon Street and Old Palace Yard at Westminster. When that bottleneck was removed, in 1914, the western route became ideal for lorries as well as motor cars, and the good people of Chelsea began to realize what havoc a modern thoroughfare could wreak upon a suburban village.

Flood control was another benefit which the completed project failed to deliver as expected. Bazalgette and his corps of engineers had predicted some recurrence of flooding and advocated embankment of the south side of the river for just that purpose. Many proprietors had suffered from flood damage in the past, and the incidence of residential damp and disease in Lambeth and Southwark was unusually high. Everyone knew that high water would eventually return. It was better, argued the engineers, to take preventive measures than to face the difficulties and expense of repairs and piecemeal embankments after disaster struck.[19] But Parliament and the First Commissioner of Works decided that the wharfingers and manufacturers in the area could build their own flood walls when needed, and at their own expense. Thomas Cubitt's privately embanked wharf at Pimlico was held up as an example. Moreover, recent experience did not seem to warrant extraordinary measures. The high tide that flooded an Embankment construction site near Westminster, in 1866, was three feet four inches above the official high-water mark at Trinity House, near London Bridge, yet it did no serious damage. Bazalgette himself noted that the completed Embankment roadway would be four feet above high water, higher than any tide on record.[20]

Yet no sooner had Webster and Grant finished the Chelsea Embankment than the Thames, as if claiming a traditional sacrifice for human interference,* rose up in an unprecedented series of floods. On 20 March 1874, the tide rose four feet four inches above the Trinity high-water mark. On 15 November 1875, it reached four feet nine inches. Another great flood tide hit in January 1877, and 1879 proved to be the wettest, most sunless year of the nineteenth century. Observers along the Thames sensed that all the tides, high and low, were more extreme than before, and that the low-lying Surrey shores were in danger of devastation.

Arguments over the cause of the flooding, based as much on pecuniary interest as on scientific understanding or technical information, broke out immediately after the first flood, and reached a crescendo by 1877.[21] Critics of the Metropolitan Board of Works, along with the Surreyside proprietors who faced damages, claimed that the new Victoria Embankment was throwing the current of the river against the opposite shore and forcing the tides farther upstream through a narrower channel. The MBW was, therefore, at fault and should be made to pay for floodwall construction

*In ancient times, people believed the construction of bridges or other impediments required some kind of appeasement of the river's natural forces, so they sacrificed criminals or, later, animals, and buried them within the structure. London Bridge was thought to hold the bones of Peter the Hermit, although when it was demolished, in 1830, only a dog skeleton was found. The Embankment, of course, claimed its share of human life during construction, though not by design.

wherever needed. Bazalgette had indeed testified, in 1860, that the Thames Conservators' river dredging, done in connection with the Embankment, would probably undermine many of the southside pilings and necessitate refacing the wharf walls to prevent flooding. John Thwaites had repeated the prediction to the Royal Commission of 1861.[22] Bazalgette, however, now claimed that this was a misleading point of evidence. The collapse of wharfage walls was really due to bad construction, he said. They were expected to wear out during the life of their lease (usually twenty years), so neither owner nor lessee bothered to keep them in good repair.[23]

Bazalgette pointed out that the Main Drainage system transferred about 600,000 tons of water and sewage from the Thames valley down to the outfalls far below London, thus actually reducing volume in the river that might contribute to flooding.[24] The stronger tidal action, he said, was a continuing reaction to the replacement of London Bridge. Other observers hypothesized, without much evidence, that by reclaiming fifty-two acres of land, the Embankment reduced the volume of the river by some 365,000 tons of water, which used to flow over the mud banks and scour out wharf foundations, so that its narrow channel actually blocked worse tides from coming upstream, much as the old London Bridge had done.[25]

Public and official reaction to the flooding followed the pattern of reaction to sewage pollution two decades previously. The perceived cause of the problem was inextricably bound up with the issues of what was to be fixed, who was legally responsible for fixing it, who had the authority and resources to fix it, and who would pay for it. Because of the indentations within and between wharf frontages on the Surrey shore, the actual distance to be improved totalled almost forty-two miles. This could be considerably shortened by a straight river wall like the Victoria Embankment, but that was not contemplated. The principals argued instead about local repairs, the raising of existing walls, and connections between them. The MBW initially argued that the Metropolis Management Act of 1855 left the provision of river walls in the hands of local vestries and districts, which had taken over the powers of former sewer commissioners in this regard. The vestries replied that they could not get wharf owners to comply with their orders, and that their share of the rates provided little revenue for such projects.

Bazalgette then made a survey of the affected shoreline and reported to the board that, although the cumulative effort was enormous, minor improvements carried out by each proprietor could accomplish the goal. The board gave him authority to send out 870 "letters of advice," which explained to individual owners what the chief engineer had determined would be necessary in each particular case. About two hundred proprietors, holding almost half the total length of shoreline, complied entirely with the "advice." One hundred thirty more complied in part, and the remainder refused.[26] Lacking the power to enforce its advice, the MBW then appealed to Parliament. It introduced a bill that would strengthen vestry authority but give the Board's engineering department control over plans and specifications, with private owners paying the bill. A select committee of Commons tried to negotiate a compromise, in the spring of 1877, but its proceedings read more like a criminal trial than a civil hearing. The member districts and vestries petitioned to use the board's money but without the board's supervision. Several wharfingers and dock companies hired their own counsel, as did the select committee itself. The board, represented by Bazalgette's barrister son Norman, relied on Bazalgette to explain their proposed remedies. Cross-examination was often hostile. The chief engineer read excerpts from the old Sewer Commission records showing that commissioners had always had the power, backed by the courts, to order repairs of the river banks at the landowner's expense. But when he was asked by what laws the board had the right to design and impose its own solution, he excused himself from answering on the ground that he was only an engineer, not a solicitor.[27] The committee chair, E. M. Grant Duff, then ordered Bazalgette to prepare copies of all the improvement plans for each of the counsellors representing the districts, vestries, and proprietors opposed to the bill. This was to be done within three weeks. Bazalgette protested that such a task was impossible. He already had his clerks working late at night on the primary set of plans and specifications. Each section of river had to be divided by individual properties, and each property had to be colored to show existing versus proposed walls. Each copy had to be traced from the master plan. He finally promised to make a second complete set of plans and give each counsel the part pertaining to his clients. More than that he was unwilling to do. The situation was admittedly unique, he said, but "I think that a greater amount of detail has been given to the opponents in this case than has ever been given before in any such case."[28]

Eventually, the select committee reported a bill that charged the expense of flood control to the metropolitan (MBW) rates. The board refused to support that provision, let the bill die, and resisted further promptings from the First Commissioner of Works until 1879. By then, the vestries and wharfingers had come to terms. The Metropolis Management (Thames River Prevention of Floods) Bill of 1879 gave the board power to order and design the necessary repairs but at the owner's expense. Owen says the engineering department was "decently liberal" toward owners who wanted to design their own frontages. Nevertheless, it pressed its advantage with vigor, and the new line of "embankments" grew up quickly. Unfortunately, the engineers acted much like the proverbial military staff, planning strategy for the previous war. On 18 January 1881, the tide at Trinity House rose six inches above the record high of 1875, and Bazalgette had to send out notices for a new round of flood protection measures.[29]

The Thames was also prone to flood along its northern shore at Millbank, the area extending upriver from the Houses of Parliament along Abingdon Street to the Horseferry Road and the old Lambeth suspension bridge. This neighborhood had never been improved like the areas on either side. It was a jumble of jerry-built tenements rising amid the warehouses, wharves, and cement works left over from the long years of building the Houses of Parliament. Abingdon Street, a narrow lane twisting westward from Old Palace Yard, led by a dogleg to the equally unimproved Millbank Street. The foreshore of the river was built up on simple earthworks reinforced with large wooden planks. Mud and gravel shoals cluttered the shoreline. Lambeth Bridge itself was too dilapidated to support vehicular traffic. The remarkable contrast between this decrepit waterfront and the New Palace of Westminster attracted painters and photographers, but most people did not share their aesthetic appreciation.

Millbank was not so low-lying as Lambeth, but low enough that, if the Thames Embankment had been designed primarily for flood control, this should have been one of the first sections completed. The neighborhood had been flooded twice in the thirteenth century, once in the seventeenth, and three times in the eighteenth, the water rising to four feet inside Westminster Hall, which was entered by boats. In 1821, the river rushed over the banks all the way from Westminster to Vauxhall "with a cataract fury," pouring into the windows of carriages that tried to brave the flood. From midcentury to 1875, floods averaged two a year.[30]

The metropolitan improvement commissions of the 1840s had noted the conditions at Millbank; plans for incorporating the area into the general scheme of the Embankment appeared from time to time in their deliberations. But the first measure to accomplish anything was produced by a Royal Commission appointed in 1863. The commissioners considered the commercial buildings south of Victoria Tower a fire hazard and judged that the New Palace could use some additional space for military parades and guard changes, pedestrian approaches and carriages. Clearing the area would provide a more aesthetic view as well. They rejected an ambitious plan to bring an arched roadway down from the new Westminster Bridge, along Parliament's river terrace and onto a new embankment to Grosvenor Road. H. M. Office of Works went ahead with a more limited project. It cleared 415 feet of property south of the Victoria Tower, extended the Parliamentary embankment 230 feet down to Great College Street, and laid out ornamental grounds with walks and carriage ways. It also recommended that the cost be defrayed by raising and leasing out some "first class housing" on the far side of Abingdon Street, where the Office of Works had taken over land for construction purposes.[31]

The Victoria Tower Gardens was an immediate success, so, in 1867, the Office of Works sought additional powers to raze the buildings further up Abingdon Street, widen and straighten the street itself, and extend the river wall another 366 feet for flood protection. The project was supported by Works Commissioner Lord John Manners and by Disraeli at the Exchequer, who may have considered the expense justified during a year when unemployment riots and Reform agitation seemed to feed each other. Under an act of 15 July 1867, the Office of Works began the tortuous business of acquiring the necessary property deeds. But either these negotiations were unusually tough or else Gladstone, elected in December 1868, was less enthusiastic about them, because the site was not fully secured until 1873.[32]

Plans and specifications for what was now designated the "Abingdon

Street Embankment" were sent to a select list of contractors drawn up by the Office of Works engineer, John Taylor, in November.[33] Only experienced Main Drainage and Embankment contractors were notified: John Aird and Sons, Benjamin Baker, William Cubitt, Lucas Brothers, Mowlem and Company, and William Webster. Webster tendered the low bid of £22,000 and began work on the contract in July 1874 (after another great flood tide in March). He encountered some opposition from the proprietors remaining near Lambeth Bridge, who complained that his cofferdam caused mud and gravel to accumulate around their wharves, but he finished the embankment on schedule in May 1876.[34]

Webster's contract did not extend all the way to Lambeth Bridge because there were still a few wharves and a cement works left at the corner. The Office of Works was responsible for laying out the ornamental flower beds and shrubs, the iron railings, and the pedestrian and carriage ways for the new Victoria Tower Gardens. It took its time getting started. The land was still vacant in November 1879 when W. H. Smith, who had led the fight to make Whitehall Gardens public, urged Works to commence digging. He was confident that the inhabitants of Westminster, "especially the children," would benefit from the recreational space, and the project would provide winter work for laborers. Smith offered to contribute £1,000 toward the cost of shrubbery and walks, which the Treasury readily accepted. The work was not completed until 1881, however, and even then Smith complained about the lack of park benches for mothers and nannies to sit on.[35]

As happened elsewhere on the Embankment, the replacement of old commercial buildings by a public garden had the effect of making the surrounding neighborhood a more desirable place to live. Although the slums persisted, their potential for redevelopment gradually became irresistible. In 1897, a private syndicate proposed to extend the Victoria Tower Gardens all the way to Lambeth Bridge and widen Millbank Street. They would, incidentally, clear out five thousand low-income residents to make way for high-rent housing. The proposal was referred to the Improvements Committee of the London County Council (successor to the MBW) and then submitted to Parliament as a private bill early in 1898. There it was attacked by the threatened residents, the Westminster vestries, many LCC members, and a number of MPs.[36] As usual, everyone thought the project worthwhile so long as it was done under their control but with someone else's money. The chair of the LCC Improvements Committee, G. J. Shaw-Lefevre, a parliamentary veteran of "extraordinary industry and public spirit,"[37] quickly produced an alternative plan which involved relocating Millbank Street to run directly west from Abingdon Street toward what everyone hoped, in vain, would be a new Lambeth Bridge. This plan also entailed the removal of many residents but would, at the same time, generate new property for the LCC to sell, thus recouping its expense for the project.[38] Shaw-Lefevre spent the next year negotiating with Prime Minister Lord Salisbury and the Westminster parishes of St. John and St. Margaret. He asked Salisbury (son of the marquis who defied the MBW over Embankment Gardens in 1867) to contribute some Office of Works property along Abingdon Street in lieu of cash, but Salisbury reminded him that the Treasury had already paid for Victoria Tower Gardens and opened it to the public without cost. The LCC had better luck with the parishes. They were in the process of being absorbed into the new City of Westminster, so they cheerfully promised £100,000, which they then charged to Westminster's account as a carry-over debt. Many LCC members thought they were still being asked to pay too much, but, on 24 October 1899 the Improvements Committee approved the scheme. Four days later, Shaw-Lefevre reported that he "triumphantly" carried it through a stormy council meeting.[39] He then introduced a parliamentary bill to authorize the plans and guided it successfully through the Commons. A select committee in the Lords, however, paid heed to the Abingdon Street working-class residents threatened with displacement (they were to move to new "Workers' Dwellings" nearby) and the loss of their view of Westminster. To appease these residents, the Lords restored Millbank Road to its original "new" route, thus undoing previous agreements and necessitating a new round of legislative fighting.[40]

The LCC (Improvements) Act of 1900 was finally signed on 6 August. It authorized widening and straightening Millbank Street, resettling the displaced residents, and converting the remaining small triangle of wharfage by Lambeth Bridge to an extension of Victoria Tower Gardens. The cost was estimated at £1,319,000. As in previous embankment schemes, the

Thames Conservators were compensated for their rights to the embanked riverbed, and the LCC was to do the work with the consent and approval of the First Commissioner of Works, in whose office the completed land would be vested.[41]

The wharves and cement works remaining by Lambeth Bridge, however, proved to be especially tenacious, and the "Workers' Dwellings" promised for relocating the nearby residents were slow to develop. The new City of Westminster insisted on approving the plans which its parishes had so generously supported, and the chief engineers at the LCC and at the Office of Works could not work out who would be responsible for the various parts of the plan. In 1909, Works undertook to build the embankment with a plain concrete wall and designed a garden to match the one at the end of the Albert Embankment across the Thames. It finally opened in the spring of 1914. By that time, the social complexion of the neighborhood had changed considerably. Two members of Parliament asked for a new gate to be installed in the fence between Westminster Palace and the gardens, so that they and "quite a number" of other members who lived in the area could walk directly to the House.[42]

Although the flood walls on the Surrey side of the Thames and the Abingdon embankment by the Victoria Tower Gardens were built in response to an unprecedented series of high tides, the inspiration of the Thames Embankment itself cannot be gainsaid. Almost before it was completed, it became a paradigm for the transformation of the Thames, a kind of template for embanking additional portions of the metropolitan riverside. In part, this was due to the publication, in previous years, of so many grandiose plans for embanking both sides of the river, from London Bridge up to Vauxhall or even Chelsea. Once the Embankment materialized as a practical and aesthetic reality, people began to perceive it as part of a more extensive design. It was at once a great achievement and a promise of future fulfillment. Suggestions for new embankments or modifications of old ones were published soon after its completion and have continued to appear right up to the present. Whenever a commercial portion of the Thames bank has actually been rebuilt, due respect is paid, both in print and in the architectural details, to the ancestral triumvirate.

A case in point was a little jumble of timber jetties called Grosvenor

Wharf, left at the southwest foot of Lambeth Bridge in Westminster after the Victoria Tower Gardens were finished. Barely 130 yards long, Grosvenor Wharf was used mainly to bring in hay and straw for local stables and paving stone for street improvements. A negligible spot for over a century, it now appeared as a blemish between the newly completed Victoria Tower Gardens and the previously embanked Grosvenor Road. Suggestions were made, in 1914, to convert the wharf into a garden while rebuilding the rickety (1862) Lambeth suspension bridge, widening the Horseferry Road approach, and carrying the newly remodelled Millbank Street down to Grosvenor Road. Nothing came of this until 1923, when the Duke of Westminster agreed to accept compensation for the property, and the London County Council applied to Parliament for the requisite authority. The *Observer's* account of the project typifies the public's view of the Embankment as both historic achievement and future task. It was "a scheme to complete the Embankment on the north side of the river," by closing "an interruption" of the river wall, where "barges still come in to the wharves there as they have done all down the ages." The newspaper's comment on the site mixed nostalgia with a disdain for river commerce that might have astonished people living back in 1860: "A Thames barge under sail is a pleasant sight, but little is to be said for the apparatus of a wharf, and it is plainly a misuse of our river that such a site should be given up to the freight of barges." The *Observer* article rehearsed the history of the Embankment, mentioning (as other articles invariably did) Sir Christopher Wren and Sir Joseph Bazalgette, and concluding with the prospect of a boulevard extending from Battersea to Blackfriars. "The south side remains less fortunate," it went on, evoking the vision of John Thwaites and W. H. Smith:

> Perhaps this generation may live to see the construction of a bridge at Charing-cross combined with a great riverside improvement to redeem the squalor of the southern mudbanks. Few of the great cities of the world have been so slow to understand and employ the natural advantages of their situation as London. It is, of course, true that the river, as harbour and highway, made the City, and Londoners have always known how to employ it in commerce and industry. But the energetic development of the utility of the river in trade has been curiously linked with neglect of what it might contribute to the comfort and beauty of a well-planned town. Let us hope that the completion of the great enterprise of the Victorians in the embankment

of the northern shore will be the first of new schemes in which our century will do as much to make London a comelier and healthier city as they did in their day.[43]

Although the LCC approved the reconstruction of Grosvenor Wharf in March 1923, actual construction did not begin until 1928. At that point, the plan to rebuild Lambeth Bridge finally came to fruition, so the two projects were combined and completed by 1932. The LCC then negotiated with the Office of Works to convert the reclaimed land into another "extension" of Victoria Tower Gardens.[44]

The *Observer* erred in claiming that this was the last piece of embankment on the northern shore between Battersea and Blackfriars. A small group of wharves handling timber and slate continued to operate near the Tate Gallery at the northeast foot of Vauxhall Bridge for many years, and the shoreline there was not really embanked until the construction of the "Riverwalk" computer and telecommunications tower in the 1950s.[45] The narrow triangle of land on the other side of the bridge road continued to house commercial wharves and factories until the 1960s. Hovis's Imperial flour mill was there, along with the London Hydraulic Power Company, which generated electricity with a tidal reservoir used in the old days to turn mechanical water wheels. In 1969, Hovis was replaced by the British Gas office tower, and, in 1985, the rest of the twenty-seven-acre parcel enclosed by the curve of Grosvenor Road up to the bridge approach was developed as an exclusive residential enclave called Crown Reach. The proprietors extended a new river wall from Vauxhall Bridge and installed replicas of the Embankment lamps, with their double-dolphin pedestals, along the top.

Farther upriver toward Chelsea, the plain river wall built by Thomas Page in the 1850s has carried on without fanfare. According to the MBW's booklet on the Chelsea Embankment (1874), Page's edifice "had no pretensions to architectural embellishment, and is devoid of many features characteristic of the more recent works." The Office of Works transferred it, in 1887, to the MBW, who gave it a new granite-faced parapet and a new ironwork fence. It still lacks the dramatic appeal of its western extension.[46]

The south side of the Thames, so lamented by the *Observer* in 1923, has of course been transformed since the "high noon" of Victorian commercial prosperity. The St. Thomas's Hospital complex added a stately presence to the Albert Embankment soon after completion, but other sections were not developed until the twentieth century. A short embankment supporting County Hall at the downstream foot of Westminster Bridge was in place by the First World War, and the London County Council moved there from the old MBW offices at Spring Gardens in 1922. It stood alone among the warehouses, breweries, and wharves of the area until after World War II.

At the end of the war, the Royal Society of Arts, which had organized the Great Exhibition of 1851, suggested a public festival to commemorate that historic celebration of art, commerce, and manufactures. Government investigated but ultimately denied the use of Hyde Park, where the Crystal Palace had been erected, because it would interfere with urban recreation at a time when public gardens were especially important to war-weary Londoners. In 1947, Sir Stafford Cripps at the Board of Trade suggested a smaller arts festival and industrial design display that would suit the conditions of postwar austerity. He turned the project over to Herbert Morrison, son of a policeman and a housemaid, who had risen from a London slum to be Home Secretary in the Churchill coalition Cabinet and was now leader of the House of Commons. Morrison reconceived the festival as "a tonic to the nation," expanded its operations to two thousand sites all over Britain, and used his extensive patronage powers to recruit influential and enthusiastic organizers. For London, the festival committee turned Battersea Park into a reincarnation of Vauxhall and Ranelagh Gardens from a century before, complete with fun fairs, fountains, and fanciful structures. The main festival site, however, was a twenty-nine-acre wedge of land between County Hall and Waterloo Bridge on the south bank of the Thames, where the London County Council had commissioned a new metropolitan concert hall. Artists, designers, and architects used the site to educate the public in the principles of modern landscape and town planning which had developed in the 1930s. Their theme was the harmony between national character and natural environment. An "Upstream Section" displayed images of the landscape and of natural resource development. The "Downstream Section" celebrated rural and urban ways of life. These exhibits may have contributed to the upsurge of popular nostalgia for a British identity rooted in countryside and village life. On the other hand, their environmental message was clearly urbanized. In addition to sculptures by Henry Moore and Barbara Hepworth, concerts of new British music, plays, and regional cuisine, the festival offered a spectacular new panorama of the Thames, dramatized by special lighting, that most people had never seen before. "For the first time," wrote John Betjeman, "Londoners would realize the river in their midst."[47]

In spite of recurrent labor strikes and incessant rain, over eight million people participated in festival activities. After the exhibits were dismantled, the gardens and outdoor cafe, the tree-lined pedestrian walkway, the Embankment extension, and Royal Festival Hall remained. As the Victoria Embankment had done in the nineteenth century, these structures served as prototypes for further development. To the east of Festival Hall rose a complex of theatres and galleries known as the South Bank Arts Centre. Occupying the great promontory overlooking Hungerford Reach, the Arts Centre was rendered in concrete slab construction that drew criticism from many quarters. It was finished in time for the Queen's Jubilee in 1977, when the nearby Festival Gardens were redesigned and renamed. The Arts Centre displaced two London landmarks: a great shot tower that stood next to Waterloo Bridge, which appears in an impression by Turner, and the Lion Brewery, whose roof sported a great stone mascot which could be seen clearly from the Savoy and Cecil Hotels across the river. The shot tower is gone, but the lion was relocated to guard the Embankment approach to Westminster Bridge near County Hall.

It is impossible to bring subsequent improvements along the Thames in London up to date, since the Victoria Embankment has been extended below Blackfriars Bridge, and the former docks and warehouses on both sides of the river there are now being reconstituted as high-rent office blocks and flats at a rapid pace. Hay's Wharf, built for the Victorian tea clippers and only closed in the 1980s, is now Hay's Galleria, full of pricey boutiques and whimsical cafes, with a new river pier catering to suburban commuter ferries and (in the public relations blurbs, at least) cross-Channel airfoils. The neighborhood of Butler's Wharf, on the west side of Tower Bridge, was eerily deserted as late as 1987. Walking through its empty streets, one sensed desolation and danger. Now it is teeming with trendy lofts, wine bars, and galleries. Even Billingsgate Market on the north side, notorious for its fishmongers' language, has been gentrified. One can lease a three-bedroom flat there with a balcony overlooking the Thames, within walking distance to the City, for only a few hundred thousand sterling.

The new embankments on the south side differ in several important respects from those completed in the nineteenth century. Except for the South Bank arts complex, they have not been built as part of a unified plan extending the length of the Thames within the metropolis. The shoreline is being developed partly by private capital for commercial enterprises, starting with the studios of London Weekend Television. Piecemeal development has produced a variegated architecture, which is quite interesting in its own way. John Major's government perceived the variety as so much "clutter." It proposed, in a 1995 white paper called "Thames Strategy," to create more "visual cohesion to establish continuity" along the river and "an appropriate urban composition." What the "Thames Strategy" might actually look like may never be known. After Mrs. Thatcher dismantled the Greater London Council, jurisdiction over the south bank of the river was divided among at least sixteen boroughs and half a dozen regional agencies. Institutionally, the region has reverted to its Victorian pattern.

Unlike the Chelsea-Blackfriars shoreline, the south side was not developed as a transportation route. In the 1860s, Government felt that the approach roads to London Bridge, Blackfriars Bridge, and Waterloo Station provided sufficient access to the area from the south and southeast, while the Albert Embankment brought traffic into Lambeth and Westminster from the southwest. Little was done to facilitate east-west routes, and the situation is not much better today. The one concession to the shoreline has been Riverside Walk, which at least opens the Thames to pedestrian recreation. The walk is being extended in front of all the "restored" warehouses and docks with a full array of cafes, boutiques, bookstalls, and buskers, just the kind of "Parisian" atmosphere the Victorians considered quite improper for respectable families.

These considerations simply point to the main difference between the two sides of the Thames. An east-west axis had developed between the City and Westminster during medieval centuries, so an orientation parallel to the river was natural. The wharves and warehouses, residences and law courts set along the shore were approached essentially from the Strand and Fleet Street, and the Embankment was viewed from the beginning as a means of duplicating that key thoroughfare. On the Surrey side, by contrast, the establishment of light manufactures (potteries, engineering firms, foundries, breweries, etc.) along the riverbank, and the need to cross the river for social, business, and political activity, meant an orientation perpendicular to the river. And that orientation has been intensified recently by gentrification and its related automobile traffic.

Farther upriver on the south side, where the Albert Embankment was brought to an undignified halt before reaching Vauxhall Bridge, a new office block will soon fill in the last piece of river wall. In the 1860s, this was the site of a gasworks and distillery, and before that the famous Vauxhall Gardens. Crosse and Blackwell's, the marmalade people, had a factory here until the turn of the century when Vauxhall Bridge was rebuilt, the approach road widened, and a petrol station and automobile garage established at the corner. By the 1980s, the site was utterly derelict, filled with trash and weeds and the debris of homeless people. In 1987, a private developer proposed to build a high-rise collection of flats, shops, and offices on the site. It was to include a river walk and a T-shaped marina, much like the tidal pools within embankments designed in the 1850s. As at Crown Reach across the bridge, a new embankment wall was to match the original, with lion-head mooring rings and the requisite dolphin lamps. The arrangement was "intended to create a dockside character and environment."[48]

"Vauxhall Cross," as the development was called, did not survive much beyond the excavation stage and has been taken over by another group, which is proceeding on more conventional lines. The idea of creating or recreating a "dockside character," however, is alive and well along the Thames below London Bridge, where bits and pieces of the old structures are incorporated into new ones like talismans. Some critics consider this the "Disneyfication" of England, a curious blend of historical preservation, nostalgia, and upscale marketing that would have astonished the Victorians who built the first Embankment. They understood the marketing aspect well enough—they were busy selling themselves to the world—but the idea of transforming old warehouses into contemporary residential and office facilities behind preserved or reconstructed facades would have violated their belief in novelty, expansion, and embellishment. And the aesthetic of leaving pieces of the original brickwork, support beams, and plumbing fixtures on display within an otherwise modern edifice could hardly have made sense to a generation dedicated to moral and civic "Improvement" as well as to "finish" in its works of art. In this respect, the South Bank has a split personality. Half of it, from Albert Embankment gardens to the National Theater at Waterloo Bridge, celebrates growth, innovation, and the future in a civic context. The other half represents retrenchment, privatization, and the exploitation of historic images.

By inspiring extensions and imitations of itself, the Embankment influenced its future both structurally and conceptually. However, as the extensions were also responses to new conditions, some of their features inevitably contrasted with the original structure. This tension between continuity and change is common to all historical development but is intensified in the case of enduring objects, because people actually reexperience the original form so long as it continues to function, while simultaneously perceiving nearby derivatives.

Such tensions also developed in the institutions, groups, and individuals mentioned earlier in this account. Of the Embankment-related institutions, the Metropolitan Board of Works gained the most benefit from the project. Starting with weak public support, questionable resources, and qualified authority, the MBW built its reputation and power with every stage of construction. The Metropolis Management Act of 1855 had given it responsibility for building and improving streets, naming and lighting them, and regulating building safety. Until the success of the Main Drainage and the Embankment was evident, however, these duties remained subordinate and often contested by vestry and City officials. The change of reputation is illustrated by the construction of Queen Victoria Street from Blackfriars Bridge to the Mansion House, together with the railway underneath it. The City put up strong opposition to what it regarded as an invasion of its territory, but, because the project was officially linked to the Embankment, the MBW retained control of it and eventually completed it on its own terms. The scope of these projects led the MBW to reform its public financing and build a bureaucratic organization, making it the largest and most resourceful metropolitan agency in London. As a result of its success, a variety of other responsibilities (and burdens) was given to it, as they had formerly been given to H. M. Office of Works. With better leadership and a bit of luck, the MBW could have evolved into a modern metropolitan government.

But, as so often happens in organizations, early success led officials to perpetuate procedures and personnel beyond their useful life. Due to the dominance and longevity of Chief Engineer Bazalgette and his coterie of assistant engineers and clerks, the MBW engineering department remained essentially a larger-than-usual mid-Victorian engineering firm, despite evolving a number of bureaucratic features. The mix of professional, kinship, and patronage relationships discussed in chapter 7 continued through the 1870s and 1880s. Although Bazalgette escaped censure for the problems that beset the board in its last years, the fact that he and his assistants Grant, Lovick, and Cooper all retired in 1889, when the board came to its undignified end, is telling. They had met the challenge of large-scale construction management for a certain period, but had not met the challenge of political evolution.

After 1874, the MBW suffered a decline in its reputation. It was accused regularly of jobbery, dullness, a tradesman's mentality, aesthetic ignorance, and self-aggrandizement. As its staff grew larger and more entrenched, its epithet changed from "The Board of Words" to "The Board of Perks."[49] Only the horrors of London prior to 1860, vividly and repeatedly recalled by the press, made it look good. Its most disagreeable and suspect function was the purchase and disposal of property in connection with public improvements. We have already seen what a time-consuming, niggling process this was. Each property had to be valued and offers made, negotiated, and sometimes taken to arbitration or to courts of law. Buildings and lots left over at the completion of a project had to be auctioned off or leased to developers. As always, the process made property owners angry and led to charges of illegitimate gain, some of which were bound to be true. George Vulliamy, the able and honest administrator of the architect's department, was in charge of the MBW's property transactions. He stayed in office as long as Bazalgette, but he aged more quickly, gradually succumbed to ill health, and finally lost control of his subordinates. Two of them engaged in questionable practices, which came to light in a series of articles by the *Financial News* in 1886. The scandal spread to other staff and to members of the board, who were then assailed with charges of a cover-up. The London Municipal Reform League, founded in 1881, took advantage of the charges to press for a full-scale investigation. With the newspapers in full cry, Parliament set up a Royal Commission on the Board of Works in 1888. After exposing the misdemeanors of an inner group, and the potential for abuse in many of the board's regulations, the commission exonerated most of the delegates and staff. But it also recommended replacing the board with a modern council government. Out of spite, the board raised its members' salaries and pensions, voted a series of expenditures for its successors to deal with, and generally behaved, wrote Owen, "as if determined to drink the cup of perversity to its depths." The newly constituted London County Council unilaterally advanced the date of its assumption of power from the first of April to the twenty-first of March, 1889, in order to prevent further damage.[50]

For a long time, the popular press conflated the ignominy of the MBW's later years with its initial successes. This was a certain logic to this, since the men and methods developed for the latter were largely responsible for the former. But it often produced confused historical images. In 1923, a newspaper compressed the public memory into one paradoxical sentence: "The Thames Embankment, as we now know it, is a Victorian enterprise, with which the defunct and discredited Metropolitan Board of Works and Sir Joseph Bazalgette, their engineer, are honourably associated."[51]

The Office of the First Commissioner of Works, which had sponsored the MBW as a solution to its own management problems and then supervised MBW projects as a compliant tool of the Treasury, gradually outgrew its Victorian limitations. In the restrictive context of London during the years of the Thames Embankment, it appeared hesitant and handicapped. In the long run, its support for an autonomous Board of Works for the metropolis was undoubtedly the right policy for it to pursue. In the twentieth century, it became a full-fledged ministry with extensive power and responsibility for projects all over Great Britain. Under Prime Minister Edward Heath, it was appropriately joined to the Ministries of Housing, Local Government, and Transport to form a Ministry of the Environment.

For the City of London and Parliament, the success of the Board of

Works was a mixed blessing. It is true that they now had a more competent body to negotiate with, and a convenient dumping ground for problems (e.g., the regulation of "baby farming" or foster care) that no one else wanted to deal with. But they also faced a new and at times very inconvenient political competitor. Today, of course, London is again faced with the problem of metropolitan administrative fragmentation. The Greater London Council, flaunting its radical populism at Mrs. Thatcher from County Hall across the Thames, has been discharged. To achieve its ends, the Ministry of Works must negotiate once again through a maze of regional, county, and local agencies, as well as powerful corporate interests and European Union regulatory departments. Whether or not a metropolitan-wide government will be reconstituted in the future, it is unlikely to be given much power over the Thames within its boundaries. The river's regime, formerly guarded by the Thames Conservators or the City of London Navigation Committee, has now been placed under regional agencies such as the Port of London Authority, which approves the design of riverside construction, and the Thames Valley Authority, one of ten agencies in the United Kingdom which regulates water quality under the aegis of the Ministry of Environment and European Union environmental directives.

The City Corporation of London continued to warn against the threat to local self-government posed by MBW ambitions. But the MBW achievements of the 1860s had the effect of dampening calls for metropolitan political reform, with the result that the City's ancient privileges were in little danger. In fact, the City fathers, perhaps guided by their Improvements Committee and their engineering department, mounted their own impressive program of public works.[52] These had little effect on the banks of the Thames, however. The St. Katherine Docks east of the Tower were still new, and the commercial importance of the river below Blackfriars Bridge still too important to be challenged.

Although the operations of the MBW engineering department inevitably reflected Bazalgette's early training and professional habits, its intimate connections with the London engineering community and with the Institution of Civil Engineers ensured a degree of modernization. The Embankment occurred during a period of gradual but important change in the profession, exemplified by the contrasting careers of Bazalgette and Thomas Page, from trial-and-error, apprentice-trained consultants and ephemeral partnerships to more systematically prepared employees working within larger firms and public offices. Both the MBW and the ICE introduced examinations for specialized grades of engineers, although both insisted that the British legacy of apprenticeship not be lost. And in both organizations, the sons and grandsons of "gentlemen engineers" assumed their social status more easily and lightly than in the past. The ICE and its first cousin, the Institution of Mechanical Engineers, still reside on Great George Street. In the 1860s, the ICE was the premier professional organization for engineers. Bazalgette submitted his report on the Main Drainage to his colleagues at one of its meetings, in 1865, and later served as president. His son Edward, as part of his membership requirement in 1878, read a complete description of the Thames Embankments.[53] Later in the century, engineering societies proliferated so that today there are hundreds of specialized metropolitan, provincial, and postcolonial institutions around the world.

The Embankment is an exceptionally enduring object, which still imposes parts of its original technological frame upon the environment, yet it has adapted easily to successive changes. Like all supposedly solid constructions, it requires constant repair and occasional modification, but the half-million cubic yards of concrete, brick, and granite show few signs of deterioration after a century and a quarter. Minor bomb damage sustained in World War II was quickly repaired. Its monumental lines, so different from most Victorian architecture, have not become cluttered with embellishment. The Egyptian obelisk dubbed "Cleopatra's Needle" arrived after several mishaps, in September 1878, and was installed on the parapet opposite Embankment Gardens.[54] The gas lamps were electrified and the roadway fitted out with electric tramways after the turn of the century. The overflow chambers or penstocks within the walls, which originally passed raw sewage into the Thames during rainstorms, were closed in response to Thames Conservancy complaints. Used only in emergencies, they gradually silted up, but they have now been cleaned out and function once again.* The Embankment appears indifferent to such minor changes.

---

*I am informed by the present staff of the Thames Valley Authority that, when sewage is discharged within the metropolis (diluted with storm water, and always at ebb tide), a boat called "The Thames Bubbler" forces oxygen into the river to bring water quality up to national and EU standards.

The massive granite-faced walls continue their relentless channeling of the Thames and their patient support for all the pipes, tunnels, roads, walks, and other utilities within and above the ground reclaimed from the river.

The Embankment, as the construction of an interface between society and its environment, was conditioned by both entities and, in turn, helped to define them. To a great extent, it continues to play that role. Due to the demographic shift from the old industrial areas of the north toward London and the southeast, population pressure on the Thames valley may be greater now than in the nineteenth century, and the river has become central to an urban and suburban society dedicated to consumerism and the commercialization of the country's heritage.

The Embankment grew, in part, out of Victorian concern for the preservation of the countryside, which included a perception of rivers as self-purifying sources of water, fresh air, fish, and spiritually renewing landscapes. That sense of environment as amenity and as heritage continued to influence development of the Thames riverside in London. Public access to the Embankment Gardens and Whitehall Gardens coincided with the founding of the Commons Preservation Society. During the same period, Victorian enthusiasm for natural history found expression through botanical field clubs whose membership, in the 1880s, amounted to 100,000. Overzealous plant collecting and widespread grouse shooting by the new bourgeoisie, among other dangers, caused a reaction in favor of countryside preservation, which carried on into the twentieth century with ever-widening support and a more complex set of goals. The countryside movement, institutionalized in a Countryside Commission in 1981, has been criticized for reviving an early-nineteenth-century hostility to industrialism that not only undermines Britain's competitive strength but also defines the rural environment as a social amenity and makes it into an object of aristocratic patronage and nostalgia. The worldwide television syndication of *All Creatures Great and Small* may be its most enduring legacy. Yet the movement has recognized the need to integrate town and rural housing, energy, transportation, recreation, forestry management, and water resources policies.[55] The reconstitution of the countryside in terms of human ecology and the built environment was adapted to urban conditions by the designers of the Festival of Britain. It has continued to guide development of the South Bank complex, with pedestrian walkways, neighborhood housing, parks, and shops mixed in with commercial and professional office blocks.

The other main impetus for the building of the Embankment was, of course, water pollution. The problems of industrial waste and sanitary sewer discharges into the Thames and other British rivers were displaced but not solved by the urban drainage projects of the 1860s and 1870s. The 1876 Rivers Pollution Prevention Act left regulation of discharges and water quality in the hands of local authorities, who were often the worst polluters. Although river contamination remained an issue through the first half of the twentieth century, no political party or government ministry saw fit to embrace a program of prevention or cleanup that would threaten established manufacturing interests. Neville Chamberlain responded to demands for action, in 1926, with the same laissez-faire economic justifications voiced a century earlier.[56] The legislation of 1876 was not modified until 1951, when the Rivers (Prevention of Pollution) Act transferred responsibility for enforcing standards to a set of River Boards, later called River Authorities. However, previously existing discharges were not covered by the legislation and the management of water supplies and sewer systems was not separated from antipollution functions.[57]

The 1960s produced a spate of pollution legislation. In fact, with the British publication of Rachel Carson's *Silent Spring*, in 1962, and the catastrophic oil spill from the tanker *Torrey Canyon*, in 1966, pollution began to dominate the whole field of conservation and ecology.[58]

The Water Resources Act of 1963 brought together a host of regulatory activities under twenty-seven River Authorities who controlled everything from the procurement of water supplies to discharges of effluent. Together with the launching of the countryside movement, this act is sometimes seen as a turning point for British environmentalism.[59] Although the Thames estuary remained afflicted with toxic waste from the rapid expansion of the new docklands development, the stretch from Teddington Lock down to the Tower improved dramatically during the 1970s and 1980s (heralded as usual by sightings of large fish), and the upper reaches of the river recovered a semblance of their preindustrial purity. Ecology was finally good politics. The Labour Party under Harold Wilson claimed to be planning for systematic antipollution action prior to the 1970 elections.

When Heath's Conservatives surprisingly won, the word "environment" actually appeared in the queen's speech for the first time, and a new ministry was organized to pursue a concerted environmental policy.[60]

Unfortunately, the institutional fragmentation that characterized Victorian river regulation did not really disappear. In 1971, an official study discovered, in addition to the twenty-seven River Authorities set up in 1963, 198 water supply companies in England and Wales and 1,300 mostly local sewerage authorities. As they had done a century before, the latter emptied their drains into the rivers and their members served as delegates to the River Authorities, so prosecution faltered and the water companies complained of contaminated supplies.[61] Britain's entry into the European Union, in 1973, only complicated the situation, since British authorities resisted the imposition of Continental techniques of water quality measurement.

Parliament finally separated urban sewerage operations from regulatory functions in 1989. Antipollution measures were consolidated under ten Water Authorities, each responsible for one or more major watersheds (the Thames is the only river system to be served by its own authority). Simultaneously, however, Government began a program of privatizing water companies. Because capital spending on sanitation facilities had fallen off during the recession of the 1970s and remained low under the Thatcher budgets, the privatized companies faced enormous outlays for new equipment and construction. Having expressed admiration for Victorian values, Mrs. Thatcher was now reconstituting Victorian institutions with many of their Victorian drawbacks. Except for the contested intervention of European Union directives, the situation is little different than it was in 1855. As one political observer concluded, "Environmental consciousness is a neat enough slogan, but it is not a political programme. Before it can be a political programme there have to be fundamental changes in attitudes and values, in the outlook of politicians, and probably in the structure of the political system itself."[62]

As I argued at the outset, the Thames Embankment was not just an engineering project, and its effects were not just physical. As John Thwaites intended, the Embankment changed Londoners' perception of the river and, by making it a focal point for the metropolis, changed their perception of London itself. A century later, the Festival of Britain designers recharged that perception by providing a new vantage point from which to view the city. Anyone standing on the balcony of Festival Hall on a clear night can follow Thwaites's vision from the gilded Clock Tower at Westminster along the majestic arc of Hungerford Reach, past the lighted parapet of Somerset House to the glowing dome of St. Paul's in the east. It is a panorama worthy of an imperial capital.

But it is only a panorama. Before the Embankment was built, the metropolitan Thames supported lives and livelihoods in physical proximity to its water and mud and wooden pilings. The great brick and granite walls sealed off that kind of relationship. By 1900, the river was virtually closed to casual boaters, to scavengers, and to other marginal characters who had no legal or bureaucratic right to work on it. It was no longer a working thoroughfare but an object of engineering reports and water quality surveys. It had ceased to be a frightening source of disease and social disorder, but it was also shut off from the life of the nearby streets. In a sense, then, it expresses a particular attitude toward the environment which may have originated in the nineteenth century but has not disappeared. Having exploited and despoiled London's natural and economic lifeline, the Victorians built a wall around it, channeled the worst of their pollution out of sight, and created a public esplanade along the river's edge. From that vantage point, they could enjoy the Thames as a cultural amenity, discerning in its waters a reflection of themselves and their improved, respectable city.

An Embankment Chronology

Embanking the Thames is a very old practice. To identify a particular period of construction as "The Embankment," one has to distinguish its principal features from the general background of continuity and change along the river. Even historians who claim to shun the modernist concept of development have to deal with this kind of problem. The usual method is to construct a chronology, sorting out sequences and fitting them together to clarify how they interact. Readers seldom see historians' chronologies, because they are overlaid by description and narrative, as the framing timbers of a house are hidden by siding, trim, and paint. But because the history of the Embankment concerns a set of overlapping and sometimes discontinuous sequences rather than an integrated process, it seems helpful to uncover the chronology that helped frame it.

Nineteenth-century engineers, like their twentieth-century successors, used chronological schedules to project the interplay of labor contracts, financing, delivery of materials to the site and the whole sequence of construction. That is why the enterprise was called a "project." Today, engineers use more sophisticated approaches—flow charts, critical path studies, project management computer software—but the purpose is the same. An historian's chronology proposes a sequence of actions that occurred in the past, whereas project schedules propose what should happen in the future. Both are exercises in composition. Visualizing the timely convergence of purposeful paths of construction in a flow chart is very much like the plotting of an historical explanation (or a Victorian novel). Both modes of discourse aim at a satisfactory working out of circumstance and character, even as they recognize the play of accident and misfortune. Authors of both modes know from experience that "on site" construction of actual buildings (or texts) requires a large dose of improvisation and may force alterations in the most carefully wrought plans.

A chronology is always selective. What is put in or left out, and the scale of the items included, are not just neutral facts but prejudgments about the nature of the subject. Chronologies are crude forms of what Jack Hexter called "bracketing," that is, sorting out the boundaries and salient characteristics of a complex historical phenomenon so that readers get a clear sense of what lies behind the label.[1]

In technology, the importance of establishing prior claim to inventions, the use of patents, and the organization of finance and production all emphasize the dating of origins, points of departure, schedules of completion. But competition among interest groups to determine the meaning of an artifact, and reconsiderations prompted by its actual use, leak through such official boundaries. So the historian has to compare the external, public boundaries marked in documents with the internal, private ones constructed by experience.

I have divided my chronology into three sections to mark off the construction of the Thames Embankment, as a project, from its past and future, and to indicate a rudimentary plot line based on my judgments about the main intersections of technology, society, and environment. The plot, essentially, is this: a long-term pattern of local responses to geologic change in the Thames River basin was brought to crisis by the rebuilding of London Bridge and the almost simultaneous appearance of cholera, perceived to be caused by the relatively sudden growth of London's population and its sewage. This crisis was met by a generation of engineers and political leaders who conceived of the riverbanks as sites for urban improvement projects supported by new kinds of public finance. The London Main Drainage and Thames Embankment projects responded to concerns about river pollution and disease, about the need for more thoroughfares, and about London's image as an imperial capital. Metropolitan railway development coincided with the need for Thames improvement, making it possible to shift commercial buildings downriver and build a solid Embankment. The project was deemed a success on most counts and became a template for future Thames-side construction.

Explaining human responses to complex environmental and social conditions means combining long-term patterns of incremental change at the structural level with the episodic interplay of cultural values, individual careers, and institutional operations. Thus the crowded, unsanitary condition of the metropolis and its river forms a background to the proposals and decisions to construct a metropolitan sewer system and lay it within a river embankment topped by a roadway. The growth of financial institutions in London and administrative expertise at Westminster are relatively long-term developments brought into sharp focus by the unprecedented scope and time constraints of the project. The decision to build the underground Metropolitan District Railway line within the half-completed Victoria Embankment, and the decision to eliminate commercial wharves along its frontage, are both responses to the relatively autonomous growth of Britain's railway system, which intersected with the Embankment project at a particular time and place. The choice of dates, therefore, is complicated.

The year 1840 is an appropriate marker for the initial thrust toward an Embankment. It marks the end of a decade of investigation into the effects of the re-building of London Bridge upon the river. The engineers' report was published by a Royal Commission concerned with a general plan for improving the metropolis, and the question of legal authority for constructing an embankment was raised in that year by the City of London's decision to raise funds on its own initiative. Al-though the railway "mania" was not yet in full swing, the revolutionary potential of this new form of transportation was already recognized. London's sanitation problems, and the threat of cholera, were fairly evident, and the state of the Thames had occasioned nasty comments for years. The year 1840 is not so much a sharp boundary or a "turning point" as it is a shift in the focus, frequency, and in-tensity of human activity with regard to London's environment. We might char-acterize it as a year when a variety of long-term developments begin to interact significantly with each other, and when people in London begin to respond to those developments in an increasingly organized way.

The most appropriate date for the "beginning" of the Embankment project proper is 1860. In that year, a House of Commons select committee initiated the sequence of effective decisions and planning for actual construction. The Victoria section was authorized by Act of Parliament in 1862, contracts were let in 1863, and construction began early in 1864. During those years also, the movement for bringing the amenities of the countryside into the city gathered momentum, not only in the design of urban parks but also in the formation of the Commons Preservation Society. The first two Rivers Pollution Commissions began investiga-tions that led to the first water quality legislation.

Why not choose to start in 1855, when Parliament created the Metropolitan Board of Works with a charge to build the new Main Drainage? The board's chief engineer proposed an Embankment to enclose the low-level sewer at that time, and the board's new authority could be seen as a *sine qua non* for the success of the project. However, Bazalgette's first plans failed, as did so many others in the two decades before 1860, and other elements of the institutional framework that made the eventual Embankment possible were yet to appear. The lengthy dispute between Crown and City over rights to the riverbed and shoreline, which had pre-cluded work below Westminster, was not settled until 1858. That year also wit-nessed the "Great Stink," which was pivotal in Parliament's decision to authorize construction of the London Main Drainage system by the MBW. By 1859, Parlia-ment was still considering a host of private bills for new railway lines into London, several of them based on embanked rights of way along the Thames, with related commercial development. And, in 1859, the only section of embankment previ-ously authorized by government, the short river wall and roadway at the east end of Chelsea, had been "in progress" for some eighteen years and was just then near-ing completion. The environmental movements noted above were still in embryo. So 1860 seems the more appropriate date.

For similar reasons, I have chosen 1874 to mark the transition to the Embank-ment's historical future. The Victoria section was formally opened in 1870 but, though it was the largest, most difficult, and most significant part of the project, the three sections really belong together. The Albert Embankment was an integral part of the original plan and, in fact, was completed before the Victoria section. The Chelsea Embankment, which seems almost a separate concern, would have been built in the 1850s, had Government not run out of money. It was included in the later scheme as a matter of course; but it was built by the MBW using new techniques of construction, and it properly belongs with the other two sections. The public celebrated 1874 as the official completion of the combined enterprise. Embankments after that date took on a different character: they were simple flood control walls, as on the Surrey side of the river, or piecemeal extensions of the original Embankment. In recent decades, developments downstream from Westminster and Blackfriars have complemented the original sections with new purposes and designs.

The way I have bracketed the construction of the Embankment is not peculiar to technological or environmental history. It is quite consistent with the general periodization of nineteenth-century England, which George Kitson Clark outlined in *The Making of Victorian England.* He argued convincingly that "the Victorian era" was given an artificial homogeneity by the accident of the queen's long life and reign. Culturally, he said, the nineteenth century developed in three broad phases: a phase of growth and social turbulence from 1815 to 1848; the "high noon" of prosperity and relative social stability that ended about 1873; and a denouement characterized by increasing class division and doubts about national purpose, which shaded into the Edwardian era. Concentrating on the first phase, Kitson Clark distinguished what was new from what was traditional in English society, and separated the conditions and typical responses of 1815–30 from those which emerged later and came to characterize the second or "high" Victorian phase, during which the London Main Drainage and Thames Embankment were built.

Kitson Clark was among the first to challenge the older practice of keying cultural history to the achievements of famous people as recounted by other famous people:

> ... to detect such changes it must obviously be necessary ... to try to disregard a little that self-conscious, self-confident minority who seem to have made history and certainly have normally written it, whose voices, unless we are careful, are the only ones we are likely to hear from the past. ... What is at issue is not the unrolling of a simple political narrative, or the logical account of successive changes in the machinery of government; it is the development of a whole community and therefore the economic growth and cultural development of large classes of men, often of obscure men, are likely to be more important for it than the behavior of individuals, even of distinguished individuals.[2]

This admonition might seem too obvious today, but, as recently as 1937, G. M. Trevelyan could still assert that the accession of Victoria a century earlier marked England's "arrival in port" after the storm of the early industrial revolution, and her death, on 22 January 1901, together with the demise of Lord Salisbury and the end of the Boer War, marked the close of the nineteenth century.[3] In a similar fashion, R. B. McCallum, filling the gaps in Elie Halevy's classic survey of the nineteenth century, tried to nail down the dividing line between the early and late Victorian periods: "The line may be taken in 1861 at the Prince Consort's death, although that is really too soon; the death of Palmerston in 1865 is another mark, as is also the passing of the second Reform Bill in 1867, or the accession of Gladstone to power in 1868. In foreign affairs, the end of the American Civil War in 1865 or the victory of Prussia over France in 1870 dates that divide. But somewhere about this time the history of Great Britain takes a turning."[4] Even after Kitson Clark led the way towards a more culturally holistic approach, some historians were still trying to graft it onto the traditional chronicle of wars and political careers. R. J. Evans prefaced the second edition (1968) of his *The Victorian Age, 1815–1914* with the following remarkable disclosure, part of which I have italicized: "I have left the main narrative untouched, *since the actual course of events does not change;* but I have thickened considerably the chapters dealing with the social and economic development of the country, which nowadays excite more general interest than formerly."[5]

Kitson Clark surveyed a cluster of features that seemed to prevail between 1848 and 1875: population growth, urban improvement, general prosperity despite several financial crises, the consolidation of gains by the new mercantile and industrial classes, liberal politics without fundamental reform, the revival of religion, the spread of literacy. He contrasted the relative social and political balance of these years with the agitations over Catholic Emancipation, the Reform Bill, Chartism, and the Corn Law Repeal of previous decades, and the more radical politics of Irish Home Rule, suffragettes, and socialists of later years. He noted a decline in agriculture and agrarian political strength after 1870, the spread of organized political constituencies, and the growth of Nonconformist influence in religion and politics. Economic problems bracket the middle years: before 1848, they stem from the dislocations of the Napoleonic Wars coupled with the chaotic genesis of industrialization; after 1875, they are associated with competition from Germany and the United States. Except for the Crimean War (a notable exception, to be sure), the midcentury years are peaceful ones at home, although fighting erupted continually in the colonies. After Disraeli's purchase of the Suez Canal shares in 1875, Britain was increasingly drawn into imperial conflict.

Although Kitson Clark's arguments were compelling, he reminded readers that any broad cultural period contains many survivals from earlier times. Midcentury Britain was influenced by the continuing power of the aristocracy and gentry, an intensity of genuine religious feeling, and a high level of crime, poverty, and violence in public life. While he contends that the masses as well as the elite had undergone fundamental changes by midcentury, they are not always evident: "the England of 1850 resembles the cruder, pre-industrial, pre-democratic, resolutely unreformed England of the eighteenth century more closely than we have been pleased to imagine, and on the surface it seemed to stay that way during the long golden hours of the Victorian interlude."[6]

Terms such as "interlude" or "high noon" are frequently used by other historians to distinguish the mid-Victorian era from its temporal neighbors, a period, says L. T. C. Rolt, "so peaceable and prosperous that we look back upon it as to some golden age."[7] Walter Arnstein agreed that "for most Britons, these two decades were, in contrast with the previous age, years of prosperity and of relative social harmony during which both talk and consciousness of class division abated. It was an age in which underlying assumptions about the necessity for a high degree of individualism at home, free trade abroad, and progress in human affairs were accepted by most with uncritical, almost religious, conviction."[8] W. L. Burn entitled his study of the years 1851–73 *The Age of Equipoise*.[9] Richard Altick, in *Victorian People and Ideas*, uses the same dates to bracket intellectual history: "It is only in the fifties and sixties that, regarding Victorianism from our vantage point of more than a hundred years, we are wholly unconscious of the eighteenth-century past on the one hand and the impending twentieth century on the other."[10] Asa Briggs used "equipoise" to denote the balance of political, economic, and social interests between the Great Exhibition and the Second Reform Bill. After 1867, he said, the middle classes lost their confidence as the Irish, the workers, the Americans, and the Germans vigorously asserted themselves. The Victorian heyday was not quite over, but "it was already late afternoon and there was a chill in the air."[11]

Walter Houghton agrees that, around 1870, the confidence and unity of Victorian public discourse began to unravel. His image of *The Victorian Frame of Mind* is that of a youth who, having survived the recklessness and dangers of childhood down to about 1830, enters a period of prolonged adolescence during which sophomoric ideas of guilt and nobility struggle with each other for expression. Eventually, the hero emerges chastened rather than reborn, mature but uncertain. Houghton uses the term "transition" for this period, and it is easy to see how it reflects the path from romanticism to realism in literature.[12]

Since England's progress in the nineteenth century was tied to industry and cities, it is not surprising to find histories of these topics following the periodization of general surveys. Checkland's *The Rise of Industrial Society in England* describes a generation of entrepreneurial inventors who came of age around 1800 in Britain, when the Napoleonic Wars gave them an advantage over Continental competitors. They were inspired by Adam Smith's theory of the free marketplace and, later, by Benthamite utilitarianism. This complex of ideas and practices was then translated into liberal political thought during the 1830s and sustained the dismantling of mercantile restrictions into the 1850s. The economic expansion of the 1850s and 1860s outgrew Smith's static theories, but Darwinian ideas of development were not unified and applied to economics until the 1880s, when British decline had already set in. Economic history thus followed the general developmental framework, but economic and political theory lagged behind.[13]

When we turn to London itself, we have to remember that what transpired in the capital was often at odds with developments elsewhere in England. The Poor Law Reform Act of 1832 and the Municipal Corporations Act of 1835 specifically exempted the metropolis, for instance, and London was not a center of heavy industrialization like Manchester or Birmingham. But London's unique experiences cannot be understood without reference to the wider framework. Donald Olsen, who has studied the city as closely as anyone, is explicit about its historical boundaries:

> The decade preceding 1837 contained enough events and novel phenomena, both symbolic and practical, to convince the most rigorous upholder of the gradualist, seamless-web interpretation, that it saw London moving from one distinct period into another: the creation of the Metropolitan Police, the introduction of the omnibus, the Great Reform Bill, the democratization of the close vestries, the coming of the railway, the invention of the Hansom cab, the retirement and death of John Nash, the burning of the Houses of Parliament, the publication of Pugin's *Contrasts*. Within the decade 1901–11 the motor car, the motorbus, and the electric tram appeared on the streets of London; underneath London electric trains ran both on the old underground and the new tubes; Norman Shaw's Piccadilly Hotel signalled the death of the old Regent Street, and the Russell Hotel announced the same for Bloomsbury; the London County Council was building suburban housing estates; Selfridge's and the first of the Lyons Corner Houses had opened. Edwardian London was already very much our own London; late Victorian London was something quite different.[14]

Within the period bracketed by Olsen, one can also discern what G. M. Young called "the mid-Victorian transition," namely "the conversion of the vast and shapeless city which Dickens knew—fog-bound and fever-haunted, brooding over its dark, mysterious river—into the imperial capital of Whitehall, the Thames Embankment and South Kensington."[15] David Owen, as we have seen, linked this transition to the establishment of the Metropolitan Board of Works, in 1855, following the unsuccessful career of Chadwick's Metropolitan Board of Health and Commissioners of Sewers in the 1840s. He also notes that the MBW's reputation peaked with the completion of the Embankment, in 1874, and declined thereafter until its virtual disgrace in 1889. Percy J. Edwards, who also focused on the MBW, identified 1832 as a boundary after which comprehensive plans for metropolitan improvement were first generated, even though sufficient funds were not forthcoming until much later. Before 1832, he observed, House of Commons select committees had dealt only with specific improvements on a piecemeal basis.[16]

After reviewing these and other histories of Victorian England, I am satisfied that the dates chosen to bracket the Thames Embankment are appropriate, not just for the specific features of its construction but for the whole cluster of elements involved in its social genesis and environmental context. I have divided the "past" into two parts: an "extended" past from Roman times to about 1840, which sets up the most basic conditions for making an embankment of the Thames a real possibility; and an "immediate" past from 1840 to 1860, in which more specific elements of the Embankment interacted to form an identifiable cluster or conjuncture.

Chronology Related to the Thames Embankment

1. TO 1840

?–1000 AD   Celts, Romans, and Anglo-Saxons build earthen dikes along the Thames riverbanks downriver from London for flood prevention. Romans embank the Thames for the first time, using the forced labor of Britons.

1100–1200   The Thames basin begins subsiding after a long period of slow rising. The tides gradually push farther upriver, creating a need for more embankments, which, in turn, channel tides even farther, causing floods.

1176–1207   Old London Bridge, first stone bridge over the Thames, built on nineteen "starlings" or thick piers that restricted tidal flows, raised upper Thames five feet above the lower side.

1272–1327   First mention of royal embankments and repairs, by Edward I and II.

1531–1847   Parish and borough sewer commissions established by Sewers Act of 1531 to oversee local sewers (open street gutters) and embankments in England; local juries could order repairs by individual property owners.

1580–1599   Water wheels constructed in the first arches of Old London Bridge to pump water supply into the City, slowing upper river currents. Thames begins to freeze occasionally in winter, allowing "frost fairs."

1667   Christopher Wren's plans to rebuild London after the Great Fire include an embanked "forty-foot way" from Temple Gardens to the Tower. Never completed.

1750   Old Parliament Houses embanked along with the first Westminster Bridge approaches for about 100 feet. Later extended eastward along the old Privy Garden in Whitehall to the Whitehall Palace Landing Stairs.

1758–1762   London Bridge renovated: houses pulled down, water wheels removed, and the central pier eliminated to form a single wide arch. Current velocity increases in the upper Thames.

1767–1769   London Common Council authorizes Robert Mylne to embank from Temple Gardens down to Paul's Wharf as part of new Blackfriars Bridge construction.

1768–1772   Embankment in front of new Adelphi buildings near Charing Cross sparks challenge to Crown rights over Thames shoreline from the Lord Mayor of London acting as Thames Conservator. The Crown wins, but London Common Council is confirmed as Thames Conservator by separate Act of Parliament, with control of the riverbed from the mouth up to Staines.

1774   "Mr. Creasy" introduces the idea of building sewers parallel to the river to intercept noxious wastes and transport them out of the city for downstream disposal or for use as agricultural fertilizer.

1776–1784	Sir William Chambers builds a high vaulted terrace over the embanked landing for Somerset House; Government leases the vaults for commercial storage.
1793	A House of Commons report says no regular survey or plan of the Thames had ever been made, nor any systematic scheme of improvement ever acted upon.
1800	Robert Mylne recommends to Parliament a continuation of Temple Gardens and Adelphi Terrace as an embankment from Westminster to Blackfriars. Opposition from wharfingers prevents further development. Population of metropolitan London reaches one million.
1802	Parliament authorizes West India Docks to relieve congestion at London Bridge and "the Pool."
1810	Another renovation of London Bridge replaces the five central arches with two. This increases the river current, scouring the channel and undermining bridge supports. It also lowers the water level in the upper Thames, exposing more of the mud banks. Parliament calls for designs for a new bridge.
1811	Weir and locks built at Teddington mark boundary between tidal and upper Thames.
1812	Millbank Penitentiary, "the English Bastille," designed by Jeremy Bentham and built with a short embankment.
1814	Last of the "frost fairs" on the Thames above London Bridge.
1814–1817	Embankment for Sir John Rennie's new Waterloo Bridge extended to meet the nearby Somerset House embankment. Quicksand encountered along the edge of the river.
1816	Vauxhall Bridge opened. Originally designed as a stone bridge, it was actually the first bridge on the Thames built of iron, by James Walker.
1818–1821	John Rennie and James Walker prepare plans for rebuilding London Bridge.
1820–1840	The flush toilet or WC, patented in 1778, spreads through middle-class London.
1824–1825	Lt. Col. Sir Frederick Trench designs a quay and raised terrace for the entire length of the Thames through London, forms a company to develop it, but fails.
1824–1832	Thomas Cubitt, largest contractor in London, uses fill dirt from new docklands to raise the ground level of Neathouse market gardens west of Vauxhall Bridge, develops Pimlico as housing estates, and builds his own works on the river.
1825–1831	Old London Bridge replaced with a longer, more open design by Sir John Rennie and sons. Stephen Leach and James Walker predict that the upper Thames current would increase. More intense scour and tidal action would lower the channel, exposing mud banks and undermining the foundations of other bridges.

Appendix: An Embankment Chronology 259

1827–1838	Painter John Martin publishes versions of his design for an embankment incorporating intercepting sewers on both sides of the river from Vauxhall to London Bridge.
1830	Millbank Penitentiary embankment extended by Sir Robert Smirke from Vauxhall Bridge east to Battersea Bridge to form the Grosvenor Road.
1830–1831	Thames Conservators (City of London) advised by two engineering teams to embank both sides of the Thames to improve river navigation, but not until the effects of the new London Bridge were better understood.
1831–1832	First cholera epidemic kills five thousand Londoners; dramatizes dangerous state of sanitation.
1832–1840	City of London appoints James Walker and others to study embankment needs. Report recommends "Walker's line," a common frontage line for wharves between Westminster and Blackfriars, to be embanked by the City as Thames Conservator.
1833	Commons Select Committee on Public Walks advocates parks for the general populace.
1834	"Sewage" first used by Parliament to denote household waste instead of rainwater drainage.
1837–1850	New Houses of Parliament built with 1200-foot embankment by James Walker southwest from Westminster Bridge; half the ground for the "New Palace" is gained from the river by embankment and fill; plans for extending the embankment, by architect Charles Barry, abandoned.
1838	The first of several Metropolitan Improvement Commissions issues a report including embankments among its other recommendations. J. Perry of Chelsea proposes the first Chelsea embankment plan.
1839	H. M. Commissioner of Woods and Forests proposes embanking both sides of the Thames through London. James Walker establishes a survey line along either side of the river.

II. 1840–1860

1840	Metropolitan Improvement Commission advocates Thames embankment along "Walker's line" from Westminster to Blackfriars, but opposition from wharfingers and coal merchants in the area prevents action. Meanwhile, the City of London as Thames Conservator petitions Parliament for the right to borrow £300,000 on its own securities to embank both sides of the river. Crown barristers deny rights to the foreshore.
1841	James Pennethorne completes Victoria Park in the East End, the first public park dedicated to working-class improvement.

1842	Metropolitan Improvement Commission recommends embanking the north side of the Thames from Chelsea to Blackfriars and opens a competition for designs. Edwin Chadwick publishes *Report on the Sanitary Condition of the Labouring Population of Great Britain*, exposing cesspools, unsanitary sewers, and river pollution, and advocating a centralized sanitary commission.
1844	Government introduces a bill to embank the Thames. Thomas Page wins the design competition and is appointed "Thames Embankment Engineer" by the Office of Works. Chancery barristers sue the City of London over rights to control development along the foreshore of the river. The suit impedes all further projects until 1857.
1845	Thames Embankment bill withdrawn by Government after protests by affected wharfingers and coal merchants. Chelsea residents offer to fund a local embankment out of property assessments. I. K. Brunel's Hungerford Bridge erected. Painter John Martin publishes a new embankment design including a railway along both sides of the Thames.
1846	First of many Parliamentary committees to investigate the application of sewage as agricultural fertilizer.
1846–1848	Government authorizes a Chelsea embankment funded through local assessments, but the funding proves unworkable.
1847	The "Thames Conservancy Bill" requires Conservators to establish a general frontage line for all future wharves and factories in the area of planned embankments.
1847–1848	Under threat of cholera, vestry and borough sewer commissions are consolidated into a Metropolitan Commissioners of Sewers (MCS), headed by Edwin Chadwick, charged to develop a metropolitan sewer system. City of London refuses to join. Chadwick urges households to eliminate cesspools and connect drains to sewers; Thames pollution increases dramatically.
1848	Thomas Page's plan for embanking both sides of the Thames is considered by Parliament but dies for lack of financing. Ordnance survey begins to establish reliable topographical data for future sewer and embankment construction. City of London enlarges its own Sewers Commission in anticipation of cholera.
1848–1849	Second cholera epidemic kills 18,000. Edwin Chadwick, embattled director of the MCS, delays laying sewer lines until the ordnance survey is completed. City appoints Dr. John Simon as its own health officer.
1850	London metropolitan population now about two million. MCS begins discussion of embankment plans. Chief Engineer Frank Forster prepares plans and estimates for an intercepting sewer grid, but it is not funded. Thomas Cubitt obtains permission to build his own embankment fronting his works in Pimlico.

1850–1854	"Pipe and sewer wars" between Chadwick and professional engineers.
1852	Metropolitan Water Act forces water companies to move intake pipes upstream and improve household supply.
1852–1857	H.M. Commissioner of Works, separated from H.M. Commissioners of Woods and Forests, resumes construction of the Chelsea Embankment, a new Chelsea Suspension Bridge, and the embankment of Battersea Park, all by Thomas Page.
1852–1859	Private bills to authorize new railway lines along the Thames banks into London introduced into Parliament. The Admiralty investigates the extent of encroachment upon the river by the proposed plans. All plans failed due to jurisdictional disputes among government agencies and demands for compensation from affected wharfingers.
1853	MCS proposes a revised plan for intercepting sewers, but omission of an embankment to control sewage along the riverbank leads to a veto by ratepayers.
1853–1854	Third cholera epidemic; Dr. John Simon's sanitary reforms in the City reduce losses to 211, while 10,500 die in MCS districts. Edwin Chadwick's concept of a central metropolitan sanitary authority is discredited.
1855	Work on sewers and embankments in London halted by controversy over the choice of small-bore or large-bore sewer pipe for the new metropolitan system. Chadwick, advocate of small-bore sewers, is forced to resign from MCS. Construction of the Royal Victoria Dock marks eastward shift of riverside wharfage.
1855–1856	First Commissioner of Works Sir Benjamin Hall sponsors legislation to create the Metropolitan Board of Works as successor to the MCS, retains MCS Chief Engineer J. W. Bazalgette. Bazalgette advocates an embankment from Westminster to Blackfriars, following previous designs, to enclose the new low-level intercepting sewer.
1857	Establishment of a new Thames Conservancy office settles the legal dispute between Crown and City over rights to the Thames riverbed and to land reclaimed by embankments. Conservancy given power of regulation, rights to dredging, but no money to finance construction. A financial crisis, caused by speculation on California and Australia gold, rapid inflation, and the Crimean War, causes 135 firms to collapse. Projected embankment railways lose their investment capital.
1858	The "Summer of the Great Stink." Unprecedented drought and heat reduce river volume. Sewage-strewn mud banks boil with toxic gases. Cholera threatens. Parliament authorizes Metropolitan Board of Works to build London Main Drainage system without interference from the First Commissioner of Works.
1859	The Metropolitan Board of Works sets guidelines for incorporating sewers into all projected embankments.

1860   House of Commons Select Committee on "the best means of providing for the increasing traffic of the Metropolis, by the Embankment of
the Thames." Founding of the Commons Preservation Society.

1860–1861   New Westminster Bridge by Thomas Page opens up river channel, increasing tidal flow and scouring currents. Page, harassed by the First
Commissioner of Works for inadequate accounting, sues to collect
back fees as official engineer and contractor.

1861   February–June: Royal Commission on the Thames Embankment
examines competing plans, recommends solid Embankment and
roadway from Westminster to Blackfriars. August–September: Embankment assigned to Metropolitan Board of Works (MBW). Parliament assigns the metropolitan coal and wine duties to the Embankment Fund and authorizes the MBW to raise rates.

1862   Whitehall tenants fail to stop plans to embank their shoreline. MBW
plans approved by Parliament, and MBW authorized to use the coal
and wine duties as security for long-term loans. MBW establishes special Thames Embankment Committee to oversee the project.

1862   Royal Commission on Surreyside Embankment considers line from
Battersea to London Bridge, eliminates all but Westminster-Vauxhall
Bridge section due to commercial considerations. Design prepared by
Bazalgette.

1862–1866   "Limited liability" law leads to company-forming mania, speculative
projects, financial instability.

1863   Coal and wine duties continued for Embankment projects to 1882.
Bazalgette publishes contract plans and specifications for first two
sections of Victoria Embankment. Parliament passes Thames Embankment (South Side) Act, authorizing MBW borrowing. William
Webster wins major contract for southern low-level sewer.

1863   Royal Commission recommends new Queen Victoria Street between
Mansion House and Blackfriars, vetoes plan for embanked road along
riverfront side of Parliament, but recommends clearing and embanking Victoria Tower Gardens. Brunel's Hungerford Bridge replaced by
Charing Cross Railway Bridge.

1864   George Furness begins Westminster-Waterloo section of the Victoria
Embankment. A. W. Ritson begins Waterloo-Temple section. Parliament approves laying the new Metropolitan District Railway line
within the projected Embankment, and a "pneumatic" railway under
the river from Whitehall to Waterloo. Parliament also authorizes
Treasury guarantee of MBW loan and advances from the Government
Consolidated Fund.

1865   January: William Webster begins northside low-level sewer. October:
Webster begins Albert Embankment. Extension of southside embank-

ment desired by MBW but vetoed by First Commissioner of Works. Chelsea Embankment Bill withdrawn due to problems in Government's budget legislation. First Royal Commission on Rivers Pollution begins investigations.

1866   Crash of Overend, Gurney, bankers and financiers; financial panic, labor strikes. George Furness saved from financial ruin by Thomas Brassey. Metropolitan District Railway stops construction along the Embankment due to financial distress. City of London Gas Works demands river access at Blackfriars Bridge location. MBW submits second bill to Parliament for Chelsea Embankment.

1867   Brassey completes Embankment-related construction for financially-distressed Whitehall-Waterloo Pneumatic Railway. MBW continues negotiations with Metropolitan District Railway and City of London Gas Company.

1868   District Railway dispute with MBW settled by arbitration. William Webster begins construction on Temple Gardens-Blackfriars section. First two sections of Victoria Embankment finished; pedestrian walkway opened to public. Chelsea Embankment Act passed. Second Royal Commission on Rivers Pollution.

1869   Joseph Cubitt's new Blackfriars Bridge opens. Arguments between bridge trustees and contractors prevent plans to continue Embankment roadway under the bridge. Albert Embankment opens. Contracts for Chelsea Embankment let. Parliament allows MBW to consolidate its loan accounts and issue its own "consols" or sixty-year stocks for revenue management.

1870   May: first train arrives at Blackfriars Railway Station via Embankment. July: Victoria Embankment formally opened by the Prince and Princess of Wales. Thames Navigation Act requires MBW to keep river free from obstructions. Arguments with Thames Conservancy over sewer outfalls and dredging continue.

1871   William Webster takes over contracts for Chelsea Embankment and sewer.

1872   Embankment and attached gardens vested in MBW for public use, over objections of City. Webster awarded contract for paving and maintenance of Victoria Embankment roadway. Public Health Act forces water companies to provide constant supplies of fresh water to households.

1872–1874   Duke of Northumberland's mansion purchased and removed to make way for new Northumberland Avenue from Charing Cross to Embankment.

1873   W. H. Smith leads public campaign to vest Whitehall Gardens in MBW rather than let Gladstone's government build public office buildings there.

1874   Chelsea Embankment opened; J. W. Bazalgette knighted.

1874   Victoria Tower Gardens site cleared by H. M. Commissioner of Works.

1874–1877   High-tide floods along south bank. MBW engineers survey forty-two miles of riverside, pressure three hundred proprietors to build or raise embankments at their own expense.

1876   Rivers Pollution Prevention Act assigns regulation of discharges and water quality to local authorities.

1878   "Cleopatra's Needle," a giant obelisk, arrives from Egypt, to be installed on the Embankment parapet below Charing Cross.

1879   Thames River (Prevention of Floods) Act gives MBW power to design and enforce embankments by owners on south side, paid for by owners and vestries.

1881–1885   Southside flood control embankments built under MBW supervision.

1884   First report of the Bramwell Commission on pollution of the Thames demands that the MBW clean up its sewage discharges at Barking Creek and Crossness.

1888–1889   The MBW is superceded by the London County Council (LCC), a popularly elected body including City of London representatives.

1897–1898   LCC Improvements Committee seeks extension of Victoria Tower Gardens to a proposed new Lambeth Bridge, opposes private development of the area.

1900–1901   Parliament authorizes revised plan for Victoria Tower Gardens extension and embankment, straightening of Millbank Street, removing and relocating residences and commercial facilities.

1911–1923   County Hall built at southeast foot of Westminster Bridge, with embankment.

1912–1914   Victoria Tower-Lambeth Bridge embankment construction completed.

1928–1932   James Walker's old Lambeth Bridge rebuilt. Remains of Doulton pottery works on the south side removed. Grosvenor Wharf at the northwest corner replaced with wedge of embankment as an extension of Victoria Tower Gardens.

1948–1950   Festival Hall built on south bank next to Hungerford Bridge.

1951   Rivers (Prevention of Pollution) Act revises act of 1876, transfers pollution control from local authorities to regional River Boards.

1951–1971   Government embanks south side from County Hall to King's Reach for Festival of Britain. South Bank Arts Centre built around Festival Hall.

1963   Water Resources Act consolidates regulation under twenty-seven River Authorities.

1968	Rehabilitation of St. Katherine's Dock. World Trade Center built on east side of Blackfriars Bridge; Embankment roadway extended under bridge to Upper Thames Street.
1970s	"Riverwalk" Telecommunications Tower and river wall replace wharves at east end of Vauxhall Bridge.
1971	Establishment of the Countryside Commission to preserve the rural environment.
1973	Britain joins the European Economic Community.
1977	Jubilee Gardens laid out next to County Hall for the queen's twenty-fifth anniversary.
1980	Tower Hotel and Embankment extension built next to Tower Bridge at St. Katherine's Dock.
1985	Local Government Act dismantles Greater London Council, which begins evacuating County Hall.
1985–	Development of Bankside-Hay's Wharf area on south side and Billingsgate Market on north side as luxury apartment and corporate office blocks, with pedestrian embankment. Another complex at northwest corner of Vauxhall Bridge with private embankment.
1990–1994	Construction of luxury apartment and office complex at southeast corner of Vauxhall Bridge completes line of Albert Embankment.
1995	Government White Paper, "Thames Strategy," proposes efforts to unify the riverside aesthetically.

Endnotes

Abbreviations:

GLRO - Greater London Record Office, Clerkenwell (now London Metropolitan Archives)

HC *Sess. Pap.* - U. K. Parliament, House of Commons *Sessional Papers*

Citations of the Parliamentary Papers vary widely in style, often erring on the side of brevity. I have cited them first by year, then Command number (except for accounts and bills, which were not Command papers), volume, and original page number. Report titles are given in the Bibliography.

ICE - Institution of Civil Engineers, London

ILN - The Illustrated London News

MBW - Metropolitan Board of Works

PRO - Public Record Office, Kew

Chapter I/Constructing the Victorian Environment

1. Sir James Fox called it "child's play." David Owen, in *The Government of Victorian London, 1855–1889* (Cambridge: Harvard University Press, 1982), 81, said the Victoria Embankment was "relatively uncomplicated" architecturally.

2. Cf. Owen, *Government*, chap. 4, "The Embankment."

3. Asa Briggs, *The Age of Improvement* (New York: Longmans, 1959).

4. John Timbs, *Curiosities of London* (London: John Camden Hotten, 1867). Timbs, born in Clerkenwell, began writing as a journalist about 1821. He became editor of *The Mirror*, in 1827, and shortly thereafter produced his first annual *Yearbook of Facts*. The success of the *Yearbook* led to an outpouring of handbooks and "miscellanea" on a variety of subjects over the next forty years. In 1855, he became a contributing editor to *Illustrated London News*, which printed a portrait and biography on 10 February 1855: 125.

5. See *The Social Construction of Technological Systems: New Directions in the Sociology and History of Technology*, ed. Wiebe Bijker, Thomas Hughes, and Trevor Pinch (Cambridge, Mass: MIT Press, 1987) and, more recently, Bijker and John Law, eds., *Shaping Technology/Building Society: Studies in Sociotechnical Change* (Cambridge: MIT Press, 1992). For a discussion of current theory in the field of technology studies, see the exchange between R. Angus Buchanan, John Law, and Philip Scranton in *Technology and Culture* 32, no. 2, part 1 (April 1991): 365–93.

6. I. G. Simmons, *Environmental History: A Concise Introduction* (Oxford: Blackwell, 1993), 77–78, 138.

7. Ibid., 53–55.

8. Alan Harris, "Changes in the Early Railway Age: 1800–1850," in *A New Historical Geography of England after 1600,* ed. H. C. Darby (Cambridge: Cambridge University Press, 1976), 215; J. B. Harley, "England circa 1850," in ibid., 229. The area of metropolitan London was defined by the Metropolis Management Act in 1855.

9. Darby, *A New Historical Geography,* 117–22, 177–78.

10. Thomas F. Glick, "Science, Technology and the Urban Environment: The Great Stink of 1858," in *Historical Ecology: Essays on Environment and Social Change,* ed. Lester J. Bilsky (Port Washington, New York and London: Kennikat Press, 1980), 124. For the lack of sunshine due to coal fires, see B. W. Clapp, *An Environmental History of Britain* (London and New York: Longman, 1994), 14, 17. The pigeons began breeding for lighter colors in the 1970s when fireplaces were converted to burn North Sea gas and London's buildings were cleaned of their soot and grime with pressure hoses.

11. I. G. Simmons, *Interpreting Nature: Cultural Consequences of the Environment* (London and New York: Routledge, 1993), 22–23.

12. Ibid., 23–24, 66; Clapp, *Environmental History of Britain,* 1–6.

13. Bill Luckin, *Pollution and Control: A Social History of the Thames in the Nineteenth Century* (Bristol and Boston: Adam Hilger, 1986), 1–2; Clapp, *Environmental History of Britain,* xi; Lawrence E. Breeze, *The British Experience with River Pollution, 1865–1976* (New York: Peter Lang, 1993).

14. Luckin, *Pollution and Control,* 4. For the American conflict between rural and urban environmental historians, see Christine Meisner Rosen and Joel Tarr, "The Importance of an Urban Perspective in Environmental History," *The Journal of Urban Studies* 20, no. 3 (May 1994): 299–310.

15. See, for instance, the delightful classic by A. E. Trueman, *Geology and Scenery in England and Wales,* revised by J. B. Whittow and J. B. Hardy (London: Pelican Books, 1971).

16. Cf. Luckin, *Pollution and Control;* Breeze, *The British Experience;* Anthony S. Wohl, *Endangered Lives: Public Health in Victorian Britain* (London: J. M. Dent and Sons, 1983); Christopher Hamlin, "Providence and Putrefaction: Victorian Sanitarians and the Natural Theology of Health and Disease," *Victorian Studies* 28 (1985): 381–411; and Hamlin, *A Science of Impurity: Water Analysis in Nineteenth Century Britain* (Berkeley: The University of California Press, 1990).

17. Robert Darnton, preface to *The Great Cat Massacre* (New York: Basic Books, 1984).

18. Wohl, *Endangered Lives,* 117. For the French, see Thomas Kselman, *Death and the Afterlife in Modern France* (Princeton: Princeton University Press, 1993).

19. Owen, *Government,* 146–52.

20. Nigel Everett, *The Tory View of Landscape* (New Haven and London: Yale University Press for the Paul Mellon Centre for Studies in British Art, 1994), introduction, 1–9, 207–8, and passim. Everett's thesis, particularly his vague use of the term "Tory," has been challenged by David Watkin (*Albion* 27, no. 2 [summer 1995]: 323–24) and other reviewers. Bill Luckin notes the important differences between London and the northern industrial cities in *Pollution and Control,* 141–43.

21. Cf. Victor Ferkiss, *Nature, Technology, and Society: Cultural Roots of the Environmental Crisis* (New York: New York University Press, 1993), 56–57, and Michael Bunce, *The Countryside Ideal: Anglo-American Images of Landscape* (London and New York: Routledge, 1994), 7–9, 80–81. For the merchant class invasion of the Chilterns, see Harley, "England *circa* 1850," in Darby, *A New Historical Geography,* 238.

22. H. C. Prince, "England *circa* 1800," in Darby, *A New Historical Geography,* 128.

23. The "levelling" influence of city life is argued (rather unevenly) by Bunce, *The Countryside Ideal,* 11–12. J. B. Harley contends the opposite: the new appreciation of the countryside, along with the continuance of royal parks in the city, "mirrored vividly the deep cleavages in contemporary society: the 'two nations' were at no time more clearly portrayed than in the human geography of 1850" ("England *circa* 1850," in Darby, *A New Historical Geography,* 227).

24. Bunce, *The Countryside Ideal*, 11–12.

25. Donald J. Olsen, review of *The People's Parks: The Design and Development of Victorian Parks in Britain*, by Hazel Conway, *Victorian Studies* 36, no. 4 (summer 1993): 491. Olsen accepts Conway's claim that parks were intended for social betterment, but dismisses her contention that they exercised social control in the Foucaultian sense. Of all such devices, he counters, parks were "surely the least coercive and least successful."

26. Bunce, *The Countryside Ideal*, 11–12. The countryside as amenity is also discussed by J. T. Coppock, "The Changing Face of England: 1850–*circa* 1900," in Darby, *A New Historical Geography*, 326, and by Simmons, *Environmental History*, 179–80.

27. Garret Hardin, "The Tragedy of the Commons," in *Pollution, Resources and the Environment*, ed. Alain Enthoven and A. Myrich Freeman III (New York: Norton, 1973), 1–13.

28. Breeze, *The British Experience*, 91.

29. Luckin, *Pollution and Control*, 9, 49, 64–65, 83. Luckin, like many other ecological historians, is here following the work of Mary Douglas, *Purity and Danger: An Analysis of Concepts of Pollution and Taboo* (1966).

Chapter II/ The Structural Background

1. My description of the Thames above Hampton Court is based on *Salter's Guide to the Thames*, 28th ed. (Oxford: Salter Bros., ca. 1926), 14–29, and Breeze, *The British Experience*, 25–32.

2. *Salter's Guide to the Thames*, 30–35, and Breeze, *The British Experience*, 25–32. For the geography of working-class housing, see W. G. Hoskins, *The Making of the English Landscape* (London: Hodder and Stoughton, 1955), 172–73.

3. Breeze, *The British Experience*, 36–38, 42.

4. Ibid., 43.

5. Timbs, *Curiosities of London*, 89.

6. For three nice photographs of Cheyne Walk taken in 1870 just before the Embankment was built, see John Betjeman, *Victorian and Edwardian London from Old Photographs* (London: B.T. Batsford, 1969), plates 90–92. The passing of the old shoreline is regretted by Elizabeth Longford in *Images of Chelsea* (Richmond-upon-Thames: Saint Helena Press, 1980), 61. Longford also includes some early prints of the Chelsea waterfront (54, 60). The improvement of Chelsea Walk into a modern roadway gave automobile commuters from the western suburbs a direct route into the West End via Grosvenor Road, creating enormous traffic problems.

7. *Crown Reach*. Promotional book produced for the Crown Reach development, London, n.d. The author visited Crown Reach in 1986 and 1987.

8. J. E. Smith, *St. John the Evangelist, Westminster: Parochial Memorials* (London, 1892), 379.

9. Timbs, *Curiosities of London*, 760.

10. Hugh Phillips, *The Thames about 1750* (London: Collins, 1951), 110.

11. Timbs, *Curiosities of London*, 74, 560.

12. Ibid., 2. For the pre-Embankment Adelphi, see GLRO, Maps collection, Westminster AH 22446. A picture and story are given in *ILN*, 20 April 1850, supplement: 278.

13. Timbs, *Curiosities of London*, 721.

14. Ibid., 737.

15. See the engravings by T. Bowles in Phillips, *The Thames About 1750*, 30–36.

16. For specific references to these ideas, see Fernand Braudel, *The Mediterranean and the Mediterranean World in the Age of Philip II*, trans. Sian Reynolds, 2d rev. ed. (New York: Harper and Row, 1973), prefaces to the first and second editions, and part 2, chap. 8.

17. The Thames basin is described in Trueman, *Geology and Scenery in England and Wales*.

18. Peter Hall, "England *circa* 1900," in Darby, *A New Historical Geography*, 380.

19. Trueman, *Geology and Scenery in England and Wales*, 78–80.

20. Ibid., 85–87.

21. Ibid., 84–85.

22. Basil Cracknell, *Portrait of London River: The Tidal Thames from Teddington to the Sea* (London: Robert Hale, 1968), 4; Timbs, *Curiosities of London*, 773.

23. A full account of the phrase and its origins is given in a footnote to a discussion of the River Thames in ICE *Minutes of Proceedings* 49 (1877): 153.

24. Trueman, *Geology and Scenery in England and Wales*, 84.

25. Cracknell, *Portrait of London River*, 51.

26. Henry Robinson, "The River Thames." ICE *Minutes of Proceedings* 15 (1856): 201–11. The tide could rise another two to four feet, leading to severe flooding, if a northwest wind prevailed off the "German Sea." This phenomenon led eventually to the construction of the Thames Barrage near Greenwich.

27. Nicholas Hawksmoor, *History of London Bridge* (1736), quoted in HC *Sess.Pap.* 1821, vol. 5: "Report of the Committee to Inquire into the present state of London Bridge and Consider Alterations and Improvements," 363.

28. HC *Sess. Pap.* 1877 (280) 17: "Report of the Select Committee on the Thames River (Prevention of Floods) Bill," 320. Pictures of old and new London Bridge side by side in 1828 are reproduced in John Pudney, *Crossing London's River: The Bridges, Ferries, and Tunnels Crossing the Thames Tideway in London* (London: J. M. Dent and Sons, 1972).

29. Cracknell, *Portrait of London River*, 32. The ice fairs are also described in Timbs, *Curiosities of London*, 362–63, and in *ILN*, 25 January 1843: 132. Cf. HC *Sess. Pap.* 1877 (280) 17: 370, testimony of Henry Law, who took sections of the river in 1823 and again in 1856. Also HC *Sess. Pap.* 1821, vol. 5: 395. For the suicides, see Timbs, *Curiosities of London*, 67.

30. Cracknell, *Portrait of London River*, 32.

31. "Third Report from the Select Committee upon the Improvement of the Port of London" (HC *Sess. Pap.* 1799–1800, "Reports and Papers" nos. 5096, 5097), in *House of Commons Sessional Papers of the Eighteenth Century*, ed. Sheila Lambert (Wilmington, Del.: Scholarly Resources, 1975), vol. 132: 239. For the engineers' agreement, see HC *Sess. Pap.* 1821, vol. 5: 286.

32. Lambert, *Sessional Papers of the Eighteenth Century*, vol. 132: 301–8, 416–17. The committee's comments on the plans are on pp. 261–70. The Telford-Douglas iron bridge featured a device that locked the arch during the night to prevent "pilferage boats" from leaving the upstream wharfage area, apparently a primary concern of port authorities.

33. HC *Sess. Pap.* 1821, vol. 5: 281, 284, 395.

34. Lambert, *Sessional Papers of the Eighteenth Century*, vol. 132: for Telford, 329–31, plus appendix B.3, 71–73, letter from Telford and Douglas; for Dodd, 416, plate 3, 417–18; appendix B.1, 301–8. Yet another design, by George Dance, is featured in Felix Barker and Ralph Hyde, *London as It Might Have Been* (London: John Murray, 1982).

35. HC *Sess. Pap.* 1821, vol. 5: 284.

36. Timbs, *Curiosities of London*, 68–69.

37. Sir John Rennie, *Autobiography* (London: E. and F. N. Spon, 1875). Also HC *Sess. Pap.* 1877 (280) 17: 371, testimony of J.W. Bazalgette.

38. HC *Sess. Pap.* 1821, vol. 5: 339–41, evidence of James Walker. Walker was the City's engineer. His father Charles Walker, "an experienced civil engineer," had prepared plans for a new London Bridge for William Pitt the Younger. Stephen Leach worked for the Admiralty, as did his son and namesake later on.

39. HC *Sess. Pap.* 1877 (280) 17: 370–77. In 1877, William Henry Barlow recalled (408) that he accompanied his father on "the occasion of the experiment . . . with reference to the possible effects of the removal of London Bridge." The "experiment" apparently measured the velocities of a set of floats moving between the various Thames bridges during flood and ebb tides, along with the water levels at each stage. See the description and table by engineers Scott and Firth in their 1832 report to the Thames Navigation Committee of the City of London, reprinted in H.C. *Sess. Pap.* 1840 (485) 12, "Second Report of the Select Committee on Metropolitan Improvements," appendix 2: 88–89.

40. "High water" means the "Trinity High Water line" marked on all bridges from London Bridge up to Westminster. It was the average spring tide high-water line at the "hermitage" entrance to the London docks, as determined by the corporation of Trinity House in August 1800 under an Act of Parliament. The Walker/Leach report is in HC *Sess. Pap.* 1821, vol. 5: 339–41; Ren-

nie's report is in HC *Sess. Pap.* 1840 (485) 12, appendices: 356–58. Water levels were reported by Captain Edward Burstal to the First Commissioner of Works, 27 January 1857 (GRLO, Records of the Metropolitan Board of Works, MBW 2326). See also HC *Sess. Pap.* 1877 (367) 17: "Report of the Select Committee . . . to inquire and report what amendments are required to deal with injuries by floods," 320 for sections, and HC *Sess. Pap.* 1837–38 (418) 16: "Second Report of the Select Committee on Metropolitan Improvements," 168, testimony of John Martin, for high-water changes.

41. Cracknell, *Portrait of London River*, 39; HC *Sess. Pap.* 1861 (2872) 31, "Report of The Royal Commission on Plans for Embanking the River Thames Within the Metropolis," appendix B: 1, testimony of George Powell.

42. HC *Sess. Pap.* 1840 (554) 12: "Report of the Select Committee on the Embankment of the Thames," 275. Similar figures were given by Henry Robinson in a description of the Thames, ICE *Minutes of Proceedings* 15 (1856): 201.

43. For Rennie and Mylne, see HC *Sess. Pap.* 1821, vol. 5: 341. Benjamin Whinnell Scott was principal clerk to the City Chamberlain; James Francis Firth was principal clerk for records concerning wharf frontages. They were reporting on a plan to extend the Port of London upriver to Blackfriars Bridge by embanking both sides of the river. Their report to the City Court of Common Council, 3 August 1832, is reprinted in HC *Sess. Pap.* 1840 (554) 12, appendix: 89.

44. These details were reported by Charles Pearson in his discourse on the Thames, *ICE Minutes of Proceedings* 15 (1856): 218–19.

45. *Minutes and Reports of the Court of Common Council,* 1839: 15–20, "Report of the Navigation Committee," 17 January. Corporation of London Record Office, Guildhall.

46. Ibid., 9 April: 150.

47. G.P. Bidder, ICE *Minutes of Proceedings* 15 (1856): 237.

48. See PRO, records of the Office of Works, WORK 6: 157/1, 5, embankment proposals dated 30 March 1836. The wooden cofferdam erected for the Parliament embankment apparently created a mess of new shoals, much to the annoyance of neighboring wharfingers and watermen. When construction was completed, it was deemed advisable to cut the cofferdam logs off at the low-water line rather than pull them out and risk erosion of the embankment foundation, and the shorn logs continued to cause buildups of mud. Both the Navigation Committee and the Office of Works argued that extending the embankment all along the river was the only solution, an argument that justified Walker's line even further. Walker himself testified, in 1840, that the mud shoals never developed as feared, in part because steamboats in the area kept the mud in suspension longer than before. The complaints are recorded on various dates in *Minutes and Reports of the Court of Common Council,* 1839; Walker's testimony is in HC *Sess. Pap.* 1840 (554) 12: 279. Later, in 1860, the Thames Conservancy asked the Board of Works to agree to dredging the foreshore to remove the shoals that, indeed, had developed there. Cf. PRO, WORK 11: 7/2, "Papers related to the Embankment of the Thames between Parliament and the Horseferry," 23 November 1860.

Chapter III/ The London Main Drainage

1. Wohl, *Endangered Lives,* 4.

2. J. R. T. Hughes, "Problems of Industrial Change," in *1859: Entering an Age of Crisis,* ed. Philip Appleman (Bloomington: Indiana University Press, 1959), 134.

3. HC *Sess. Pap.* 1882 (809) 52: "Average number of corpses per year in the Thames, 1877–1882," gives an average of 240 corpses a year in the river from 1877 to 1882, "from all causes, including stillbirths, murders, suicides, and accidents." See the *Edinburgh Review* 91 (1850): 381 for a contemporary description of all the effluvia going into the Thames. More recent inventories are provided by Wohl, *Endangered Lives,* 80–84, 233–56; Luckin, *Pollution and Control,* 15–16; and Breeze, *The British Experience,* 50–52. Wohl quotes official reports that some twenty thousand tons of animal manure accumulated on London streets each year.

4. Henry Robinson, "On the Past and Present Condition of the River Thames." ICE *Minutes of Proceedings* 15 (22 January 1856): 195.

5. Breeze, *The British Experience,* 56–57. Clapp, *Environmental History of Britain,* 74, reproduces a letter of complaint about river pollution written in 1868 by a resident of Wakefield, using Calder River water as ink; it is still legible.

6. Luckin, *Pollution and Control,* 141–43; Breeze, *The British Experience,* 7. Breeze reminds us (52), however, that London-area industries could be very toxic. The residents of Luton, upstream on the Lea, specialized in bleaching and dying straw plait for hats and bonnets. They dumped large amounts of sulphur, dyestuffs, oxalic acid, and other byproducts into the river. Some of this discharge drained into the parallel New River aqueduct, from which the East London Waterworks drew drinking water for over a million metropolitan residents.

7. For a concise history of sewers, cesspits, and sanitation in London, see the HC *Sess. Pap.* 1884 (3842-I) 41: "First Report of the Royal Commission on the Discharge of the Sewage of the Metropolis," appendix B.

8. W. H. G. Armytage, *A Social History of Engineering* (Boulder, Colo.: Westview Press, 1976), 139.

9. For cholera's shock value compared to its actual impact, see Wohl, *Endangered Lives,* 120, and Luckin, *Pollution and Control,* 69–70, 81–82.

10. Donald J. Olsen, "Introduction," in Owen, *Government,* 11–13. Snow's designation of polluted water as the carrier of cholera, in 1849 and 1854, was integrated into the prevailing miasma theory. Bad water was said to create conditions for organic decay, which was then transmitted through the atmosphere to humans.

11. Bramah's legacy, which also includes the hydraulic pump used on beer barrels, is detailed in the *Dictionary of National Biography,* s.v. "Bramah, Joseph." The design problems of WCs are detailed with gusto by Wallace Reyburn in his tribute to the eponymous engineer Thomas Crapper, *Flushed with Pride* (London: Macdonald and Jane's, 1969).

12. The prevalence of sewage discharge is often exaggerated, however. Thomas Glick contends that, by 1853, only one-tenth of the homes in London were drained by sewer pipes. Cf. "Science, Technology and the Urban Environment: The Great Stink of 1858," in Bilsky, *Historical Ecology,* 129. Most households used the "dry conservancy" method: human waste was collected on an ash-filled tray inside a "dry closet" and removed periodically by municipal or private agents, much as cesspits were. According to Wohl (*Endangered Lives,* 101), dry closets remained the predominant method of removing excrement in Britain (if not in London) until 1890.

13. Owen, *Government,* 28. Owen gives the figure of seventy-one sewers for the metropolis as a whole; J. W. Bazalgette, chief engineer for the Metropolitan Board of Works, said there were sixty main outlets above London Bridge (ICE *Minutes of Proceedings* 15 [January 1856]: 224.) For the theory of epidemic atmosphere, a variation of the "miasma" hypothesis common to the early nineteenth century, see S. E. Finer, *The Life and Times of Edwin Chadwick* (New York: Barnes and Noble, 1952).

14. HC *Sess. Pap.* 1854–55 (1980) 21: "Report of the Royal Commission on Metropolitan Improvements," 36.

15. HC *Sess. Pap.* 1857–58 (442) 11: "Report from the Select Committee Appointed to take into consideration Mr. Gurney's Report on the River Thames," 6, 37, 65.

16. HC *Sess Pap.* 1854–55 (1980) 21: 27.

17. Timbs, *Curiosities of London,* 238–39, 353. London's atmosphere is further detailed in Wohl, *Endangered Lives,* 210, 213, 222, and in P. Brimblecombe, *The Big Smoke* (London: Methuen, 1987).

18. Wohl, *Endangered Lives,* 147–48. Wohl suggests that Chadwick deliberately ignored the effects of poverty on the health of the laboring population because government agencies could not or would not take action against it. Joseph Childers, however, has argued recently that Chadwick constructed a literary representation of "the Poor" designed to appeal to middle-class reformist sentiments. The *Report* emphasized their susceptibility to disease without displaying habits and living conditions their social betters would find repulsive. See Childers, "Observation and Representation: Mr. Chadwick Writes the Poor," *Victorian Studies* 37, no. 3 (spring 1994): 405–32.

19. G. M. Young, *Victorian England: Portrait of an Age* (London: Oxford University Press, 1936), 60.

20. The *Times,* 20 March 1855, quoting the First Commissioner of Works, Benjamin Hall. David Owen (*Government,* 33) gives Hall's estimate as 10,550 commissioners. Sewer commissioners were Crown appointments, often distributed as political favors without expectation of active service, the urban equivalent of rural justices of the peace. Many took no part in commission business, and some listed in a survey of the early 1840s were dead. Cf. Christopher Hamlin, "Edwin Chadwick and the Engineers, 1842–1854: Systems and Antisystems in the Pipe-and-Brick Sewers War," *Technology and Culture* 33, no. 4 (October 1992): 680–709.

21. Breeze, *The British Experience,* 15. Wohl, *Endangered Lives,* 169, argues that London was no different from other large towns in the percentage of manufacturers and builders who made up sewerage boards. Nor were they any less parsimonious.

22. Hamlin, "Edwin Chadwick and the Engineers." Chadwick's system is also described by Breeze, *The British Experience,* 11. Chadwick's search for fresh water sources led to the Metropolis Water Act of 1852, which required the water companies to move their intake pipes upriver above Teddington and filter the water before delivery.

23. The MCS members appointed in 1847 resigned in 1849. The new (third) commission included Robert Stephenson, Samuel Morton Peto, and other distinguished engineers. They appointed Stephenson's assistant, Frank Forster, to synthesize the best features of all previous plans. Forster came up with the principle of using pumping stations to provide the necessary flow rate in low-lying intercepting sewers. Cf. HC *Sess. Pap.* 1884 (3842-I) 41, Appendix B.

24. Hamlin, "Edwin Chadwick and the Engineers," 689.

25. Ibid., 709.

26. Ibid., 698–700.

27. Luckin (*Pollution and Control,* 4–5) argues that Chadwick's integrated water supply and sewerage strategy was little more than "a canny and, in some respects, commonsensical yoking together of generalised medical and environmental doctrines long current among those with an interest in public health." It had little influence among medical and scientific experts of the day, but the doctrines with which it was associated remained potent long after Chadwick disappeared from the scene.

28. Owen, *Government,* 48–49.

29. Cf. ibid., chap. 2: "The Creation of the Metropolitan Board of Works."

30. Ibid., 37–39.

31. John Martin (1789–1854) specialized in huge apocalyptic Biblical paintings, two of which can be seen in the Tate Gallery. His proposed embankment would have reached from Pimlico down to Limehouse, where the Regent's Canal opened into the Thames. In one of its many manifestations, his plan included railway lines. In 1844, a parliamentary committee found it to be technically deficient both in structural and in sanitary terms, although it was thought to be architecturally magnificent. The original plan is in the Corporation of London Record Office (Guildhall), Noble Collection B, V group 8; it was published as an appendix to the report of the Metropolitan Improvement Committee, HC *Sess. Pap.* 1844 (15) 15. In 1861, when a Royal Commission was evaluating plans for the Thames Embankment, Martin's daughter appeared with counsel to demand compensation for her late father's ideas, which everyone else had borrowed. Cf. HC *Sess. Pap.* 1861 (2872) 31, "Report of the Royal Commission on plans for Embanking the River Thames," appendix: 84. The origin and examples of intercepting sewers are discussed in ICE *Minutes of Proceedings* 15 (1856): 203, 227, 230.

32. HC *Sess. Pap.* 1857–58 (442) 11. Gurney was already ventilating the House of Commons with a steam-jet furnace set up in the Clock Tower, and had applied it to sewers in the neighborhood during the cholera epidemic of 1854.

33. T. R. Harris, *Sir Goldsworthy Gurney, 1793–1875* (Cornwall: The Trevithick Society, 1975), 57–59. For the "animalculae," see Hamlin, *A Science of Impurity,* 104. In 1850, a journalist published drawings of the "animalculae" found in London's water supplies, representing a crowded circle of repulsive creatures as if seen through a microscope. It was reprinted by the *Illustrated London News* and, although later exposed as a composite invention, became what Hamlin calls "one of the most effective appeals to sensibility in the history of public health."

34. Hamlin, *A Science of Impurity*, 3–5. See also Breeze, *The British Experience*, 29–30, and Luckin, *Pollution and Control*, 95.

35. Hamlin, *A Science of Impurity*, chap. 4.

36. Owen, *Government*, 55.

37. The *Times*, 1 July 1858, quoted by Glick, "Science, Technology, and the Urban Environment," in Bilsky, *Historical Ecology*, 138. Glick argues that both the engineering solution and the miasmatic pollution theory which supported it were typically retrograde. He implies that earlier acceptance of the germ theory would have prompted alternative solutions. But this begs the question of how public officials might have responded. Lacking a system of on-site household waste disposal, they would still have favored cleaning up the river and disposing of the sewage through a system of closed drains. The treatment facilities at the outfalls might have been designed more like today's city wastewater plants, but the Main Drainage would still have been constructed.

38. HC *Sess. Pap.* 1846 (474) 10, "Report of the Select Committee on the application of sewage to agricultural purposes." Testimony of company director Augustus Granville, M.D., questions 965–1029.

39. Ibid., questions 830–97. Wicksteed's name is also spelled Wicksted (in 1846) and Wickstead in some records. His memorialist in the ICE *Minutes of Proceedings* (33: 241–46) remarked, "there were no half measures with him!"

40. HC *Sess. Pap.* 1846 (474) 10, "Report of the Select Committee on the application of sewage to agricultural purposes." For the Edinburgh experiments, see Christopher Hamlin, "Environmental Sensibility in Edinburgh, 1839–1840: The 'Fetid Irrigation' Controversy," *Journal of Urban History* 20, no. 3 (May 1994): 311–39.

41. Breeze, *The British Experience*, 80–81; Clapp, *Environmental History of Britain*, 87; Wohl, *Endangered Lives*, 101. Advocates and opponents of sewage recycling knew no more about the real properties and dangers of sewage than the scientists who were pondering the question of water quality during the same period. The arguments might be couched in terms of "pollution" and "purification," but these had more social than medical meaning. See Hamlin, "Environmental Sensibility in Edinburgh."

42. *ILN*, 29 October 1853: 367 showed the cross section of a sewer designed by William Austin of Westminster for Edwin Chadwick at the MCS. One of a variety of plans offered at that time, it is of interest because, following Continental designs, it featured subways under main streets carrying gas and water pipes as well as a drainage ditch flanked by service platforms. Austin proposed to discharge his system into holding tanks at the river's edge, where the solid manure would be filtered out and transported at night by rail or barge to outlying districts.

43. Cf. HC *Sess. Pap.* 1852–53 (629) 26: "Report of the Select Committee on the London Drainage Bill"; 1857–58 (2372) 32: "Preliminary Report of the Royal Commission on the Sewage of Towns"; 1861 (2882) 33: "Second Report of the Royal Commission on the Sewage of Towns"; 1862 (160) (469) 14: "First and Second Reports of the Select Commitee on the Application of Sewage to Agricultural Purposes"; 1864 (487) 14: "Report of the Select Committee on Metropolis and town sewers."

44. HC *Sess. Pap.* 1865 (171) 8: "Special Report of the Select Committee on the Metropolis Sewage and Essex Reclamation Bill."

45. Owen, *Government*, 62–65.

46. Cf. Luckin, *Pollution and Control*, 13–14.

47. The tangled dispute between Hall and the MBW is covered in detail, with relevant quotations from the correspondence, in HC *Sess. Pap.* 1884 (3842-I) 41, appendix B.

48. HC *Sess. Pap.* 1857–58 (442) 11. Gurney had used his steam-jet furnace to clean out sewers during the cholera epidemics of 1848 and 1854. Cf. HC *Sess. Pap.* 1850 (480) 33: "Correspondence between the Commissioners of Sewers and Mr. Goldsworthy Gurney, etc.," and HC *Sess. Pap.* 1854–55 (105) 53: "Mr. Gurney's Report to the Office of Works, on his experiment for withdrawing and decomposing the noxious effluvia from the Sewers in the neighbourhood of the Houses of Parliament."

49. The fact that two former members of Young England acted decisively to clean up the

Thames gives credence to Nigel Everett's thesis, in *The Tory View of Landscape,* that the Tories' traditional view of landscape inclined them toward greater stewardship of the environment. In 1876, Disraeli's ministry passed the weak, ineffective, but at least visible Rivers Pollution Prevention Act, which remained in force until 1951. However, it is always possible to discern less idealistic motives in Disraeli's legislative initiatives. Moreover, MBW control over the Embankment design was supported by Lord Palmerston and his stepson William Cowper, than whom no one could be more Whiggish.

50. Letters to the *Times,* 6 July 1858 :10; 26 June 1858 : 9; 25 June 1858: 9.

51. *Hansard Parliamentary Debates,* 3d ser., 149 (1858): 436.

52. Amid the traumatic conditions of 1858, all sorts of diseases and fevers were reported and rumored to be spreading. In terms of actual cases recorded, however, the summer was no more unhealthy than other years. Cf. Luckin, *Pollution and Control,* 17, 109.

53. Letter to the *Times,* 18 June 1858: 9.

54. *Times,* 26 June 1858: 9.

55. *Hansard Parliamentary Debates,* 3d ser., 151 (1858): 575.

56. *Times,* 18 June 1858: 9.

57. Owen, *Government,* 49.

58. ICE *Minutes of Proceedings* 24 (1865): 352–53.

59. Owen, *Government,* 55–58. Of course, if sewage was accidentally discharged at *low* tide, it could be carried by the next wave of incoming sea water all the way up to Vauxhall Bridge.

60. Ibid., 55–62.

61. HC *Sess. Pap.* 1860 (494) 20, "Report of the Select Committee on the best means of providing for increasing the traffic of the Metropolis by the Embankment of the Thames," Bazalgette testimony question 1089 and Thwaites testimony question 12. F. C. Penrose and Thomas Page corroborated the evidence on quicksand. See also Bazalgette's letter to the MBW, 4 July 1856, in GLRO, MBW 2411 batch #5, Miscellaneous papers on the Thames Embankment, item 32. Further evidence about quicksand was given to the Royal Commission on the Thames Embankment, in HC *Sess. Pap.* 1861 (2872) 31, minutes of evidence 1 (Powell) and 4 (Baxter).

Chapter IV/ Before Its Time: The First Thames Embankment

1. Cf. Darnton, *The Great Cat Massacre,* preface, and Bijker and Law, *Shaping Technology/ Building Society,* introduction.

2. Longford, *Images of Chelsea,* tends to conflate the two projects and even connects the earlier embankment work with the excavation of the new main drainage lines in South Kensington, a separate project.

3. PRO, *Current Guide to the PRO,* Part 1 [116.1, sections 1–4], Works Department, "The Office of Works"; Sir Harold Emmerson, *The Ministry of Works* (London: George Allen and Unwin, 1956), 15–17.

4. The Ministry of Works, *History of the King's Works* (6 volumes, London: Ministry of Works, n.d.), vol. 6, 209. For the incorporation of the Board of Works on each project, see Hall's memoir of 1857, "Returns by the First Commissioner of Works to the House of Commons on Improvements under his department," PRO, WORK 6: 15.2/3.

5. Emmerson, *Ministry of Works,* 17.

6. *History of the King's Works,* vol. 6: 209, 244–47.

7. Owen, *Government,* chaps. 5, 6, 8.

8. Trenham Walshman Phillipps (1795–1855) "made himself indispensable to successive First Commissioners," according to the *History of the King's Works,* vol. 6, 188. He was hired as a clerk in 1814, became private secretary to the First Commissioner, in 1827, and then head clerk in 1835. He remained as secretary after the reorganization of 1851 and, like most good civil servants, wielded huge authority by knowing more than his political chief about the whole operation.

9. *Dictionary of National Biography,* s.v. "Ponsonby, John William." He was the eldest son and eventual (1844) heir of the Earl of Bessborough. He was, however, elevated to the Lords, in July 1834, as Baron Duncannon of Bessborough, and John Cam Hobhouse (1786–1869) took over as

First Commissioner until Melbourne's government fell in December. Lord Granville Somerset (1792–1848) was Peel's appointee.

10. *History of the King's Works,* vol. 6, 210–11.

11. *Dictionary of National Biography,* s.v. "Howard, George William Frederick." He was the eldest son of the sixth Earl of Carlisle and Lady Georgiana Cavendish. Morpeth was the family seat, which Lord Morpeth represented in Parliament from 1826 to 1830. His father had been First Commissioner of Works under Canning in 1827. Cf. *History of the King's Works* , vol. 6, 213.

12. *History of the King's Works,* vol. 6, 191.

13. *History of the King's Works,* vol. 6, 191. PRO, *Current Guide to the PRO,* Part 1 [116.1, section 3], Works Department, "The Office of Works"; Emmerson, *The Ministry of Works,* 15–17.

14. *Dictionary of National Biography,* s.v. "Seymour, Edward Adolphus Seymour." He married Jane Georgiana Sheridan, youngest of the playwright's three beautiful granddaughters, who presided as Queen of Beauty over the famous tournament of chivalry at Eglinton castle in 1839. Lord Seymour represented Totnes until it was disenfranchised as a pocket borough, in 1855, and then succeeded his father as twelfth Duke of Somerset.

15. *Dictionary of National Biography,* s.v. "Manners, Lord John James Robert." Tall and dark-complected, Lord John had a "rather unpleasant" tone of voice which improved considerably when he became excited. See *ILN,* 27 April 1844: 268. He was the second son of the seventh Duke of Rutland and eventually succeeded to the Duchy when his older brother died. His mother was the daughter of the fifth Earl of Carlisle, making him cousin to Lord Morpeth. Although firmly aristocratic and conservative, he was devoted to labor reform, supporting the factory acts and the Ten Hours Act (1847). He entered Parliament as a colleague of Gladstone but later joined Disraeli and the Young England movement. He had a "chivalrous and romantic" cast of ideas, which led Disraeli to cast him as "Lord Henry Sydney" in *Coningsby.*

16. *Dictionary of National Biography,* Second Supplement, s.v. "Molesworth, Sir William." Cf. his obituary in the *Times,* 23 October 1855, 7a. Sir William responded to an unhappy childhood by attacking authority of every form. He was returned to Parliament, in 1832, from East Cornwall, a seat held by others in his family, and became a philosophical radical supporting the Chartists. He was an adequate speaker in the House, but not a strong debater.

17. Owen, *Government,* 30–31.

18. Emmerson, *The Ministry of Works,* 17, attributes the discovery of mismanagement to Sir Benjamin Hall, but puts the date at 1851; Hall did not take over the Office of Works until 1855.

19. *The Houses of Parliament,* ed. M. H. Port (New York and London: Yale University Press, 1976), 177–78.

20. Cf. Michael Bunce's discussion of Mill and Shaw-Lefevre in *The Countryside Ideal,* 179.

21. *ILN,* 25 February 1860: 176.

22. Ibid., 24 July 1858: 75.

23. Owen, *Government,* 40.

24. Ibid., 40–41. See also Clifton, *Professionalism, Patronage and Public Service,* 19–20.

25. GLRO, MBW 966, Works Committee reports, 27 March and 1 May 1860. Sir Joseph Paxton reportedly concluded from a survey of the riverside, in 1838, that the bridges and commercial properties made an embankment impossible. HC *Sess.Pap.* 1854–55 (415) 10: "Reports of the Select Committee on Metropolitan Communications," minutes of evidence 93.

26. GLRO, MBW 2365, "Reports 1860–61"; Owen, *Government,* 76.

27. GLRO, MBW 2365: Thwaites to Cowper, 8 October and 2 November 1860.

28. *ILN,* 25 June 1842: 100. The independent commission was proposed by George Parker Bidder Jr., who, despite consulting for the MBW on engineering matters, had a deep suspicion of all government agencies and a professional interest in keeping the project open to private contractors. He was probably supported by Lord Robert Grosvenor, a long-time president of the Metropolitan Improvement Society. As early as 1842, Lord Robert had urged Robert Peel to put London's improvements in the hands of a special commission of expert engineers to prepare a "scientific report" on drainage, streets, railway termini, and "the most practicable means of throwing open the banks of the Thames," because his society "did not wish to see the subject entrusted to a

board of virtuosi, of amateur architects, or of parties who might be interested in any of the plans [being considered by Parliament]." In 1864, he formed another committee of Commons that, according to Owen, went on a "fishing expedition" for MBW improprieties in the hope of discrediting its engineers (*Government,* 85). Lawrence Breeze (*The British Experience,* 227 n. #9) agrees that Lord Robert was considered lacking in common sense, but notes that he, more than any other environmental advocate, realized that sewage disposal was the "loose end" of the MBW Main Drainage system.

29. Owen, *Government,* 79.

30. Acton Smee Ayrton (1816–1886), the Liberal member for Tower Hamlets, was made First Commissioner by Gladstone in November 1869. Cf. *Dictionary of National Biography,* First Supplement, s.v. "Ayrton, Acton Smee."

31. HC *Sess. Pap.* 1839 (136) 13, "First Report of the Select Committee on Metropolitan Improvements." The efforts were described to the Select Committee of 1860 (HC *Sess. Pap.* 1860 [494] 20, minutes of evidence 188).

32. PRO, WORK 6, 139/4: 62, memo by Thomas Page on the history of Chelsea embankments. In March 1844, Page wrote to the First Commissioner of Works, Lord Lincoln, that a sewage-infested mud bank lying near the Chelsea Water Works intake pipe was covered by "blood worms" that turned the whole bank red. PRO, WORK 6: 157/1: 30.

33. HC *Sess. Pap.* 1844 (15) 15: "Report of the Royal Commission on Metropolitan Improvements," 9. Capitalized in the announcement. Members of the commission are listed in a Return to Parliament, PRO, WORK 6, 139/4: 62, 324.

34. HC *Sess. Pap.* 1844 (15) 15: 1–21, "Report of the Royal Commission on Metropolitan Improvements." Thomas Page later testified (HC *Sess.Pap.* 1860 [494] 20, minutes of evidence 149) that the commission referred the plans to a panel of engineers and public officials.

35. ICE *Minutes of Proceedings* 49 (1877): 263, memoir of T. Page. Henry Law, a member of the ICE, helped Page with the survey. He had apprenticed under Sir Marc Brunel and was another assistant engineer for the Thames Tunnel project. HC *Sess. Pap.* 1877 (280) 17: 369.

36. See the map and description of "The Great Thames Embankment" in *ILN,* 30 December 1843: 432. A bill to authorize the Westminster-Blackfriars embankment was presented to Parliament at the end of the 1843 session but died at adjournment.

37. HC *Sess. Pap.* 1844 (15) 15: 22, 42–45. The plans are on pp. 362–63. Page's plans and maps for the Chelsea portion are located in an appendix, another copy of which is in the Guildhall Library, Noble MS Collection B, V group 8. A description of the pre-Embankment shoreline written by Page and Pennethorne is in the PRO, WORK 6, 139/4: 324.

38. *ILN,* 30 December 1843: 432.

39. In 1855, the chairman of a select committee hearing on metropolitan improvements remarked of the lawsuit, "The lawyers have got a good annuity out of that question!" HC *Sess. Pap.* 1854–55 (415) 10, minutes of evidence 31.

40. PRO, WORK 6, 157/1: 33–36.

41. ICE *Minutes of Proceedings* 49 (1877): 263–64.

42. MBW, *Chelsea Embankment,* issued 5 August 1871 on the occasion of laying the foundation stone for the later addition. Copy in GLRO, MBW 2411, batch #5.

43. PRO, WORK 6, 138/4: 324, Page's report of 1856, and correspondence pp. 120–21. The amendment is given in HC *Sess. Pap.* 1852–53, vol. 7: Bill for the Chelsea Bridge and Embankment 415–17.

44. PRO, WORK 6, 139/4: 114, 128, 330. The 1852 act was 15 & 16 Vict. c. 71; the 1853 act was 16 & 17 Vict. c. 87. A schedule of the property owners and occupiers in the line of embankment and bridge building is given with the bill in HC *Sess. Pap.* 1852–53, vol. 7: 384–93. The agreements necessary for a westward extension of Page's embankment along Cheyne Walk never materialized. The Chelsea vestry board pleaded with Sir Benjamin Hall, in 1856, and with Lord John Manners, in 1858, to extend the embankment anyway, arguing that while a few owners were holding out, all other deeds, leases, rents, and local rates were in limbo. But the Act of 1853 had a four-year limitation on the search for agreements, and it lapsed before construction was authorized. Cf. MBW,

Chelsea Embankment (1871), GLRO, MBW 2411, batch #5. For the relevant Works agreements and correspondence, see PRO, WORK 6, 139/4: 290–94, 328, 406.

45. PRO, WORK 6, 139/4: 4, 48–49, 103–4, 324.

46. The Whigs took over in December; Page's plans are dated 17 November. Drawings are in the Corporation of London Record Office, Guildhall, under "Maps" #H.2.6. The tide, one of a whole series that flooded and damaged London until the building of the Thames Barrage at Greenwich in the 1970s, is mentioned by Page in PRO, WORK 6, 139/4: 129.

47. PRO, WORK 6, 139/4: 135. In his specifications, Page stated the embankment would require 3,291 cubic yards of excavation, 62,490 yards of gravel and other fill dredged from the river, 3,100 yards of brickwork for walls and buttresses, and 8,100 yards of concrete for foundations. He argued that contractors should make bids on the basis of these quantities rather than on a lump sum general contract, so that it would be harder to inflate their calculations.

48. Cf. Port, *The Houses of Parliament*, 157.

49. PRO, WORK 6, 139/4: 135, 140.

50. Ibid., 161–62. Creegan's letter of application is a concise account of the work expected from such a position and a reminder of the difference between a consulting engineer and a site supervisor: "I have been for a period of 6 years acting in the general capacity of Engineering & Architectural draughtsman, surveyor, accountant, etc. under W. G. Mylne Esq., F.R.S. the Engineer of the New River Company and have been accustomed during that period to see the various works connected with that extensive Establishment carried out in strict accordance to the tenor of the Plans, Specifications, and Contracts. [He then gives other references.] In conclusion I beg to say that I can when necessary be working Mason, Carpenter, Bricklayer or Pipedriver, and set out their work for them, showing the workmen by occular demonstration practically how the work should be carried out."

51. Ibid., 165–68, 185. The Smith distillery was probably situated next to Vauxhall Bridge, where the new Crown Reach condominiums have now been developed. Proceeding westward from Smith's, one would have come across the Trollope Wharf; a large, messy opening for the King's Scholars' Pond sewer; the Gas Light and Coke Company wharf; and, finally, some coal wharves next to Cubitt's Westminster Gardens. Across the road, on either side of the King's Scholars Pond Sewer, were Thomas Grissell's large workshops for construction of the Houses of Parliament, finally demolished in the 1970s, and the "Spread Eagle" tavern, still in business today. There is still a short line of wharves along the shore today, without their docks, used for miscellaneous commercial storage. It is the only unembanked section left on the north side of the Thames.

52. Ibid., 165–68, 185, 190, 229–38, 252, 326–30, 429, 443. Cf. the pictures and report of the work in progress in *ILN*, 24 January 1857. Page's Chelsea Bridge was rebuilt in 1936. Elizabeth Longford, in *Images of Chelsea*, 63–67, reprints the *ILN* pictures and a print of Chelsea Bridge at its opening, showing the nearby embankment "as the artist thought it would look." Page certified the embankment finished in 1859 but, in July 1860, told the Works secretary that roadway details and some of the iron railings, along with a set of river stairs, were still to be added to the west end when and if it was extended to Battersea Bridge. Meanwhile, the projections were causing new mud shoals and sewage buildups. PRO, WORK 6, 139/4: 449.

53. GLRO, MBW 2411, batch #5, *Chelsea Embankment Opened*. In 1857, the MBW, hoping to economize on the western portion of its low-level sewer line, proposed to intercept the local drains west of Battersea Bridge with a catchment basin, deodorize the sewage, and dump it into the river rather than carry it all the way through the metropolis to Barking Creek. Horrified, the people of Chelsea rallied behind the idea of an embankment to extend the main sewer line out to their neighborhood. Led by Sir William Tite, the Chelsea vestry asked that the £38,000 returned to the Treasury from the failed embankment project of that year be reappropriated regardless of property agreements.

Page drew up plans to rebuild the old Cheyne Walk river wall seventy feet out into the river, providing room for the sewer and a new roadway. He wrote to Gardiner that the new thoroughfare would fulfill the old vision of a direct route from London to Chelsea and, incidentally, would increase toll revenues on the new Chelsea Bridge. Estimating the cost at £30,000, he suggested naively that the newly created Thames Conservancy could authorize the work, thus avoiding another appeal to Parliament. Neither Gardiner nor the MBW was impressed, and the project was allowed to die once again. Cf. PRO, WORK 6: 139/4: 403–13.

54. Ibid., 324, 403; WORK 6, 149/1: 40–41, 71–72, 78. The new First Commissioner, Sir Benjamin Hall, was not amused by Pennethorne's argument. In the same report of 1857 in which he attacked long-standing abuses in the Office of Works staff operations, he reviewed the history of the Chelsea and Battersea embankments with typical scorn, citing the difficulties encountered by solicitors and contractors and the excessive costs of construction. Pennethorne, who had written a preliminary draft of this report, responded with a long complaint about Hall's mistaken figures and critical remarks. Distancing himself from Page's operations, he charged that the report was harmful to his professional reputation and demanded that his reply be forwarded to the Treasury and the House of Commons. Hall's report of 1857 is in PRO, WORK 6, 152/3; Pennethorne's complaint is in 149/1: 195–204. Pennethorne was a society architect, a relative of John Nash's wife, who raised him in her own household. Nash (who consulted for the Office of Works, 1815–1832) transferred his considerable practice to Pennethorne in 1834. *History of the King's Works*, vol. 6, 193 n. 2.

55. Page encased each of the seven pier foundations in long iron cylinder piles fitted with grooves on each side so that great ribbed sheets of cast iron could be slid down like shutters between them. This arrangement could be installed at low tide, and the cylinders and iron sheeting could be used over again for each pier. There is some evidence that this method had been tried elsewhere, but, so far as I can tell, Page was the first to use it in London. He had to battle with Robert Stephenson and other conservative engineers in a select committee of Commons to get his design approved, but it worked. Later, Bazalgette and the Embankment contractor Ridley experimented with a similar system. The bridge and its construction were described in *ILN*, 13 January 1856: 38, and 3 February 1855: 100; by Timbs, *Curiosities of London*, 71; and in the memoir of Page in ICE *Minutes of Proceedings* 49 (1877): 264–65. The minutes of various commissions and select committees, plans and correspondence on the bridge are in PRO, WORK 6, 175/1–6. Since work on the piers could only proceed during low tide, which often occurred at night, Messrs. Mare the contractors hired a new electric light, powered by a seventy-two-cell battery, to illuminate the work site.

56. PRO, WORK 6, 166/1: 1–3, Molesworth to Treasury, 27 May 1854.

57. For Hall's criticism, see PRO, WORK 6, 152/3, "Returns by the First Commissioner of Works on Improvements under his department"; Hall's correspondence with Page is in WORK 6, 149/1: 1–5. The Chelsea Bridge was praised by *ILN*, 10 April 1858: 364. Page bought several rounds of good English ale for the two-hundred-odd workmen to celebrate the opening. Another engraving of the "Chelsea New Bridge" appeared in *ILN*, 25 September 1858: 281, along with an engraving of Page's unfinished Westminster Bridge.

58. PRO, WORK 6, 166/1: 78, 85, 86, 111, 117, 159, 162, 172, 232–45.

59. Ibid., 111, 159, 162, 172, 232–45.

60. Ibid., 247–50. The solicitor for the Union Bank wrote to Works asking for payment of his fees, as he had defaulted on a loan and would soon be bankrupt. Apparently, Page had cosigned a loan for a friend, a "well-known writer for *The Times*," who then defaulted and was thrown into debtor's prison when Page himself could not pay. In asking for an advance "without prejudice to the court case," Page quoted the bank manager to the effect that Works' refusal to pay was "one of the most disgraceful cases within his knowledge." The bank manager later denied this statement: if he had made remarks on the case, he said, they would have emphasized Page's disgraceful delay in sending in his accounts. Cowper actually advanced Page £750 to get out of this mess.

61. Ibid., 400, 430–31, 442. The Office of Works hired F. I. Bramwell, an "engineer of eminence," to counter Page's technical testimony.

62. Ibid., 164. Barry's problems are mentioned in *History of the King's Works*, vol. 6: 198.

63. PRO, WORK 6, 166/1: 87.

64. HC *Sess. Pap.* 1860 (494) 20, minutes of evidence 59–68. Bazalgette was asked about Page's plan by William Tite, the member from Chelsea who was leading the effort to extend the Chelsea Embankment.

65. Owen, *Government*, 156–57; Clifton, *Professionalism, Patronage, and Public Service*, 163.

66. Gloria C. Clifton, "The Staff of the Metropolitan Board of Works, 1855–1889: The Development of a Professional Local Government Bureaucracy in Victorian London" (Ph.D. dissertation, University of London, 1986), 426–29.

67. Professor Denis Smith of Northeast London Polytechnic has been investigating the Bazalgettes for some years. His findings were summarized in "Sir Joseph William Bazalgette (1819–1891), Engineer to the Metropolitan Board of Works," *Transactions of the Newcomen Society* 58 (1986–87): 89–111. A full-scale biography will probably never be written, as Bazalgette left no personal letters or diary. Gloria Clifton's thesis and subsequent monograph on the MBW staff are cited above.

68. Smith, "Bazalgette," 90. Mental or physical breakdowns were fairly common among engineers and other professionals during the mid-Victorian years. The conditions of work were often dangerous—open sewers laid low dozens of MCS and MBW staff—and men tended to drive themselves to the limit on projects where time was money, accidents always intervened, and constant attention was the only defense against failure. However, M. Jeanne Peterson notes that men in less strenuous professions also tended to have breakdowns, including bouts of depression, in early or midcareer, and at a higher rate than their supposedly fragile, emotionally overwrought Victorian wives. *Family, Love and Work in the Lives of Victorian Gentlewomen* (Bloomington: Indiana University Press, 1989), 117–20.

69. Smith, "Bazalgette," 91.

70. Ibid., 93.

71. Ibid.

72. Ibid., 95. Clifton, *Professionalism, Patronage, and Public Service*, 78, says that men trained in civil engineering were also employed in the architect's department.

73. Ibid., chap. 10, "Work Ambitions and Social Aspirations." For engineering wages, see GLRO, MBW *Minutes of Proceedings*, 25 June 1869; for clerk's salaries, see the MBW *Pocket Book and Diary for the year 1868* (London: John Smith, 1868), copy in GLRO, MBW 673c, 78–79. Other wage data are in the *Builder*, 1 November 1862: 794. Francis Sheppard connects wages to class in *London, 1808–1870: The Infernal Wen* (London: Secker and Warburg, 1971), 354.

74. Clifton, *Professionalism, Patronage and Public Service*, chap. 7, "Methods of Recruitment." On Bazalgette's pupils, see Smith, "Bazalgette," 99.

75. George Willoughby Hemans (1814–1886), had been raised in northern Wales by his mother Felicia Hemans, a well-known sentimental poet, then schooled in Italy and France. Like Bazalgette, he was articled as a pupil to Sir John McNeil, in 1834, and did railway work in Ireland before coming to London. He had a strong reputation among the "Parliamentary" engineers of Great George Street. *ILN*, 10 August 1851, and ICE *Minutes of Proceedings* 85 (1885–86): 394–99.

76. HC *Sess. Pap.* 1860 (494) 20, minutes of evidence 468.

77. The conflict of interest involved George Furness, contractor for the Victoria Embankment. It is most thoroughly covered in Clifton, *Professionalism, Patronage and Public Service*, 131–34.

78. See the biographical note and portrait in *ILN*, 12 March 1859: 253.

79. Smith, "Bazalgette," 102.

80. Ibid., 93.

81. Compare the *ILN* view of the Chelsea Embankment site, 24 January 1857: 67 with their view of the Victoria site, 22 April, 1865: 369 (figure 3).

Chapter V/ The Genealogy of the Thames Embankment

1. GLRO, "Records of the Commissioners of Sewers for the London Area, 1531–1847," folio 2 (1608).

2. GLRO, "Records of the Metropolitan Commissioners of Sewers: Surrey and Kent Commissioners of Sewers," Roll 2 (1608).

3. Cf. Barker and Hyde, *London As It Might Have Been*, chap. 8, "The Thames Quay," 81. The 1670 Act of Parliament and the 1671 letters patent authorizing the embankment line are in the Corporation of London Record Office (Guildhall), chap. 98 (296B) and PD 83.18.

4. Plans for Mylne's bridge and embankment are in the Corporation of London Record Office (Guildhall) under "Maps—Thames" H.2.b. See also GLRO, History Library, 27.51 (BLA), "Plan for Raising Three Hundred Thousand Pounds for the Purpose of Compleating the Bridge at Black-friars, and Redeeming the Toll Thereon; Embanking the North Side of the River Thames between Paul's Wharf and Milford Lane, etc.," 1767.

5. The original Somerset House was constructed by Edward Seymour, the "Protector" Somerset, in 1549. Upon his treason and hanging, it was forfeited to the Crown and used as a residence by various domestic and foreign royalty. It was extensively remodeled, in 1662, by Inigo Jones, with a broad set of landing stairs leading to the river's edge; but the soft building stone decayed, and Somerset House was used as a barracks before being rebuilt in 1774. For Chambers's design, see GLRO, Maps, Westminster AD and AE. There is some evidence that the arches were originally intended as storage areas for commercial lease, as was later contemplated for the Thames Embankment and actually done with the arches under the Adelphi buildings and metropolitan railway viaducts. But the leases were apparently never taken up, and contractor G. W. Hemans testified, in 1861, that the Somerset House foundations were "dead arches" (HC *Sess. Pap.* 1861 [2872] 31: 3–4).

6. Timbs, *Curiosities of London*, 658. A drawing of the New Palace under construction, showing the cofferdam sunk into the river on the west side of Westminster Bridge, is in *ILN*, 25 June 1842: 104.

7. Previous embankments near Parliament are shown on a map and detailed in a letter from Thomas Page to the Earl of Lincoln, 1 June 1843, printed in HC *Sess. Pap.* 1844 (15) 15: 190–91. Others were described by Edward Bazalgette in his history of the Thames Embankment, ICE *Minutes of Proceedings* 54 (1878): 33. A general ground plan of the Old Houses of Parliament, showing the river wall running up to Westminster Bridge in 1751, is in PRO, WORK 29/21. After the new Houses of Parliament were completed, in 1858, John Jay received a contract to extend the foundation embankment up to the new Westminster Bridge (PRO, WORK 11/6/7). The Grosvenor-Millbank embankment licenses are mentioned in *Minutes and Reports of the Court of Common Council*, 1813, vol. 31, in the Corporation of London Record Office, Guildhall. More licenses are indicated in the index to the *Minutes*.

8. Walker's plans are described in HC *Sess. Pap.* 1840 (554) 12: 284–342 and appendix. Another copy of his plan is in the Corporation of London Record Office, Guildhall, MS, Noble Collection B, V Group 8, #111.

9. Barker and Hyde, *London As It Might Have Been*, 81–94.

10. John Lacy, "The River Frontage between Westminster and Blackfriars Bridges with a proposed cattle market in Lambeth, London." Drawing L 55.4, Guildhall Map Room, London, under "London Improvements." Listed in the Guildhall documents catalogue #23 (1980).

11. Cf. Barker and Hyde, *London As It Might Have Been*, 82–85.

12. *The Autobiography of Sir John Rennie* (London: E. and F.N. Spon, 1875), 194–97.

13. Trench's prospectus, accompanied by a set of watercolor plates by Thomas Allom, can still be seen in the British Museum Library, catalogue item 1787.aa.50, *Prospectus of a Proposed Improvement of the Thames Banks [by a Public Company]*.

14. Trench's design is reprinted and discussed in Barker and Hyde, *London as it Might Have Been*, 81–94. For Palmerston's role, and the attempt to finance the scheme through government, see *The Autobiography of Sir John Rennie*, 194–97.

15. Barker and Hyde, *London As it Might Have Been*, 84–85. A similar type of company, headed by the Duke of Rutland (rather than the Duchess), the Duke of Grafton, and other notables, formed plans, in 1841, for an eighty-five-foot embankment from Millbank (near Westminster) down to Queenhithe, with wharves, a rope-pulled railway, etc. Sir Frederick Trench was still alive then, but his involvement in this project is not clear. See the testimony of John Fordham Stanford, HC *Sess. Pap.* 1861 (2872) 31, minutes of evidence 35–37.

16. S. G. Checkland, *The Rise of Industrial Society in England, 1815–1885* (New York: St. Martin's Press, 1964), 283.

17. PRO, WORK 6, 157/1, 21–22.

18. GLRO, MBW *Minutes of Proceedings* 1856, 26 June and 3 July.

19. PRO, WORK 6, 157/1: 44, Lionel Gisborne to Sir William Molesworth, 18 August 1853.

20. In 1856, in reply to a talk on the Thames by Henry Robinson of the Admiralty, engineer Charles B. Vignoles denied that bad air could arise from the river's mud banks if they were periodically washed by the tides. Further, he argued that the tidal changes stirred up the nearby atmosphere, cleared the air, and helped keep the metropolis healthy. A lively debate ensued, in which the majority of professional engineers present agreed with Vignoles. Cf. *ICE Minutes of Proceedings* 15 (1886): 210.

21. Michael Robbins, *The Railway Age in Britain and its Impact on the World* (London: Penguin Books, 1965), 53–54. For Albert's ride, see L. T. C. Rolt, *Victorian Engineering* (London: Penguin Books, 1970), 21.

22. Donald Olsen, *The Growth of Victorian London* (Harmondsworth: Penguin Books, 1979), 314–15. Rolt (*Victorian Engineering,* 156) remarks that, when the price of admission to the Crystal Palace Exhibition of 1851 dropped to one shilling, the railways laid on dozens of excursion trains and gave thousands of rural subjects their first trip to the metropolis.

23. Barker and Hyde, *London as it Might Have Been,* 89.

24. *Stanford's Map of London,* British Museum Map Room, 11-c-4. The Greater London Record Office has a folio of maps [Westminster AK] with fanciful lithographs of proposed railways and roadways along the Embankment. One shows a roadway sunk between the outer embankment wall and a set of inner gardens, which are raised on top of a level of shops fronting the roadway. The shops look exactly like the present Temple Underground Station, and I suspect the station design is derived from those early plans.

25. ICE *Minutes of Proceedings* 15 (1856): 237–38.

26. HC *Sess. Pap.* 1861 (2872) 31, minutes of evidence.

27. Barker and Hyde, *London As it Might Have Been,* 84–85.

28. The 1844 plan was published as an appendix to the report of the Metropolitan Improvement Committee, HC *Sess. Pap.* 1844 (15) 15. The original is in the Guildhall Library, Noble Collection B, V group 8.

29. GLRO, MBW 2365, Engineer's monthly report, 3 October 1861. Prince Albert died December 14; Queen Victoria was said to blame the effluvia from a large sewer line then under reconstruction outside Kensington Palace.

30. HC *Sess. Pap.* 1861 (2872) 31: 23, evidence of Thomas Weller.

31. Olsen, *Growth of Victorian London,* 299.

32. J. R. Coppock, "The Changing Face of England: 1850–*circa* 1900" in Darby, *A New Historical Geography,* 348.

33. Timbs, *Curiosities of London,* 777.

34. See the critical editorial about steamboat facilities in *ILN,* 27 July 1844: 1.

35. Pudney, *Crossing London's River,* 86.

36. GLRO, MBW 1444, 20 January 1857, quoted in Owen, *Government,* 75. Another version of the report was published in GLRO, MBW *Minutes* for 6 February 1857, where the same statement is given with different wording.

37. Owen, *Government,* 75.

38. HC *Sess. Pap.* 1860 (494) 20: iii, "Report."

39. HC *Sess. Pap.* 1861 (2872) 31: 338. A "Tabular Summary of Plans and designs for the Embankment" can be found in appendix A, 278–89, and many of the plans follow this section.

40. Ibid., minutes of evidence 20–22.

41. Ibid. I suppose one can imagine the effect by walking through the new steel and glass Hay's Galleria in the Surrey Docks redevelopment.

42. Ibid., minutes of evidence 35–37. Stanford's 1841 plans involved the Duke of Rutland and the Duke of Grafton, which suggests they were connected with Sir Frederick Trench's revival, mentioned by other officials at that time.

43. Ibid., minutes of evidence 53–54.

44. HC *Sess.Pap.* 1862 (3043) 28: "Report of the Royal Commission on the Thames Embankment, on Plans for Embanking the Surrey Side of the Thames within the Metropolis," 168. Hen-

man appeared before the commission when it was considering the embankment of the south side of the river, after having approved the northside plans.

45. H. H. Bird's testimony is in HC *Sess. Pap.* 1861 (2872) 31, minutes of evidence 38–47. Lionel Gisborne's project was described in testimony before the Select Committee on Metropolitan Communications, HC *Sess. Pap.* 1854–55 (415) 10, minutes of evidence, questions 969–1029.

46. Gisborne's plans are abstracted in PRO, WORK, 157/1: 41–42, and in GLRO, MBW 2411, batch #5, bundle 1, "Report of Works Committee on Thames Embankment Plans," 6 February 1857. The published MBW *Minutes of Proceedings* show that Bird, his partner Edward Loder, and their solicitor Robert E. Jackson appeared before the board to promote their plans, on 11 April 1856, after sending a memorial to the Committee on Works.

47. For descriptions of the Fowler, Fulton and Hemans plans, see HC *Sess. Pap.* 1860 (494) 20: 613–20, and 1861 (2872) 31: 3–8.

48. Development maps of London in the 1850s and 1860s regularly showed the railways, streets, and docks planned or projected by various companies. The Stanford map of 1863 (British Museum Map Room, 11.c.4) lists H. H. Fulton and others for the North and South London Railway Junction line and for extensions of the London, Brighton and South Coast Railway.

49. The Royal Commission chair, Mayor William Cubitt, doubted that the picturesque Thames sailing barges, called "billy boys," could sail across an embankment at low tide without hitting an overhead archway with their masts. He was later corrected by the boatmen themselves, who said the masts were hinged and lowered frequently when sailing under bridges. HC *Sess. Pap.* 1861 (2872) 31: 35–37, evidence of John Stanford.

50. HC *Sess. Pap.* 1860 (494) 20: 398, 595–604.

51. Roughly in a line from Whitehall downriver, these were: Yool and Judkins Coal Yard and Wharf; Stewart and Company's Portland stone wharf; the Cannon, Great Scotland Yard, Percy, Northumberland, Adelphi, Brown's, and Salisbury wharves; the large Beaufort complex; the Somerset, Waterloo, and Duchy of Lancaster wharves; Arundel wharf and stairs; the Milford complex: Gwynne and Co. mechanical engineers; Barraud's draw dock, the Grand Junction wharf, and the Phoenix wharf, all at Whitefriars; and the City of London Gas Works. These are shown in MBW, *Thames Embankment: Plans and Specifications*, 1861: contract #1, drawing 1 and contract #2, drawing 1.

52. G. L. Gomme, *London in the Reign of Victoria, 1837–1898* (London: Blackie and Son, 1898), 9. See also HC *Sess. Pap.* 1861 (2872) 31, question 354, evidence of Stephen Leach, Sr. Leach had made the same observation to a select committee on the Embankment in 1840.

53. HC *Sess. Pap.* 1861 (2872) 31, minutes of evidence 4, R. Baxter, agent for John Fowler.

54. HC *Sess. Pap.* 1860 (494) 20, minutes of evidence 23, testimony of Captain Vetch.

55. Robbins, *The Railway Age in Britain*, 31–40.

56. Sheppard, *The Infernal Wen*, 195.

57. The *Builder* 51 (1886): 656, quoted in Olsen, *Growth of Victorian London*, 277.

58. Elspet Fraser-Stephen, *Two Centuries in the London Coal Trade: The Story of Charrington's* (London: privately printed, 1952), 52–53.

59. HC *Sess. Pap.* 1861 (2872) 31, minutes of evidence 6, 13, 16. One radical suggestion for solving the problem of wharf access along the Thames was to cut a six-hundred-foot-wide canal straight across the Surrey side of the river from Vauxhall to London Bridge, straightening out the northern "loop" and eliminating the mud banks along with everything else. This proposal was made, in 1860, by Mr. Brown, Registrar General of Seamen and Shipping, who argued that the project would gain valuable land for traffic roadways and new buildings in the former Thames riverbed. It is similar to a proposal made, in 1799, to accomodate a new set of docks for the Port of London. Cf. PRO, WORK 6,157, 80 and Lambert, *Sessional Papers of the Eighteenth Century*, vol. 124 (1799, "Reports and Papers"), "Second Report of the Committee on the Port of London," 424.

60. Letter to the *Nautical Magazine*, April 1838: 260–63, from "E.B." in reply to a previous essay on the need to improve the Thames in London. For national and imperial perspectives on the Thames pollution crisis of 1858, see Luckin, *Pollution and Control*, 18–19.

61. HC *Sess. Pap.* 1860 (494) 20, minutes of evidence 19–22.

62. The *Builder*, 22 (1864): 95, reported by Olsen, *Growth of Victorian London*, 60.

63. HC *Sess. Pap.* 1861 (2872) 31, minutes of evidence 80, questions 1068–69.

64. Ibid., "Report," iii.

Chapter VI/ The Financial and Institutional Environment

1. The secretary of the Duchy of Lancaster, which controlled Crown land along the river, would not even read his enabling bill until it passed through the committee stage. For this and other such experiences, see HC *Sess. Pap.* 1860 (494) 20, minutes of evidence 93.

2. GLRO, MBW *Minutes of Proceedings*, 5 October 1860. For H. H. Bird's version of the "circumlocution offices" mentioned in *Bleak House*, see his testimony in HC *Sess. Pap.* 1854–55 (415) 10, minutes of evidence 127–29, and 1861 (2872) 31, minutes of evidence, questions 676–77.

3. The navigation of the upper Thames from Cricklade to Staines was under the purview of a body called the Thames Commissioners, established by Parliament in 1751. Over six hundred men held the title of Thames Commissioner, from members of Parliament for Thames Valley constituencies to town officials, the Lord Mayor and aldermen of London, Oxford University dons, vicars of various churches, landowners, millowners, and bondholders. A small steering committee handled the business of the Thames Commissioners with traditional caution and ineptitude. Breeze, *The British Experience*, 27–28.

4. Timbs, *Curiosities of London*, 1.

5. According to Henry Robinson in his remarks on the history of the Thames River, ICE *Minutes of Proceedings* 15 (1856): 198.

6. Corporation of London Record Office (Guildhall), "Thames Embankment: Report to Court of Common Council, from the Committee for Improving the Navigation of the River Thames," 3 August 1832.

7. HC *Sess. Pap.* 1840 (554) 12. Sir Frederick Trench was one of the members of the select committee.

8. *The Thames Conservancy, 1857–1957*, ed. G. E. Walker, secretary (London: Published by the Thames Conservancy, 1957), 1–2. See also Charles Pearson's account of the chancery suit in ICE *Minutes of Proceedings* 15 (January 1856): 219–20.

9. Ibid.

10. HC *Sess. Pap.* 1844 (15) 15, minutes of evidence 145, testimony of Green, Horne, and Todd.

11. In a letter to the Commissioner of Works, 25 July 1846, in PRO, WORK 6: 139/4: 26, the Admiralty acknowledged that the Thames Conservators might object to its review of embankment plans. Some observers thought the Corporation of London was at fault in the issue, but, in 1853, the *ILN* (29 October 1853: 367) excoriated the Conservators for failing to embank the river and provide parallel sewer lines as a natural system of drainage, as had already been done along the Mersey at Liverpool.

12. Walker, *The Thames Conservancy*, 1, 2, 6, 10. For the running dispute between the Conservancy and the MBW over the latter's discharge of sewage into the river, see Luckin, *Pollution and Control*, 143–46.

13. Walker, *The Thames Conservancy*, 6, 10.

14. For the Conservancy's reclaimed land rights, see HC *Sess. Pap.* 1860 (494) 20, minutes of evidence 335, 345, testimony of J. Thwaites. For the argument over dredging, and the denial of MBW representation, see Owen, *Government*, 87, 228.

15. HC *Sess. Pap.* 1860 (494) 20, minutes of evidence 359, testimony of Captain Vetch.

16. According to a new statistical study by John A. Phillips and Charles Wetherell, "The Great Reform Act of 1832 and the Political Modernization of England," *American Historical Review* 100, no. 2 (April 1995): 411–36.

17. Bernard Cracroft, "The Analysis of the House of Commons," in *Essays on Reform* (1867), quoted in Briggs, *The Age of Improvement*, 406.

18. Walter Bagehot, *The English Constitution*, quoted in Briggs, *The Age of Improvement*, 415.

19. Briggs, *The Age of Improvement*, 416.

20. The status of Molesworth's appointment is remarked by Norman McCord, *British History, 1815–1906* (New York: Oxford University Press, 1991), 244.

21. Jasper Ridley, *Lord Palmerston* (London: Constable, 1970), 439.

22. R. J. Evans, *The Victorian Age, 1815–1914,* 2 ed. (New York: St. Martin's Press, 1968), 124.

23. Ridley, *Lord Palmerston,* 503.

24. According to Sidney Low and Lloyd C. Sanders, *The History of England During the Reign of Queen Victoria, 1837–1901* (London: Longmans, Green, 1907), 180–82.

25. McCord, *British History, 1815–1906,* 292.

26. Owen, *Government,* 45.

27. Ibid., 78.

28. Ibid., 79–80; *The Builder,* 20 (12 July 1862): 494–95. Montague House can be seen on plans for the Embankment in records of the First Commissioner of Works, PRO, WORK 6, 160/7, and in a drawing for *ILN,* 24 September 1864: 310.

29. Owen, *Government,* 84.

30. Ibid., 98–99.

31. In 1872, Buccleuch's neighbor, the sixth Duke of Northumberland, was persuaded to sell his mansion for the good of the metropolis. It was demolished over public protest to make way for Northumberland Avenue, built to connect the Embankment with Charing Cross, very near the route suggested by the Whitehall residents in 1862. Ibid., 80.

32. Ibid., 80.

33. Sheppard, *The Infernal Wen,* 50–54, 62–67. For the bank's effects on the market, see W. T. C. King, *A History of the London Discount Market* (London: George Routledge and Sons, 1936), 138, 156–57.

34. Sheppard, *The Infernal Wen,* 54; Olsen, *Growth of Victorian London,* 158–60. Cf. H. J. Dyos, "The Speculative Builders and Developers of Victorian London," *Victorian Studies* 11 (1968): 668–69.

35. McCord, *British History, 1815–1906,* 219–20.

36. R. K. Webb, *Modern England From the Eighteenth Century to the Present* (New York: Dodd, Mead and Company, 1968), 267.

37. King, *London Discount Market,* 193–99; Checkland, *The Rise of Industrial Society,* 38–39.

38. Sheppard, *The Infernal Wen,* 74–78; King, *London Discount Market,* 231–38.

39. Webb, *Modern England,* 324.

40. For Thwaites's testimony, see HC *Sess. Pap.* 1860 (494) 20, minutes of evidence 343. Thwaites argued, perhaps with tongue in cheek, that the government had been collecting £80,000 a year since 1831 in London hackney carriage fees, a charge not levied elsewhere in the kingdom, and that the total sum of around £2 millions should be returned to the MBW for the Embankment and other projects (341). *The Builder,* a leading trade journal, repeated the suggestion on 2 June 1860: 354. Resolutions demanding national funding, sent from Chelsea, Kensington, Islington, St. Martins in the Fields, and Whitechapel, in the summer of 1860, are in GLRO, MBW 2365. As early as 1856, the MBW clerk, E. H. Woolrych, had advised members that the proposed Embankment was so large, expensive, and legally complex that an application to Parliament for funds and authority was clearly required. GLRO, MBW *Minutes,* 29 July 1856.

41. PRO, WORK 6, 158/1, 1862–63: 5–48 give Cowper's expenses for the act of 1862. The bill from Parliamentary agents Baxter Rose Norton and Company, which runs 233 folio pages, includes charges for reading and writing letters, for legal conferences and counsel's attendance at parliamentary meetings, for having clerks take down testimony in parliamentary committees and prepare multiple copies, and for researching property likely to be affected by the Embankment. For the MBW's permission to charge parliamentary expenses against the Thames Embankment Fund, see GLRO, MBW *Minutes,* 13 November 1868.

42. Briggs, *The Age of Improvement,* 490.

43. Sir John Rennie, *Autobiography,* 194–97.

44. Given the antiquity of the coal and wine duties and the variety of uses to which they were put, it is not surprising that different sources give conflicting dates, amounts, and applications. I have presented what seems to be the most reasonable data, based on contextual materials as well as the manuscript sources. For the First Commissioner of Works accounts, see PRO, WORK 6,

150/3; For the London Bridge Fund legislation, see HC *Sess. Pap.* 1837–38 (418) (661) 16, "First and Second Reports of the Select Committee on Metropolitan Improvements." Other historical details are recounted in the *Builder*, 2 June 1860: 354, and in Percy J. Edwards, *A History of London Street Improvements* (London: London County Council, 1898), 15, paraphrasing the MBW annual report of 1888. Owen (*Government*, 382 n. 8) says the coal duty was 8d. But that was only one portion of the whole coal duty. Another 1d. levy was added to the 8d., in 1861, and yet another 4d. which remained with the City Corporation. Cf. Clifton, "Staff," 81–82.

45. PRO, WORK 6, 150/3: 58, Lord John Manners to John Thwaites, 31 March 1859.

46. See, for example, *ILN*, 9 March 1844: 150 and 26 March 1853: 229. The 1853 article was accompanied by an abstract from Parliamentary Reports showing the increase in coal tonnage brought in by canals and railways.

47. HC *Sess. Pap.* 1859 (Session 2), vol. 26: 275–80, "Annual Report of the MBW," 30 July 1859.

48. Available records indicate that the new Embankment fund received a £32,136 balance from the Office of Works' "Metropolis Improvement Fund," backdated to March 1861, and £5,485 from the coal and wine duties that would have accrued to that fund from March through July. It also received £137,992 from the "London Bridge Approaches Fund," i. e., the revenue from the original fund not assigned to the Office of Works. Since the MBW's accounting year was dated from 31 March, it then reported additions to the fund from July 1861 through March 1862 of £121,072, £169,337 the following year, £176,326 in 1864, and £186,773 in 1865, with accrued interest on investments of £21,628. Fund revenues were typically invested in consols and other government securities, earning about 3 percent. HC *Sess. Pap.* 1867, vol. 58: 735, Documents relating to Metropolitan coal duties and MBW loan guarantees: Accounts from the MBW. In 1863, Parliament extended the coal and wine duties to 1882.

49. Clifton, "Staff," 81–82.

50. Wohl, *Endangered Lives*, 174.

51. HC *Sess. Pap.* 1864, vol. 50, Correspondence between Her Majesty's Treasury and the Metropolitan Board of Works: 179–80, letter of 24 March 1864.

52. HC *Sess. Pap.* 1867, vol. 58: 739, Accounts from the MBW, appendix listing parliamentary acts. For the Thwaites-Gladstone correspondence, see HC *Sess. Pap.* 1864, vol. 50: 181–82. The relevant legislation is the Thames Embankment and Metropolitan Improvement (Loans) Act, 27 & 28 Vict. c. 61.

53. HC *Sess. Pap.* 1867, vol. 58: 737, Accounts from the MBW. Owen, *Government*, 165, says the board was allowed to borrow £4,200,000 altogether on the strength of a minimum threepenny rate assessment.

54. Bazalgette's report for 6 November 1882 is in GLRO, MBW 2411, batch #5, Miscellaneous papers on the Embankment, item #21. Another drain on the budget was the MBW's practice of repaying loans in equal installments of capital, plus interest on the remaining balance. This meant higher initial payments because the interest was higher at first, even though it reduced the overall amount of interest paid on the loan. Cf. Clifton, "Staff," 82–83. The debt figure is given by Owen, *Government*, 165.

55. GLRO, MBW *Minutes of Proceedings*, 19 January, 9 February, and 23 February 1866. For the establishment charges, see the *Minutes* for 2 February 1866.

56. Clifton, "Staff," 84.

57. Owen, *Government*, 162–68.

58. HC *Sess. Pap.* 1836, vol. 20: 3, "Report of the Select Committee on Metropolitan Improvements."

59. For Henman's testimony, see HC *Sess. Pap.* 1862 (3043) 8, "Report of the Royal Commission on the Thames Embankment, on plans for Embanking the Surrey Side of the Thames within the Metropolis," minutes of evidence 73.

Chapter VII/ Relevant Interest Groups

1. Walmsley's testimony is in HC *Sess. Pap.* 1861 (2872) 31, minutes of evidence 20–22.

2. Aytown Ellis, *Three Hundred Years on London River: The Hay's Wharf Story, 1651–1951*

(London: Bodley Head, 1952). Humphery was Lord Mayor of London, in 1842, and later an MP.

3. Owen, *Government*, 82–83. An agent for John Fowler, the railway engineer, testifying before the Royal Commission in 1861, guessed that the amount would run to £200,000, but he said that the law was such a mess that estimates for compensation were never correct. See HC *Sess. Pap.* 1861 (2872) 31, minutes of evidence 4.

4. For the sale to St. Thomas's Hospital, see Owen, *Government*, 91.

5. For the steam ferry facilities and reputation, see Harold P. Clunn, *The Face of London* (London: Simpkin Marshall, 1933), quoted in Pudney, *Crossing London's River*, 81–82. The belief that railways could not compete with the steam packets was expressed, *inter alios*, by engineer William Alexander Brooks to the Royal Commission on the Thames Embankment, HC *Sess. Pap.* 1861 (2872) 31, minutes of evidence 19. James Walker, who surveyed the river in the late 1830s, observed that the flotilla of steamers at least churned up the river enough to reduce the formation of mud shoals (HC *Sess. Pap.* 1840 [554] 12, minutes of evidence 279).

6. MBW, *Chelsea Embankment Opened*, 45.

7. Material on mudlarks and other riverbank denizens occurs in the fourth volume of Henry Mayhew, *London Labour and the London Poor* (London, 1862, 1865; reprinted, New York: Dover Publications, 1968), 366–70. Lumpers are mentioned by Peter Cracknell, *Crossing London's Rivers*, 95. Telford's testimony on pilferage on the Thames is in the "Third Report of the Select Committee on the Port of London, 1799–1800": 329–31, reprinted in Lambert, *Sessional Papers of the Eighteenth Century*, vol. 132, appendix B.3, 71–73. Bonnet's *The Mudlark* (Garden City, New York: Doubleday, 1949), made into a film version with Alec Guiness, featured a spunky waif who dug up a locket containing a miniature of the young Queen Victoria. He was so enthralled with this vision of loveliness that he ventured out into the great worlds of London and Westminster, stole into Kensington Palace, and made friends with Disraeli.

8. Ridley, *Lord Palmerston*, 509.

9. R. Angus Buchanan, "Gentleman Engineers: The Making of a Profession," *Victorian Studies* 26, no. 4 (summer 1983): 427–29; and Buchanan, *The Engineers: A History of the Engineering Profession in Britain, 1750–1914* (London: Jessica Kingsley, 1989), 90. *The Engineers* incorporates some of the material from the 1983 essay but does not develop the essay's promising discussion of the profession's social contexts, centering instead on the institutional development of the ICE and related bodies.

10. Buchanan, *The Engineers*, 192. Buchanan argues (195–96) that engineers consciously fashioned their institution into a gentleman's club as a sign of their social status. They also worked so hard and long at their jobs, whether out of satisfaction, anxiety, or ambition, that they had little time or inclination to develop political, social, or intellectual interests, or even take holidays with their families.

11. Buchanan, "Gentleman Engineers," 427–29, and *The Engineers*, 69–90. See also J. G. Watson, *The Institution of Civil Engineers: A Short History* (London: Institution of Civil Engineers, 1982). Fowler's ruling came during discussion of a paper by Henry Robinson, "The Past and Present Condition of the River Thames," ICE *Minutes of Proceedings* 15 (1856): 233.

12. For the Bazalgette presentations, see ICE *Minutes of Proceedings* 24 (1865): 280–354 and 54 (1878), 1–60.

13. Checkland, *The Rise of Industrial Society*, 91. As George Kitson Clark wrote, "A *professional* man was not necessarily an *educated* man. He was by definition a *trained* man, but from a 19th-century point of view a professional training was not necessarily an education, certainly not the education of a gentleman." *The Making of Victorian England*, (Cambridge, Mass.: Harvard University Press, 1962), 263.

14. Armytage, *A Social History of Engineering*, 238; Watson, *A Short History*, 11. See also *The Education and Status of Civil Engineers, in the United Kingdom and in Foreign Countries; Compiled from Documents Supplied to the Council of the Institution of Civil Engineers, 1868 to 1870* (London: ICE, 1870). For the establishment of chairs in civil and mechanical engineering, see Amytage, *A Social History of Engineering*, 149–50.

15. Michael Sanderson, *The Universities and British Industry, 1850–1970* (London: Routledge and Kegan Paul, 1972), 24; R. Angus Buchanan, "Institutional Proliferation in the British Engineering Profession, 1847–1947," *Economic History Review*, 2d ser., 38 (1985): 57. For a recent comment on the empirical tradition among British engineers, see R. B. Hill, "Address," ICE *Minutes of Proceedings*, n. s., 68 (1980), part 1: 161.

16. Checkland, *The Rise of Industrial Society*, 91.

17. R. Angus Buchanan, "The Rise of Scientific Engineering in Britain," *British Journal for the History of Science* 18 (1985): 218–33.

18. Sir John Rennie, *Autobiography*, 422.

19. ICE *Minutes of Proceedings* 33 (1871): 264. George Furness was one of the great railway contractors and his rebuilding of the Odessa railways after the Crimean War was his most ambitious undertaking. Furness's Odessa contract had been won through the agency of Bazalgette, a fact that brought charges of collusion against the MBW engineer when Furness was awarded the Embankment contract. But Furness encountered a series of financial and technical problems and had to be saved from bankruptcy by the intervention of wealthy contractor Thomas Brassey (Owen, *Government*, 84–89). He was no doubt glad to turn his section of Embankment over to Gardner in 1868. Not a licensed engineer, Furness was made an associate of the ICE. He later earned a reputation for philanthropy and served in a variety of local government positions.

20. As remarked in another memoir, ICE *Minutes of Proceedings* 65 (1880): 368.

21. Clifton, "Staff," 180–81, 320; GLRO, MBW 2321, Engineer's Monthly Reports for 1870–71; MBW *Minutes*, 23 October 1868.

22. Clifton, "Staff," 169–71. For a more detailed description of clerks of works' duties on the Embankment, see Bazalgette's explanation in GLRO, MBW *Minutes*, 23 October, 1868.

23. Cf. Cubitt's challenge to James Frank Smith of Leicester, who claimed to be both architect and engineer: HC *Sess. Pap.* 1862 (3008) 28, "Report of the Select Committee on the Thames Embankment (North) Bill," minutes of evidence, question 1804.

24. GLRO, MBW *Minutes*, 17 June 1870; Clifton, *Professionalism, Patronage and Public Service*, 96.

25. Buchanan, "Gentlemen Engineers," 425.

26. Clifton, *Professionalism, Patronage and Public Service*, chap. 9, "Salaries and Promotion Prospects." See also GLRO, MBW *Minutes*, 25 June 1869. For clerks' salaries, see the MBW *Pocket Book and Diary for the year 1868* (London: John Smith and Co., 1868), GLRO, MBW 673c. Other wage data are in the *Builder*, 1 November 1862: 794.

27. Sheppard, *The Infernal Wen*, 354. Clifton, *Professionalism, Patronage and Public Service*, has a more complete discussion of the servant-keeping standard in relation to MBW wages. For other examples of successful engineers and their social pretensions, see Buchanan, "Gentleman Engineers."

28. Watson, *A Short History*, 4.

29. Sir John Rennie, *Autobiography*, 432.

30. HC *Sess. Pap.* 1861 (2872) 31, appendix A, "Tabular Statement of Plans."

31. HC *Sess. Pap.* 1861 (2872) 31: 292, (a list of witnesses) and minutes of evidence, passim.

32. P. J. Booker, *A History of Engineering Drawing* (London: Chatto and Windus, 1963), 133. Clifton, *Professionalism, Patronage and Public Service*, 103, reports that, in 1871, the technically trained draughtsmen at the MBW asked that they be given a more specific job title to distinguish them from mere copyists.

33. The association is listed in the *Post Office London Trades Directory* for 1866. For the salary of the superintending draughtsman, see GLRO, MBW *Minutes*, 17 June 1870.

34. Sheppard, *The Infernal Wen*, 331–32; Armytage, *A Social History of Engineering*, 142–43; Kitson Clark, *The Making of Victorian England*, 62. William Newton was one of the Thames-side proprietors affected by construction of the Embankment, according to MBW records in the GLRO, MBW 676: 169.

35. Briggs, *The Age of Improvement*, 405.

36. According to Sir John Clapham, *Free Trade and Steel*, quoted by Kitson Clark, *The Making of Victorian England*, 137.

37. Sheppard, *The Infernal Wen*, 331–36; Webb, *Modern England*, 318–20.

38. GLRO, MBW 2365, MBW *Reports* nos. 47–70 (1860–61), Engineer's Monthly Report, 1 February 1861. Six years earlier, a late February frost had stopped work along the Thames and forced thousands of coalwhippers and dockworkers to seek outdoor relief from local workhouses, while the queen and Prince Albert enjoyed sleigh rides in Hyde Park. Cf. *ILN*, 24 February 1855: 189–90 and 3 March 1855: 210.

39. Briggs, *Age of Improvement*, 493. Depositing the Engineers' funds in the Post Office meant that they would end up being loaned to the MBW under the arrangements for financing the Embankment worked out in 1869.

40. Checkland, *The Rise of Industrial Society*, 414. Ricardo had argued that the supply of capital for industry and wages was fixed, so that an increase in wages for one group would not only depress wages elsewhere but shrink the supply of capital for investment. Workers could see for themselves that Ricardo's premise was false, but it took a critique from John Stuart Mill, in 1867, to dispose of the theory.

41. Webb, *Modern England*, 320.

42. GLRO, MBW *Specifications* for Victoria Embankment Contract #1 (1863), 64; for Albert Embankment Contract #4 (1865), 60; for Chelsea Embankment (1871), 55.

43. For a discussion of butty gangs, see Arthur Helps, *Life and Labours of Mr. Brassey, 1805–1870* (London: Bell and Daldy, 1872), 102, and George Pauley, *The Chronicles of a Contractor* (London: Constable, 1926). The members of a butty gang were supposed to divide the group wages equally, but foremen were notorious for keeping back a portion for themselves.

44. See Peter Mathias's discussion of union labor in London and in large industry, which was not always typical of the rest of the country, in *The First Industrial Nation: An Economic History of Britain, 1700–1914*, 2d ed. (London: Methuen, 1983), 338–39.

45. For Brassey and Furness, see GLRO, MBW 2321, Engineer's Monthly Reports for 27 September 1866 and 1 October 1869. For Webster, see his obituary in the ICE *Minutes of Proceedings* 92 (1888): 410–11.

46. The effect of large-scale railway, shipyard, and engineering works on the organization of the skilled trades is described in Mathias, *The First Industrial Nation*, 338.

47. Hermione Hobhouse, *Thomas Cubitt, Master Builder* (London, Macmillan, 1971), 527; E. W. Cooney, "The Origins of the Victorian Master Builders," *Economic History Review* 8 (1955): 167–76; see also Cooney's "The Speculative Builders and Developers of Victorian London: A Comment," *Victorian Studies* 13, no. 3 (March 1970): 355–57. In the 1820s, Cubitt discovered that the "Five Fields" of wet wasteland that then lay between Westminster and Chelsea were composed of deep layers of clay upon a sound base of gravel. He contracted with the aristocratic landlords to remove the clay, manufactured housing bricks out of it, and stabilized the gravel base for the great housing estates around Belgrave Square and Eaton Square. Later, he did the same for south Belgravia down through St. George's Square. Timbs, *Curiosities of London*, 43.

48. Helps, *Life and Labours of Mr. Brassey*, 102; for other contractors, see the *Post Office London Trades Directory* for 1866 and 1872.

49. Locations listed in *Post Office London Trades Directory*, 1866 and 1872, and in various tables and minutes of evidence of HC *Sess. Pap.* 1861 (2872) 31.

50. Helps, *Life and Labours of Mr. Brassey*, 98–99.

51. For Brassey and Harrison, see Helps, *Life and Labours of Mr. Brassey*, 101–2. On Bidder, see the memoir in ICE *Minutes of Proceedings* 57 (1878): 294–309, and *George Parker Bidder, The Calculating Boy*, ed. G. F. Clark (Bedford, England: KSL Publications, 1983), 3–9, 301–26. Bidder's lecture to the ICE on mental calculation was reported, with a portrait, in *ILN*, 15 March 1856: 268.

52. HC *Sess. Pap.* 1861 (2872) 31, minutes of evidence 12, testimony of Henry Wyndham Sich.

53. Owen, *Government*, 80. The difficulties are described in the MBW's commemorative booklet, *Chelsea Embankment Opened*. Problems with the MBW's core samples were noted by Furness

in discussion of Thomas Ridley's paper on cofferdams used for the Embankment, ICE *Minutes of Proceedings* 31 (1870): 24. Among other things, the samples apparently missed several pockets of quicksand, which was common in this area. The ordnance survey maps were crucial for two reasons. They indicated the incidence of decline in the Thames valley that made gravity-fed drainage works possible, and they showed contractors where to find the dozens of neighborhood sewer pipes, of varying shapes and sizes, installed over the decades by vestry boards.

54. H. Peto, *Sir Morton Peto: A Memorial Sketch* (London: privately printed, 1893), 10, 29, 36–56.

55. For Aird et al., see GLRO, MBW 676, 218; for Rendel, HC *Sess. Pap.* 1861 (2872) 31, minutes of evidence 306. Fox and Sons appear in the *Post Office London Trades Directory* for 1866, and Pearson is listed in GLRO, MBW 2321, Engineer's Monthly Reports, 6 March 1862, 1 August 1878, and 1 May 1883. The Leaches are mentioned in Walker, *The Thames Conservancy*, 3; see also Stephen Leach's evidence in HC *Sess. Pap.* 1840 (554) 12: 297.

56. *Post Office London Trades Directory*, 1862, 1866.

57. Pauley, *The Chronicles of a Contractor*, 6–7.

58. Clifton, *Professionalism, Patronage and Public Service*, chap. 6, "Methods of Recruitment."

59. Ibid., 62, 83.

60. Edward Bazalgette, "The Victoria, Albert, and Chelsea Embankments of the River Thames," ICE *Minutes of Proceedings* 54 (1878): 1–60.

61. Clifton, "Staff," 136, 217.

62. Grant memoir, ICE *Minutes of Proceedings* 92 (1888): 389–90.

63. Stefi Weisburd, "Hard Science," a report on innovations in cement, *Science News* 134 (9 July 1988): 24. Cf. Norman Davey, *A History of Building Materials* (London: Phoenix House, 1961), 106.

64. Grant memoir, ICE *Minutes of Proceedings* 92 (1888): 392–93. His chief paper on the experiments is "Portland Cement: Its Nature, Tests, and Uses," ICE *Minutes of Proceedings* 62 (1880): 98–248, but he had declared the reliability of the material as early as 1865 (ICE *Minutes of Proceedings* 24 [1864–65]: 182), and there is an excellent discussion on its use in the Albert Embankment in ICE *Minutes of Proceedings* 32 (1870–71): 266–328. One of the commentators in the last discussion said the savings of concrete over brickwork was 40 percent, and that Portland cement resisted the high ammonia content of sewage that tended to leach out ordinary mortar. Before Grant demonstrated the reliability of Portland cement concrete, foundations below water level had to be laid in dry conditions created by sheet piling and cofferdams, which was tedious, hazardous, and expensive.

65. Owen, *Government*, 90–91.

66. Letter from Neil Rhind, 19 August 1987. Mr. Rhind has studied the Webster family because William Jr., a polymath and patron of the arts, built the fine Blackheath music hall, now almost completely restored under the aegis of the Blackheath Preservation Trust.

67. ICE *Minutes of Proceedings* 92 (1888): 389–90, Webster obituary. Other obituaries were printed in the *Builder* and the *Building News*. For his contracts, cf. GLRO, MBW Engineer's Monthly Reports, MBW 2321 and MBW 2411, batch #5. Webster's Embankment roadway contract included paving, periodic cleaning of slops and horse droppings, watering for dust, snow removal, gully cleaning, and cleaning and repairing the pedestrian footways, landing steps, and public urinals (Engineer's Monthly Report, 21 November 1877).

68. See Owen, *Government*, 57; GLRO, MBW *Minutes*, 27 July and 10 August 1860. Information on the Airds' houses (father and son) from Neil Rhind.

69. GLRO, MBW 676, Engineer's Monthly Reports.

70. Obituary, ICE *Minutes of Proceedings* 92 (1888), 410–11.

71. His associate's award is noted in his obituary, ICE *Minutes of Proceedings* 92 (1888): 410–11. The seal appears on a letter he sent to the First Commissioner of Works, in April 1875, complaining that a Works payment for £9,900 had been misdirected and lost for a week. PRO, WORK 11.7/2.

72. Grant memoir, ICE *Minutes of Proceedings* 92 (1888): 392–93. I have been informed by Lewisham Archivist C. W. Harrison that Webster owned a cement works, "The Burnham Cement

Works on the Medway" in partnership with another neighbor, civil engineer William Porter (I claim no relation). The partnership is noted for 1881; whether it was formed earlier is unknown.

73. ICE *Minutes of Proceedings* 92 (1888): 392–93; also letter and documents from Lewisham Borough archivist C. W. Harrison.

Chapter VIII/ Contingencies

1. Owen, *Government*, 81, and Smith, "Bazalgette," 96. For Bird's comment, see HC *Sess. Pap.* 1861 (2872) 31, minutes of evidence 43.

2. The Thames Embankment contracts specified 10 percent as the standard allowance for contingencies, but Sir John Rennie, writing in 1875, advised engineers to add 15 percent (*Autobiography*, 454). These rates are still common today.

3. Flushing the riverbanks was not a new idea. In 1843, Thomas Page described how Thames wharfingers closed their dock gates at high tide and then released the pent-up water at low tide to keep the adjacent banks clear of mud and debris. PRO, WORK 6: 157/1, Thames Embankment proposals p. 6, letter to T. W. Phillips, 22 March 1843. One of the old tidal flushing reservoirs was located next to the present headquarters of London Weekend Television on the South Bank, and the discharge pipes are still there.

4. Owen, *Government*, 81. Owen creates a bit of confusion by giving the length of the "first section" as 3,740 feet, and then describing other features of the Victoria Embankment as a whole. But the "first section" was merely the first of the three contracts drawn up for the Victoria. For details, see the MBW *Thames Embankment: Plans and Specifications*, 1861. The penstocks were key elements in the system, since all the sewage flowing down the lower valley toward the river would drain into them, often at considerable volume and velocity. Bazalgette wrote, "The Contractor is to note that in these penstocks more than ordinary care, both in construction and fixing, will be requisite, in consequence of the pressure to which they will be subjected" (Contract #1, p. 12). The penstocks were allowed to silt up in the twentieth century but are now being cleaned out and allowed to discharge overflows periodically into the Thames.

5. HC *Sess. Pap.* 1877 (280) 17, minutes of evidence 423.

6. The MBW *Thames Embankment: Plans and Specifications*, 1861, gives a good picture of construction prices over a twelve-year period in London. For example, the specs for contract #1 allow 2s. per cubic yard for excavations, 7s. per yard for concrete, £16 per rod for brickwork, 2s. 4d. per yard for Portland stone, and 3s. per yard for Devon or Cornwall granite. The contracts specified Portland cement, and it is a measure of how uncertain this material still was that all cement had to be stored on the site for at least three weeks before use and then tested for another week to five hundred pounds pressure. The specifications also mandated that all "coins and other antiquities" found during excavations were to be turned over to the engineer and the Board. Bazalgette later reported that elephant and whale bones, deer and ox horns, flint war tools, skulls, coffins, and Roman coins of various dates were uncovered. Despite the contract sanctions, workers kept or sold a good many artifacts; others were donated to the British Museum. ICE *Minutes of Proceedings* 24 (1865): 314.

7. Smith, "Bazalgette," 93.

8. ICE *Minutes of Proceedings* 24 (1865): 357. The idea for a subway apparently came from John Fowler, who had seen it in Paris (HC *Sess. Pap.* 1863 [344] 12: "Report of the Select Committee on the Thames Embankment Bill," minutes of evidence 445.) The Embankment cut across the district lines of four gas light companies, and each wanted to lay its own mains within the subway. Bazalgette recommended (GLRO, MBW *Minutes*, 3 July 1868: 873) that one company lay a common main for all of them rather than complicate his project further. The Equitable Gas Light and Coke Company got the job.

9. Iron facing was featured on the Embankment proposed by Bazalgette and G. W. Hemans in 1860. They claimed it was cheaper than marble and could be lowered into place through water, thus saving the cost of building cofferdams. But Lionel Gisborne convinced the select committee that iron would turn brittle and crack in cold weather, so that marble would be cheaper in the long run. HC *Sess. Pap.* 1860 (494) 20: 397, 437.

10. Letter from Bazalgette to the Thames Embankment Committee, 17 November 1862 (GLRO, MBW 706). The Board was correct in its judgment. The first railway to build in the metropolis, the London and Greenwich Railway Company, tried to rent the arches of its viaducts as dwelling houses, without success (Sheppard, *The Infernal Wen*, 125). By 1867, Bazalgette was opposed to building urban railways above ground on leasable archways. He claimed that previous attempts to do this had failed. The income to be gained from such arrangements was very small, so that underground railways, despite the initial expense, were cheaper in the long run. *ICE Minutes of Proceedings* 27 (1867–68): 439.

11. Letter to the *Builder*, 14 November 1863: 813.

12. Comments from the *Builder*, 9 May 1863: 327, and 7 May 1864: 326, noted in Owen, *Government*, 81.

13. Timbs, *Curiosities of London*, 761. The Duke of Buckingham obtained the estate from James I after its previous occupant, Sir Francis Bacon, fell into disgrace. The York Waterworks and buildings are shown, terraced down to the gate, on eighteenth-century maps: see the GLRO maps Westminster AJ #2353 and #2363. Further details are given on a plaque mounted on the side of the gate.

14. Cf. GLRO, MBW, *Thames Embankment: Plans and Specifications*, 1861, Contract #1, 19–22. The landing stairs intended as the site for York Gate were to be fitted with a tidal reservoir for flushing the area, but this, too, was omitted, and new stairs designed, when the gate was abandoned. Cf. GLRO, MBW *Minutes*, 7 May 1869: 565.

15. GLRO, MBW 2321, Engineer's Monthly Reports, 4 October 1865.

16. GLRO, MBW 676: 218, 288.

17. Owen, *Government*, 84–86.

18. Ibid., 82–83. Cf. GLRO, MBW *Minutes*, 5 May 1864, Engineer's Monthly Report.

19. GLRO, MBW 676: 388–89. A stone was laid, on 10 October 1863, with board members, the First Commissioner of Works, and other officials attending. The party then took a steamer to Richmond for a celebratory dinner charged to public funds—an extravagance that one board member, William Carpmael, complained was improper. For the official photograph, see Stella Margetson, *Fifty Years of Victorian London* (London: Macdonald, 1969), 87. Another "foundation stone" was laid at Whitehall Steps, in June 1864, by Thwaites, twenty-six feet below the low-water line, within the iron cofferdam erected to keep out the river during construction.

20. The formation of the Westminster pier encroached upon the Embankment roadway as originally planned, so Bazalgette had to ask Cowper at Works and Gore at Woods and Forests for permission to move the roadway northward onto Crown property to maintain its width. GLRO, MBW *Minutes*, 23 February 1866: 250.

21. Thomas D. Ridley, "Description of the Cofferdams used in the Execution of the Thames Embankment," ICE *Minutes of Proceedings* 31 (1870): 2. The sluice gates were discussed by Edward Bazalgette, Sir Joseph's son and assistant, in his ICE paper on the Thames Embankment (*Minutes of Proceedings* 54 [1877–78]: 13–15). The gates could release natural runoff from the Thames valley that seeped into the work site from its land side, regardless of the cofferdam. Also, the gates were sometimes opened to allow river water into the site for construction purposes. The clay puddle between the two rows of dam pilings could be dangerous. One of William Webster's men fell in and drowned while working on contract #3.

22. One of the first papers on cofferdams was presented, in 1870, at the ICE by Thomas Dawson Ridley (not Samuel Ridley, the discharged contractor for the first section of Embankment, but the works engineer for A. W. Ritson on contract #2 cited in note 21). Introducing Ridley, ICE President Charles Vignoles "said it had been truly remarked . . . that there were at present no textbooks upon cofferdams. He was afraid there were but few upon any branch of engineering for systematic adoption in tuition [i.e., academic] courses. Almost all experience, particularly of cofferdams, was what was gained by contractors, who had not published it, and each in succession was obliged to form his own conclusion as to the best method of effecting his object. That was one of the misfortunes of the present mode of carrying on engineering. A work was first of all completed, and then rules deduced from it. This had often been considered and discussed, and it

presented a remarkable contrast between the engineering practice of this country, and the theoretical researches of other countries." Ridley himself had consulted an experienced hydraulic engineer, R. P. Brereton, before building his cofferdam. ICE *Minutes of Proceedings* 31 (1870): 31. Even the esteemed William Webster had to learn from experience. He had just got his cofferdam in place for the Temple-Blackfriars section in December 1868 when a high tide and strong winds destroyed two hundred feet of it. Two months of "active measures" repaired the damage, and the new dam withstood record tides. GLRO, MBW *Minutes*, 1 May 1869: 567.

23. HC *Sess. Pap.* 1860 (494) 20, minutes of evidence 54, question 771. For Gisborne, see HC *Sess. Pap.* 1854–55 (415)10, minutes of evidence, questions 969–1029. Gisborne said that the iron cylinder system had been adopted elsewhere before 1854.

24. ICE *Minutes of Proceedings* 54 (1878):16–31. There is some evidence (ibid., 13) that Bazalgette and Furness also tried iron H-beams joined at the legs to form a hollow dam, but it didn't work. The use of iron caissons as floats for the steamboat platforms is mentioned also in Bazalgette's annual report, GLRO, MBW *Minutes*, 12 June 1866: 775. Floating ferry platforms, which rose and fell with the tides, were first proposed by Captain James Walker in 1840; see HC *Sess. Pap.* 1840 (554) 12, minutes of evidence 347. Although Edward Bazalgette said that iron caissons were not used elsewhere on the Embankment, *ILN* (25 January 1868: 94) reported that William Webster was using them at Lambeth for the Albert Embankment, and they are listed in the MBW *Specifications* for contract #4.

25. Owen, *Government*, 86–87.

26. Ibid., 86. The contract in MBW *Specifications* (#1, p. 32) lists "Haytor" or "Dartmoor" granite "or other approved granite of uniform color, free from stains, flaws, etc. and equal to samples at the Board offices." The granite slabs were to be fastened with slate dowels and joints with a one-eighth-inch tolerance.

27. The "Metropolitan District" was a consolidation of the "District" with the "Metropolitan," the latter being the first underground railway to open, in the spring of 1860. The underground was opposed by local authorities, condemned as a "road to hell" by religious extremists, and sued by landowners claiming their foundations were undermined by the excavations. It opened to the public, in January 1863, and carried an average of 29,196 passengers per day in its first three weeks of operation. Opposition was largely gone by the time Fowler proposed the "Metropolitan District" line in 1864. The two lines separated again, in 1871, when the "District" carried over eight million passengers. W. J. Passingham, *The Romance of London's Underground* (New York: Benjamin Bloom, 1972), 13–16, 34–36.

28. GLRO, MBW *Minutes*, 18 March 1864. There is some evidence that the roadway had to be narrowed almost twenty feet where it joined the Westminster Bridge Road opposite the Clock Tower. It certainly looks narrow today.

29. Cf. Howard Clayton, *The Atmospheric Railways* (Oxford: published by the author, 1966), 128; Derek Bayliss, *The Post Office Railway, London* (Sheffield: Turntable Publications, 1978), 10; and Passingham, *The Romance of London's Underground*, 48–50. The Post Office Railway ran initially from Euston Station to the General Post Office up Eversholt Street at Mornington Crescent. The PO publicized their new line with free rides for celebrities like Gladstone and Prince Jerome Bonaparte. Timbs, *Curiosities of London*, 706–7, describes the would-be Whitehall and Waterloo Railway in 1867, although it had already gone bankrupt. The London Transport Office archives at South Kensington used to have (1985) a plan of the W and W in their documents collection, but it vanished when the records were transferred to microfiche. All of the commercial pneumatic lines eventually failed, because the technical problem of maintaining a vacuum power sufficient to haul passengers was not solved; but the Post Office Railway ran until after World War II.

30. GLRO, MBW *Minutes*, 16 March 1866: 342; 15 June 1866: 811; 12 October 1866: 1255; 16 November 1866: 1418; and 1 February 1867: 145.

31. Brassey's biographer observed that the start of the great harbor and dock works in the late 1850s, with their massive concrete elements, made for difficult, complicated contracts: "There was not a great deal of previous experience to fall back on." Charles Walker, *Thomas Brassey: Railway Builder* (London: Frederick Muller, 1969), 150–51.

32. GLRO, MBW *Minutes,* 23 October 1868: 1184. Owen, *Government,* 89, gives an earlier progress report based on the MBW *Annual Report* for 1867–68.

33. For pictures of construction in progress, see *ILN,* 22 April 1865: 369, and 25 January 1868: 88. Bazalgette's comment on Furness came during construction of the northern outfall works in October 1861. GLRO, MBW 2321, Engineer's Monthly Report.

34. GLRO, MBW *Minutes,* 23 February: 255, and 28 September 1866: 1174.

35. The measuring error is reported in GLRO, MBW *Minutes,* 16 November 1866: 1416. For the ornaments, see the London County Council General Purposes Committee, *Principal Contracts Completed by the Metropolitan Board of Works* (London: LCC, May 1906), 6, contract #1. I assume that the ornamental lamp standards were not included in the tender because Bazalgette's original design was abandoned for a later one.

36. HC *Sess. Pap.* 1889 (5705) 39: "Final Report of the Royal Commission appointed to Inquire into Certain Matters connected with the working of the Metropolitan Board of Works," 322, Bazalgette's evidence of 10 July 1888.

37. Cf. HC *Sess. Pap.* 1862 (3043) 28, appendix A, tabular statement of plans. A newspaper advertisement attracted some twenty designs. The commission praised all of them but accepted none.

38. HC *Sess. Pap.* 1863 (367) 12: "Report of the Select Committee on the Thames Embankment (South Side) Bill," minutes of evidence 7. Lambeth officials also objected that a viaduct would mask all the buildings behind it.

39. The Doulton dry dock is shown on the ordnance survey map for 1894–96. Ferry Street no longer exists, of course; it can be seen on the Albert Embankment contract drawings #2, GLRO, MBW 2532. For the White Hart Dock, see HC *Sess. Pap.* 1863 (367) 12, minutes of evidence 14. The flooding problem with the old White Hart Dock was mentioned by William Lawrence, a wharfinger, in testimony to the Royal Commission. Cf. HC *Sess. Pap.* 1862 (3008) 28, minutes of evidence 125.

40. The double tramway docks are shown in the MBW *Thames Embankment: Plans and Specifications,* 1861, for contract #4, drawings #1 and 2. The mud still accumulates by the remaining dock openings (creating a popular fishing site) and is still dredged periodically.

41. Different sources give the name of the gas company as the "City of London" or the "Chartered Gas Light and Coke Company." Given the consolidation of the industry shortly thereafter, this is not surprising. The viaduct was discussed by David Watson of the City of London Gas Company in testimony before the Royal Commission, HC *Sess. Pap.* 1862 (3008) 28, minutes of evidence 148. It is named on the ordnance survey map of 1862 as the "London." It is also mentioned by Stirling Everard, *The History of the Gas Light and Coke Company* [the "Chartered"] (London: Ernest Benn, 1949), 213. Everard adds that the new site at Nine Elms was completely wrecked by a tremendous explosion, in the autumn of 1865, just after the company had relocated there. The MBW's memorial booklet, *Chelsea Embankment Opened,* says that the viaduct was "abandoned for lack of funds," but other sources contend that the approach road to Vauxhall Bridge curved so far away from the river at that point that it was useless to embank. In a preliminary report to the board, 6 November 1862, Bazalgette had already suggested postponing the extension and linking up to the bridge approach roads, effecting a savings of almost 50 percent in the estimated cost (GLRO, MBW 2411 batch #5, bundle 1). For the river widening, see HC *Sess. Pap.* 1863 (367) 12: minutes of evidence 2. For the little garden (still there), see the MBW *Minutes,* 31 October 1879.

42. Owen, *Government,* 91. For Webster, see the Engineer's Monthly Reports in GLRO, MBW 2321, February 1863 and August 1866.

43. ICE *Minutes of Proceedings* 41 (1874–75): 118.

44. See Grant's comments in the discussion following Bazalgette's paper on the Main Drainage scheme, ICE *Minutes of Proceedings* 24 (1864–65): 343–44. He and Webster also devised a tray for holding activated charcoal in the sewer vents that opened to street level. This addressed a problem common to all the city's sewers, which was particularly noticable to pedestrians along the Embankment walkways. Charcoal filters were subsequently installed in ventilation shafts

added to the Victoria Embankment sewer line (GLRO, MBW 2321, Engineer's Monthly Reports, 7 January 1870).

45. Specifically, the MBW specifications for contract #1, drawing #1, show the City Gas Works between Whitefriars Dock (where Sion College is now) and Bridewell Wharf (at the end of what was once Bridewell Street, leading down from St. Bride's Church). A large waste-disposal pipe leads out into the Thames. An *ILN* panorama of London (10 August 1861: 150) shows the large storage tank just behind the company's headquarters.

46. See the "Report of the Royal Commission," HC *Sess. Pap.* 1861 (2872) 31, appendix, map facing p. 277; also drawings #6 and #7, MBW *Thames Embankment: Plans and Specifications,* 1861,: contract #1.

47. GLRO, MBW 2411 batch #5, bundle 2, "Works Committee Report on the Board's actions with respect to the Metropolitan District Railway works," 1868.

48. Ibid.

49. King, *London Discount Market,* 243–44; Checkland, *The Rise of Industrial Society,* 43–44.

50. The board's compromises are detailed in the Works Committee Report, GLRO, MBW 2411, batch #5, bundle 2. For Hawksley, who produced a plan almost exactly like the one worked out by Bazalgette and Fowler the year before, see the MBW *Minutes,* 6 March 1868: 404–5, and 13 March 1868: 427.

51. Clifton, *Professionalism, Patronage and Public Service,* 42.

52. Everard, *Gas Light and Coke Company,* 230–36. Beck's associate in the consolidation was the superintendent of the Horseferry works, Frederick John Evans (1818–1880), a civil engineer. See his memoir in ICE *Minutes of Proceedings* 63 (1880–81): 312. Owen (*Government,* 90) suggests that Hawksley's arbitration plan was responsible for the solid Embankment, but I find no evidence for this. Bazalgette had already worked it out. The double subway is shown on later maps but disappears before the end of the century. I have not been able to discover exactly when it was filled in. The MBW "Engineer's Monthly Reports" (GLRO, MBW 2321) show that a new hotel (the "Royal") was built next to the bridge, in 1875, and William Webster reconstructed part of the site when the Royal was replaced by Dekeyser's larger hotel in 1884–85. Sion College and the City of London School also moved into the site, in the 1880s, but, although the Gas Light and Coke Company removed its old discharge pipes, the subways were still shown on the site plans. Cf. MBW *Minutes,* 13 August 1880, and MBW 2492, Embankment plans and sections in 1879.

53. Cf. GLRO, MBW 2411 batch #5, bundle #11: 52, 55. The report by Bazalgette and the board's architect, George Vulliamy, dated 18 October 1865, is in MBW 2371, bundle #215. The plans were described by Timbs, *Curiosities of London,* 774. Some idea of how the row of shops might act as a retaining wall and provide a terrace for the park may be gained by viewing Temple Underground Station from the back. The flat roof of the station, bordered by the crescent Temple Place, could easily be converted into a terrace garden. Another plan for the Embankment Gardens site, encompassing a smaller crescent flanked by imposing multipurpose buildings in the prevailing eclectic style, was published by architect Charles Forster Hayward in the *Architect,* 13 March 1869.

54. Although a deputation from the parishioners of the Queen's Royal Chapel in the Savoy petitioned the board to change the road planned to run through their corner, arguing that it could cut through their burial ground and the traffic would ruin the chapel itself. GLRO, MBW *Minutes,* 20 April 1866: 482.

55. Lady Gwendolyn Cecil, *The Life of Robert, Marquis of Salisbury.* Vol. 1, 1830–1868 (London: Hodder and Stoughton, 1921), 1, 5, 171.

56. Low and Sanders, *The History of England,* 202–10.

57. Six-inch hardwood blocks were laid on end in a prepared bed. Northumberland Avenue, laid out from Charing Cross to the Embankment, in 1876, was also paved with wood blocks. Bazalgette said wood paving lasted five to ten years and needed less maintenance than other materials, although he admitted "there is but little reliable information on the subject." GLRO, MBW 2411, bundle #5, item #19: 5–6. Cf. Osbert Henry Howarth's paper, "Wood as Paving Material Under Heavy Traffic," ICE *Minutes of Proceedings* 57 (1879): 31.

58. Savoy Lane was authorized by the Metropolitan Board of Works (Various Powers) Act, 1875. See GLRO, MBW 2321, Engineer's Monthly Report, 1 December 1876; the MBW *Minutes,* 15 October 1880; and Edwards, *A History of London Street Improvements.* The planned crescent roads were officially abandoned in the Thames Embankment (North) Act, 1870 (33 & 34 Vict. c. 92), which vested the gardens in the MBW.

59. For a more extensive treatment of the developing public appreciation of open spaces, see Hazel Conway, *The People's Parks: The Design and Development of Victorian Parks in Britain* (Cambridge and New York: Cambridge University Press, 1991).

60. Owen, *Government,* 96. Charles Gore seems to have been overshadowed in this episode by his superior, Lowe. Acton Smee Ayrton (1816–1886), the Liberal member for Tower Hamlets, was made First Commissioner by Gladstone in November 1869. Cf. *Dictionary of National Biography,* First Supplement, s.v. "Ayrton, Acton Smee." Among other reforms, he changed the system of contracting public works so that, when unforeseen "extras" were negotiated at the end of the job, the contractor was compensated at the rates stipulated in the original contract. Under the old system, contractors typically put in low bids with the expectation that they could extort high prices for the "extras" and thus turn a profit. PRO, WORK 11, 7/2, letter of Douglas Galton to Works, November 1873.

61. Owen, *Government,* 95–97. The act is 36 & 37 Vict. c. 40. John Timbs (*Curiosities of London,* 553–54) said that the ninety-nine-year lease secured by the Duke of Buccleuch, in 1856, to rebuild Montague House was another obstacle in the way of erecting public buildings on the reclaimed land.

62. Owen, *Government,* 98. The act is 35 & 36 Vict. c. 66.

63. GLRO, MBW 2411 batch #5, Bazalgette's report for 21 November 1877. Webster also won the contract for finishing the Embankment intersections with Whitehall Place, Surrey and Arundel Streets around Temple Underground Station, and the extension of Villiers Street next to Charing Cross Station. He was also awarded £10,000 for linking the Temple-Blackfriars sewer line with the new Queen Victoria Street sewer on the other side of Blackfriars Bridge. At this time, he held about £240,000 worth of MBW contracts and was getting ready to bid on the £120,000 Western Pumping Station. GLRO, MBW 2321, Engineer's Monthly Reports.

64. Quoted in Olsen, *Growth of Victorian London,* 108, 217. The board's comment on Chelsea is from its booklet, *Chelsea Embankment Opened . . .*

65. Owen, *Government,* 94. A year before the formal opening, the MBW held a giant "rummage" sale of used construction materials and miscellaneous objects left over from the Embankment work sites and from the property purchased along the way. They realized £685 from the sale. GLRO, MBW *Minutes,* 4 June 1869.

66. MBW, *Chelsea Embankment Opened . . .* , 8–9. Edwards, *A History of London Street Improvements,* 127.

67. Longford, *Images of Chelsea,* 65, prints nos. 178 and 190, shows the Chelsea waterfront before and after the Embankment to buttress her claim that the "improvement" destroyed Chelsea's charm. She also says (67) that residents "fought off a scheme to extend the Embankment westwards to Cremorne," where the low-level sewer was laid in 1874. But that scheme resulted from a petition by Chelsea vestry, in 1880, which the MBW refused.

68. Edwards, *A History of London Street Improvements,* 127; Owen, *Government,* 99–100. Cf. MBW *Thames Embankment: Plans and Specifications,* 1861, contract #5. Chelsea was one of the new boroughs created by the second Reform Act, 1867.

Chapter IX/ The Historical Future

1. Wohl, *Endangered Lives,* 116. W. P. D. Logan, "Mortality in England and Wales, 1848–1947," *Population Studies* 4 (1951): 134, table 1, cites data showing a 22 percent decline for infectious diseases for men and 26 percent for women in the same period. The death rate from cholera dropped to .006 per million people in the decade 1901–10.

2. For a reminder of the bad old days, see the Bazalgette obituary in the *Times,* 16 March 1891:

4. The swimming baths are shown on plans and sections of the Victoria Embankment prepared in 1879 (GLRO, MBW 2492) and on the ordnance survey map for 1895.

3. Wohl, *Endangered Lives*, 111; Luckin, *Pollution and Control*, 24–25. For the continuing debates on water quality measurement, see Hamlin, *A Science of Impurity*.

4. Wohl, *Endangered Lives*, 74; Checkland, *The Rise of Industrial Society*, 317–18.

5. Luckin, *Pollution and Control*, 100, 111; Wohl, *Endangered Lives*, 95, 107–8. Wohl chants a litany of horrors experienced by the small town of Franham in Surrey when it tried to install a new sewer system. The backup and overflow of sewerage in stormwater drains continues in both British and American cities because officials can't bring themselves to finance separate systems; the image of economical flushing is stronger than the evidence of long experience. The damage done by rat tunnels is described by Glick, "Science, Technology and the Urban Environment," in Bilsky, *Historical Ecology*, 125–26.

6. Breeze, *The British Experience*, 77, 200.

7. Luckin, *Pollution and Control*, 21, 172–73. Breeze adds that a third Royal Commission on Sewage Disposal sat from 1898 to 1915, when it was replaced by a standing committee. When advocates brought pressure directly on Balfour and Neville Chamberlain, in 1926, they were told that Government needed more time to study the problem because it involved a threat to manufacturing prosperity. Chamberlain did set up a Joint Advisory Committee of experts from the Health and Agriculture ministries, which made practical recommendations. Enabling legislation, however, was not prepared and passed until World War II. Breeze, *The British Experience*, 200–208.

8. Luckin, *Pollution and Control*, 22.

9. Ibid., 141–43.

10. J. T. Coppock, "The Changing Face of England: 1850–*circa* 1900," and Peter Hall, "England *circa* 1900," in Darby, *A New Historical Geography of England*, 355–59, 440–45. Wohl, *Endangered Lives*, 234, gives the sewage volume.

11. A full account of the MBW-Conservancy battle is given by Owen, *Government*, 69–73. The Bramwell Commission report, which contains an excellent history of the Main Drainage project, is in HC *Sess. Pap.* 1884 (3842-I) 41. About the barge *Bazalgette* , the *Times* lamented, "it is not as the eponymous hero of a sludge ship that Londoners ought to remember the name of Sir Joseph Bazalgette" (obituary, 16 March 1891: 4).

12. A mudlarking guide is quoted in the *Washington Post*, 16 May 1996: A16, col. 2.

13. Owen, *Government*, 123.

14. For the pre-Embankment Adelphi, see GLRO, Maps collection, Westminster AH 22446. The arches were described in a letter (with drawings) to the *Daily Graphic*, 16 December 1896: 10, by a former artist who recalled venturing into the netherworld as a student to study the effect of sunlight reflected from the Thames onto the vaulted ceiling. Some of the arches are apparently still extant and under restoration. Those under the Royal Society of Arts in John Adam Street nearer the Strand have recently been excavated and restored for use as a dining hall.

15. Sheppard, *The Infernal Wen*, 123, claims the new railway gave the *coup de grâce* to the steamboats after other railways reduced their traffic, but he does not factor in the freeing of tolls or the new Embankment streets.

16. Owen, *Government*, 119–23.

17. *Building News* 27 (1874): 481. Olsen (*Growth of Victorian London,* 303 n. 13) cites the *Builder's* description of "the loneliness and desertion that reign after dark" along the Embankment in 1882, a description that could easily apply in the present. Checkland, in *The Rise of Industrial Society,* claims that the financial crises of 1866 and 1873 led gentlemen and would-be gentlemen of marginal means to give up their carriages, which might account for some drop in traffic, but it does not explain the contrast between the empty Embankment and the crowded Strand.

18. GLRO, MBW 2411, batch #5, item 17, tabular statement of traffic.

19. Cf. the *Builder* 20 (1862): 564. Ellis, *Three Hundred Years on London River: The Hay's Wharf Story, 1651–1951* (London: Bodley Head, 1952), 87, describes an occasion when high tides flowed over the company dock and soaked the year's supply of tea, just arrived on the new clipper ships.

Usually the warehouse doors were simply left open to allow tides to flow through. Frank Forster, the MCS chief engineer in 1850, had estimated that 100,000 residents would be affected by flooding, damp, and disease without proper controls. Cf. John Thwaites, "Points for Consideration of the Thames Embankment Commission in reference to the Embankment of the South Side," 10 December 1861, in PRO, WORK 6, 158/3: 1.

20. GLRO, MBW *Minutes,* 30 November 1866, 1473. The river wall, of course, rose four feet above the roadway.

21. Aside from Owen's brief account, the main sources for the Thames flooding and subsequent embankments are the "Report of the Select Committee on the Thames River (Prevention of Floods) Bill," HC *Sess. Pap.* 1877 (280) 17; a discussion led by hydrographer J B. Redman at the ICE, in *Minutes of Proceedings* 49 (1877): 67–93; and the papers of the First Commissioner of Works in PRO, WORK 11. Redman's measurements of the river showed that, at high tide, the water averaged six inches higher, from Blackfriars to Battersea, than it had before the Embankment was built. It had been hoped that the Embankment, by increasing the river's current, would scour out a deeper channel, lowering the overall water level. Redman reported (87–93) that this had happened in the lower reaches of the river but, around Waterloo Bridge, the tons of ballast poured in the river bed to secure the bridge's shallow foundations negated the effect of the scour.

22. HC *Sess. Pap.* 1860 (494) 20, minutes of evidence 56, question 793.

23. Cf. HC *Sess. Pap.* 1862 (3008) 28: 115, 120, minutes of evidence, testimony of Thomas Hawksley and William Carpmael.

24. HC *Sess. Pap.* 1877 (280) 17, minutes of evidence 319.

25. ICE *Minutes of Proceedings* 49 (1877): 82–83.

26. HC *Sess. Pap.* 1877 (280) 17: 323–24. The "letter of advice" form, with blanks to be filled in for specifications, is dated 10 January 1877. One of the letters was sent to the First Commissioner of Works, advising him to raise the top of the stairs at Speaker's Landing next to Westminster Bridge—a nice touch. Because some of the cooperating owners were large companies, such as the London and South Western Railway, the area improved was almost half the total length, more than the number of respondents would indicate.

27. HC *Sess. Pap.* 1877 (280) 17: 306–17. In fact, the old Metropolitan Commission of Sewers had contemplated serving flood-wall repair notices to proprietors on the south side after high tides in 1850–52. "Points for Consideration of the Thames Embankment Commission," PRO, WORK 6, 158/3. Norman Bazalgette was involved with one of the several sewage reclamation companies and presented a paper on "The Sewage Question" to the Institution of Civil Engineers in 1876. Cf. ICE *Minutes of Proceedings* 48 (1876): 105–59.

28. HC *Sess. Pap.* 1877 (280) 17: 419, 423.

29. PRO, WORK 11, 171.

30. HC *Sess. Pap.* 1877 (280) 17, minutes of evidence 316–17.

31. HC *Sess. Pap.* 1863 (3093) 26, "Report of the Royal Commission to consider plans for making a Communication between the Embankment at Blackfriars Bridge and the Mansion House, and also between the Embankment at Westminster Bridge and the Embankment at Millbank," 431–36.

32. For the Houses of Parliament (Victoria Tower Extension) Bill, brought in by Lord John Manners, see HC *Sess. Pap.* 1867, vol. 3: 14. The relevant correspondence is in PRO, WORK 11, 63 but the plans are bound with Webster's contract in WORK 29/2364. The Office of Works purchased the bed and foreshore of the Thames at Millbank from the Thames Conservators, as the MBW had done for the Victoria Embankment. Works also compensated St. John Parish, Westminster, for the lost rates.

33. PRO, WORK 11, 7/2: 5. The Office of Works' solicitor demanded an open competition to pick the lowest bidder. Taylor replied that a reputable contractor was needed and refused to carry out the project unless he had a free hand to choose the firm he thought best. Douglas Galton, the First Commissioner, backed him up.

34. PRO, WORK 11, 7/2: 5–37. Engineer Taylor noted that Parliament had appropriated £35,000 for the contract, but that sum included all the foot and carriage ways and the ornamental

iron railings, which were done later. The MBW's booklet *Chelsea Embankment Opened . . .* said that the Abingdon Street Embankment was made necessary by the flood tide of 20 March, but this is obviously wrong; it had been in design stages since 1867.

35. PRO, WORK 11, 63. This folio also contains the layout of the grounds. The land cleared for the gardens is shown as "vacant land" on the 1879 plans and sections for the MBW's flood control measures. GLRO, MBW 2492.

36. GLRO, London County Council *Minutes* 9 November 1897: 1189, and Report of Improvement Committee, 27 June 1899.

37. George John Shaw-Lefevre, Baron Eversley (1831–1928) was an egocentric traveller, politician, and keen observer of human behavior who had been in Parliament since 1863. His father was a long-time clerk of Parliament who had served on the Select Committee for Metropolitan Improvements, in 1837, when the Thames Embankment was first planned. The son had photographed the Crimean War, ridden horseback from Vienna to Constantinople, and visited slave markets in the United States. Along with William Cowper, he founded the Commons Preservation Society. Under Gladstone, he filled a number of secondary posts and was First Commissioner of Works, in 1880–81, when the Victoria Tower Gardens were completed. The DNB entry notes his extraordinary memory, "which was especially retentive of curious and often grotesque stories about strange characters and persons of note whom he had met." *Dictionary of National Biography,* twentieth-century supplement 1922–1930 (Oxford, 1937).

38. GLRO, London County Council *Minutes* 27 June 1899: 912.

39. Ibid., Reports of the Improvements Committee, 27 June, 4 July, 17 October, 24 October 1899; PRO, WORK 11:63 and 11:191.

40. Ibid., 18 July, 24 July 1900; 29 January 1901.

41. London County Council (Improvements) Act, 1900, 63 & 64 Vict c. 264.

42. PRO, WORK 11, 63, letters of 1 April and 10 July, 1914 . The building and opening of the embankment are described in PRO, WORK 16, 826.

43. The *Observer,* 3 March 1923. The mention of Charing Cross is a reference to another perennial idea for linking Trafalgar Square with Waterloo Station via a new bridge, replacing the railway-cum-pedestrian bridge at Hungerford. One version, offered by architect Richard Rogers, in 1986, called for removing Charing Cross Station to Waterloo East and covering the length of Embankment Boulevard with a pedestrian walkway. Cf. *ILN,* October 1986: 52.

44. PRO, WORK 16, 1148, correspondence of July–November 1932. The Office of Works agreed to maintain the park and, reluctantly, to build a public lavatory. Because the slope down to the river is fairly steep there, the Royal Human Society donated life-saving equipment. The new Lambeth Bridge was opened by King George V and Queen Mary on 19 July 1832.

45. Smith, *St. John The Evangelist, Westminster,* 379.

46. For the Page embankment fortunes under the Office of Works, see PRO, WORK 6, 140/2. For the improvements by MBW, see GLRO, MBW 2321, Engineer's Monthly Report, 8 January 1888. The comment is from the MBW's *Chelsea Embankment Opened . . . ,* 3.

47. F. M. Leventhal, "'A Tonic to the Nation': The Festival of Britain, 1951," *Albion* 27 (fall 1995): 448–53.

48. Vauxhall Cross: planning application to Lambeth Council, 21 April 1987, Anthony Bowhill and Associates. For previous site plans, see GLRO, MBW 2456, "Deposited Plans for Metropolitan Improvement," 1889, and the 1916 ordnance survey map.

49. In *Punch,* quoted by Owen, *Government,* 176.

50. Owen, *Government,* chap. 8, "The Odor of Corruption," and chap. 9, "The Twilight of the Metropolitan Board of Works."

51. The *Observer,* 3 March 1923.

52. Owen, *Government,* 255.

53. J. W. Bazalgette, "The Main Drainage of London, and Interception of the Sewage of the River Thames," 14 March 1865, ICE *Minutes of Proceedings* 24 (1865): 280–315, with 3 appendices and 9 plates. E. Bazalgette, "The Victoria, Albert, and Chelsea Embankments of the River Thames," 9 and 16 April 1878, ICE *Minutes of Proceedings* 4 (1878) part 4: 1–60.

54. Cf. GLRO, MBW *Annual Report* for 1878, 23–24. The MBW architect, George Vulliamy, designed the bronze sphinxes and cornices to hide the obelisk's dilapidated base.

55. For critical accounts of the countryside movement, see Martin Wiener, *English Culture and the Decline of the Industrial Spirit* (New York: Cambridge University Press, 1981); Bunce, *The Countryside Ideal*; and Stanley P. Johnson, *The Politics of Environment: The British Experience* (London: Tom Stacey, 1973).

56. In 1824, a parliamentary committee responded to complaints from fishermen about river pollution by stating "In those rivers on which large commercial cities are situated and on which the interests of manufacturers have led to the expenditure of vast capital, it is not to be looked for that the salmon fishery should flourish; and while it may be from those causes nearly extinct, it would be chimerical to expect that it should ever be restored." HC *Sess. Pap.* 1824 (427) 7, appendix 3, quoted in Clapp, *Environmental History of Britain*, 93. Chamberlain told advocates, in 1926, that government needed "more time" to study the problem of river pollution because it involved a threat to manufacturing prosperity.

57. Clapp, *Environmental History of Britain*, 95.

58. Johnson, *The Politics of Environment*, 82–83.

59. Ibid., 17, 82–83.

60. Clapp, *Environmental History of Britain*, 88, 92–93.

61. Johnson, *The Politics of Environment*, 170.

62. Ibid., 10. Lawrence Breeze bemoans Mrs. Thatcher's privatization policy as an abandonment of "that thread of Tory social consciousness found in Disraeli and his Conservative successors" in favor of "the surging impulses of greed and grab. . . ." *The British Experience*, 208.

Appendix/An Embankment Chronology

1. J. H. Hexter, *The History Primer* (New York: Basic Books, 1971), 266.

2. Kitson Clark, *The Making of Victorian England*, 62–63.

3. G. M. Trevelyan, *British History in the Nineteenth Century and After, 1782–1919* (1937; reprint, New York: Harper, 1966), 423.

4. Elie Halevy, *The History of the English People in the Nineteenth Century*, vol. 4, *Victorian Years (1841–1895)*, part 2, 1852–1895, by R. B. McCallum (New York: Barnes and Noble, 1961), 434–35.

5. Evans, *The Victorian Age*.

6. Kitson Clark, *The Making of Victorian England*, 206.

7. Rolt, *Victorian Engineering*, 13.

8. Geoffrey Best, *Mid-Victorian Britain, 1851–1875* (1971), quoted in Walter L. Arnstein, *Britain Yesterday and Today, 1830 to the Present*, 6th ed. (Lexington, Mass: D. C. Heath, 1992), 71. Arnstein actually uses the date 1873, but the difference is negligible.

9. W. L. Burn, *The Age of Equipoise* (New York: Norton, 1964).

10. Richard D. Altick, *Victorian People and Ideas* (New York: Norton, 1973), 12.

11. Briggs, *The Age of Improvement*, 394–445, "The Balance of Interests." The quotation is a paraphrase from G. M. Young's *Victorian England, Portrait of an Age*.

12. Walter E. Houghton, *The Victorian Frame of Mind, 1830–1870* (New Haven: Yale University Press, 1957).

13. Checkland, *The Rise of Industrial Society*.

14. Olsen, *Growth of Victorian London*, 34.

15. Young, *Victorian England*, 82.

16. Edwards, *A History of London Street Improvements*, 10.

Bibliography

Note: The most complete bibliography on London government administration and construction in the Victorian period is the one prepared for David Owen's *The Government of Victorian London* (q.v.), 416–54.

I. Public Records, Periodicals, and Archival Documents

British Museum Library:

> Trench, Frederick William. *Prospectus of a Proposed Improvement of the Thames Banks [by a Public Company].* London, 1825. Accompanied by a set of watercolor plates by Thomas Allom. Catalogue item 1787.aa.50.

> Stanford, Edward. *Stanford's Library Map of London.* Special edition, showing proposed metropolitan railways and improvements deposited in Parliament by 30 November 1863 for the 1864 session. London: Stanford, 1863. Map Room #11-c-4.

Corporation of London Record Office, Guildhall:

> *Minutes and Reports of the Court of Common Council.*

> Manuscripts Collection, Noble Collection B, V group 8: plans for London-area improvements.

> Guildhall Map Room, "London Improvements."

Greater London Record Office and History Library, Clerkenwell (GLRO):

> "Plan for Raising Three Hundred Thousand Pounds for the Purpose of Compleating the Bridge at Black-friars, and Redeeming the Toll Thereon; Embanking the North Side of the River Thames between Paul's Wharf and Milford Lane," 1767. History Library, 27.51 (BLA).

> London County Council General Purposes Committee. *Principal Contracts Completed by the Metropolitan Board of Works.* London, 1906.

> Maps collection: Westminster AD, AE, AH, AJ, and AK.

> MCS series, Records of the Commissioners of Sewers for the London Area, 1531–1847.

> MBW series, Records of the Metropolitan Board of Works.

> Metropolitan Board of Works. *Chelsea Embankment Opened by their Royal Highnesses the Duke and Duchess of Edinburgh on May 9th, 1874.* London: Judd and Co., 1874 (MBW 2411, batch #5).

—*A Descriptive Account of the Victoria Embankment Opened by Her Most Gracious Majesty Queen Victoria on the 13th July 1870.* London, 1871.

—*Thames Embankment: Plans and Specifications,* 1861.

—*Minutes of Proceedings, 1855–1889.*

—*Pocket Book and Diary.* Annual. London: John Smith and Co. (MBW 673c).

U.K. House of Commons, *Sessional Papers:*

1799, vol. 124, "Reports and Papers," and 1799–1800, vol. 132, Reports and Papers: "Reports from the Select Committee upon the Improvement of the Port of London." In *House of Commons Sessional Papers of the Eighteenth Century,* edited by Sheila Lambert. Wilmington, Del.: Scholarly Resources, 1975.

1821, vol. 5, "Report of the Committee to Inquire into the present state of London Bridge and Consider Alterations and Improvements."

1836, 20: "Report of the Select Committee on Metropolitan Improvements."

1837–38 (418) (661) 16, "First and Second Reports of the Select Committee on Metropolitan Improvements."

1839 (136) 13, "First Report of the Select Committee on Metropolitan Improvements."

1840 (410) (485) 12, "First and Second Reports of the Select Committee on Metropolitan Improvements."

1840 (554) 12, "Report of the Select Committee on the Embankment of the Thames."

1844 (15) 15; 1845 (4348, 619, 627) 17; 1846 (682) 14; 1847 (861) 16, "Reports of the Royal Commission on Metropolitan Improvements" [Chelsea Embankment]."

1846 (474) 10, "Report of the Select Committee on the application of sewage to agricultural purposes."

1850 (480) 33, Correspondence between the Commissioners of Sewers and Mr. Goldsworthy Gurney, etc.

1852–53 (629) 26, "Report of the Select Committee on the London Drainage Bill."

1852–53, vol. 7, Bill for the Chelsea Bridge and Embankment, with a schedule of the property owners and occupiers in the line of embankment.

1854–55 (105) 53, "Mr. Gurney's Report to the Office of Works, on his experiment for withdrawing and decomposing the noxious effluvia from the Sewers in the neighbourhood of the Houses of Parliament."

1854–55 (415) 10, "Report of the Select Committee on Metropolitan Communications."

1854–55 (1980) 21, "Report of the Royal Commission on Metropolitan Improvements."

1857–58 (21) 48, "Report of Mr. Gurney to the First Commissioner of Works on the State of the Thames in the Neighbourhood of the Houses of Parliament."

1857–58 (442) 11, "Report from the Select Committee appointed to take into consideration Mr. Gurney's Report on the River Thames."

1857–58 (2372) 32, "Preliminary Report of the Royal Commission on the Sewage of Towns."

1859 (Session 2) 26: 275–280, Annual Report of the Metropolitan Board of Works, 30 July 1859.

1860 (494) 20, "Report of the Select Committee on the best means of providing for increasing traffic of the Metropolis by the Embankment of the Thames."

1861 (2872) 31, "Report of the Royal Commission on plans for Embanking the River Thames within the Metropolis."

1861 (2882) 33, "Second Report of the Royal Commission on the Sewage of Towns."

1862 (160) (469) 14, "First and Second Reports of the Select Committee on the Application of Sewage to Agricultural Purposes."

1862 (369) 47, "Correspondence relating to works under the Thames Embankment Bill."

1862 (3008) 28, "Report of the Select Committee on the Thames Embankment (North) Bill."

1862 (3043) 28, "Report of the Royal Commission on the Thames Embankment, on Plans for Embanking the Surrey Side of the Thames within the Metropolis."

1863 (344) 12, "Report of the Select Committee on the Thames Embankment Bill."

1863 (367) 12, "Report on the Thames Embankment (South Side) Bill."

1863 (3093) 26, "Report of the Royal Commission to consider plans for making a Communication between the Embankment at Blackfriars Bridge and Mansion House, and also between the Embankment at Westminster Bridge and the Embankment at Millbank."

1864 (487) 14, "Report of the Select Committee on Metropolis and town sewers."

1864, vol. 50, "Correspondence between Her Majesty's Treasury and the Metropolitan Board of Works."

1865 (171) 8, "Special Report of the Select Committee on the Metropolis Sewage and Essex Reclamation Bill."

1867, vol. 3, "Houses of Parliament (Victoria Tower Garden extension) Bill."

1867, vol. 58, "Documents relating to metropolitan coal duties and MBW loan guarantees."

1871 (411) 12, "Report of the Select Committee on whether land reclaimed from the Thames between Whitehall Place and Whitehall Gardens should be appropriated to the advantage of inhabitants of the Metropolis."

1872 (287) 12, "Report of the Select Committee on the Thames Embankment (Land) Bill."

1877 (280) 17, "Report of the Select Committee on the Thames River (Prevention of Floods) Bill."

1877 (367) 17, "Report of the Select Committee . . . to inquire and report what amendments are required to deal with injuries by floods."

1878–79 (178) 13, "Report of the Select Committee on the Thames River (Prevention of Floods) Bill."

1882 (809) 52, Average number of corpses per year in the Thames, 1877–1882.

1884 (3842-I) 41, "First Report of the Royal Commission on the Discharge of the Sewage of the Metropolis."

1889 (5705) 39, "Final Report of the Royal Commission appointed to Inquire into Certain Matters connected with the working of the Metropolitan Board of Works."

Hansard Parliamentary Debates, 3d series.

Institution of Civil Engineers (ICE):

Minutes of Proceedings.

Metropolitan Board of Works, *Thames Embankment Plans and Specifications,* 1861.

London Transport Office, South Kensington archives: metropolitan railway plans.

Newspapers and Periodicals:

The Architect
The Builder
Building News
The Edinburgh Review
The Illustrated London News
The Nautical Magazine
The Observer
Times (London)

Public Record Office, Kew (PRO):

Office of Works papers (WORK).

Current Guide to the PRO, Part I.

II. Secondary Works

Altick, Richard D. *Victorian People and Ideas.* New York: Norton, 1973.

Anthony Bowhill and Associates, *Vauxhall Cross.* Planning application to Lambeth Council, 21 April 1987.

Armytage, W. H. G. *A Social History of Engineering.* Boulder, Colo.: Westview Press, 1976.

Arnstein, Walter L. *Britain Yesterday and Today, 1830 to the Present.* 6th ed. Lexington, Mass.: D. C. Heath, 1992.

Appleman, Philip, ed. *1859: Entering an Age of Crisis.* Bloomington: Indiana University Press, 1959.

Barker, Felix, and Ralph Hyde. *London as It Might Have Been.* London: John Murray, 1982.

Barker, Felix, and Peter Jackson. *London: 2,000 Years of a City and Its People.* New York: Macmillan, 1974.

Bayliss, Derek. *The Post Office Railway, London.* Sheffield: Turntable Publications, 1978.

Betjeman, John. *Victorian and Edwardian London from Old Photographs.* London: B.T. Batsford, 1969.

Bijker, Wiebe, Thomas Hughes, and Trevor Pinch, eds. *The Social Construction of Technological Systems: New Directions in the Sociology and History of Technology.* Cambridge, Mass.: MIT Press, 1987.

Bijker, Wiebe, and John Law, eds. *Shaping Technology/Building Society: Studies in Sociotechnical Change.* Cambridge, Mass.: MIT Press, 1992.

Bilsky, Lester J., ed. *Historical Ecology: Essays on Environment and Social Change.* Port Washington, New York and London: Kennikat Press, 1980.

Bonnet, Theodore. *The Mudlark.* Garden City, New York: Doubleday, 1949.

Booker, P. J. *A History of Engineering Drawing.* London: Chatto and Windus, 1963.

Braudel, Fernand. *The Mediterranean and the Mediterranean World in the Age of Philip II.* Trans. Sian Reynolds. 2d rev. ed. New York: Harper and Row, 1973.

Breeze, Lawrence E. *The British Experience with River Pollution, 1865–1976.* New York: Peter Lang, 1993.

Briggs, Asa. *The Age of Improvement.* New York: Longmans, Green, 1959.

Brimblecombe, P. *The Big Smoke.* London: Methuen, 1987.

Buchanan, R. Angus. *The Engineers: A History of the Engineering Profession in Britain, 1750–1914.* London: Jessica Kingsley, 1989.

— "Gentleman Engineers: The Making of a Profession," *Victorian Studies* 26, no.4 (summer 1983): 407–29.

— "Insitutional Proliferation in the British Engineering Profession, 1847–1947," *Economic History Review,* 2d ser., 38 (1985): 46–60.

— "The Rise of Scientific Engineering in Britain," *British Journal for the History of Science* 18 (1985): 218–33.

— "Theory and Narrative in the History of Technology," *Technology and Culture* 32, no. 2, part 1 (April 1991): 365–76. "Response" by John Law, 377–84; "Comment" by Philip Scranton, 385–93.

Bunce, Michael. *The Countryside Ideal: Anglo-American Images of Landscape.* London and New York: Routledge, 1994.

Burn, W. L. *The Age of Equipoise.* New York: Norton, 1964.

Cecil, Lady Gwendolyn. *The Life of Robert, Marquis of Salisbury.* Vol. 1 (1830–1868). London: Hodder and Stoughton, 1921.

Checkland, S. G. *The Rise of Industrial Society in England, 1815–1885.* New York: St. Martin's Press, 1964.

Childers, Joseph, "Observation and Representation: Mr. Chadwick Writes the Poor." *Victorian Studies* 37, no. 3 (spring 1994): 405–32.

Clapham, J. H. *An Economic History of Modern Britain.* Vol. 2. New York: Macmillan, 1932.

Clapp, B. W. *An Environmental History of Britain.* London and New York: Longman, 1994.

Clark, E. F., ed. *George Parker Bidder, The Calculating Boy.* Bedford, England: KSL Publications, 1983.

Clayton, Howard. *The Atmospheric Railways.* Oxford: privately published, 1966.

Clifton, Gloria C. *Professionalism, Patronage and Public Service in Victorian London: The Staff of the Metropolitan Board of Works, 1855–1889.* London: The Athelone Press, 1992.

— "The Staff of the Metropolitan Board of Works, 1855–1889: The Development of a Professional Local Government Bureaucracy in Victorian London." Ph.D. dissertation, University of London, 1986.

Conway, Hazel. *The People's Parks: The Design and Development of Victorian Parks in Britain.* Cambridge and New York: Cambridge University Press, 1991.

Cooney, E. W. "The Origins of the Victorian Master Builders," *Economic History Review* 8 (1955): 167–76.

—"The Speculative Builders and Developers of Victorian London: A Comment." *Victorian Studies* 13, no. 3 (March 1970): 355–57.

Cracknell, Basil. *Portrait of London River: The Tidal Thames from Teddington to the Sea.* London: Robert Hale, 1968.

Crown Reach (Promotional book). London: The Crown Reach development, n.d.

Darby, H. C., ed. *A New Historical Geography of England after 1600.* Cambridge: Cambridge University Press, 1976.

Darnton, Robert. *The Great Cat Massacre and Other Essays on French Cultural History.* New York: Vintage Books, 1985.

Davey, Norman. *A History of Building Materials.* London: Phoenix House, 1961.

Dictionary of National Biography. Various editions.

Dyos, H. J., "The Speculative Builders and Developers of Victorian London." *Victorian Studies* 11 (1968): 668–690.

Edwards, Percy J. *A History of London Street Improvements.* London: London County Council, 1898.

Ellis, Aytown. *Three Hundred Years on London River: The Hay's Wharf Story, 1651–1951.* London: Bodley Head, 1952.

Emmerson, Sir Harold. *The Ministry of Works.* London: George Allen and Unwin, 1956.

Enthoven, Alain, and A. Myrich Freeman III, eds. *Pollution, Resources and the Environment.* New York: Norton, 1973.

Evans, R. J. *The Victorian Age, 1815–1914.* 2d ed. New York: St. Martin's Press, 1968.

Everard, Stirling. *The History of the Gas Light and Coke Company.* London: Ernest Benn, 1949.

Everett, Nigel. *The Tory View of Landscape.* New Haven and London: Yale University Press for the Paul Mellon Centre for Studies in British Art, 1994.

Ferkiss, Victor. *Nature, Technology and Society: Cultural Roots of the Environmental Crisis.* New York: New York University Press, 1993.

Finer, S. E. *The Life and Times of Edwin Chadwick.* New York: Barnes and Noble, 1952.

Fraser-Stephen, Elspet. *Two Centuries in the London Coal Trade: The Story of Charrington's.* London: privately printed, 1952.

Gomme, G. L. *London in the Reign of Victoria, 1837–1898.* London: Blackie and Son, 1898.

Halevy, Elie. *The History of the English People in the Nineteenth Century.* Vol. 4, *Victorian Years (1841–1895),* part 2, 1852–1895, by R. B. McCallum. New York: Barnes and Noble, 1961.

Hamlin, Christopher. "Edwin Chadwick and the Engineers, 1842–1854: Systems and Antisystems in the Pipe-and-Brick Sewers War." *Technology and Culture* 33, no. 4 (October 1992): 680–709.

—"Environmental Sensibility in Edinburgh, 1839–1840: The 'Fetid Irrigation' Controversy." *Journal of Urban History* 20 (1994): 311–39.

—"Providence and Putrefaction: Victorian Sanitarians and the Natural Theology of Health and Disease." *Victorian Studies* 28 (1985): 381–411.

—*A Science of Impurity: Water Analysis in Nineteenth Century Britain.* Berkeley: The University of California Press, 1990.

Harris, T. R. *Sir Goldsworthy Gurney, 1793–1875.* Cornwall: The Trevithick Society, 1975.

Helps, Arthur. *Life and Labours of Mr. Brassey, 1805–1870.* London: Bell and Daldy, 1872.

Hexter, J. H. *The History Primer.* New York: Basic Books, 1971.

Hobhouse, Hermione. *Thomas Cubitt, Master Builder.* London: Macmillan, 1971.

Hoskins, W. G. *The Making of the English Landscape.* London: Hodder and Stoughton, 1955.

Houghton, Walter E. *The Victorian Frame of Mind, 1830–1870.* New Haven: Yale University Press, 1957.

Institution of Civil Engineers. *The Education and Status of Civil Engineers, in the United Kingdom and in Foreign Countries; Compiled from Documents Supplied to the Council of the Institution of Civil Engineers, 1868 to 1870.* London: ICE, 1870.

Johnson, Stanley P. *The Politics of Environment: The British Experience.* London: Tom Stacey, 1973.

King, W. T. C. *A History of the London Discount Market.* London: George Routledge and Sons, 1936.

Kitson Clark, G. *The Making of Victorian England*. Cambridge, Mass.: Harvard University Press, 1962.

Kselman, Thomas. *Death and the Afterlife in Modern France*. Princeton: Princeton University Press, 1993.

Layton, Edwin Jr. Review of *The Engineers*, by R. Angus Buchanan, *Technology and Culture* 32, no. 1 (January 1991): 143–44.

Leventhal, F. M. "'A Tonic to the Nation': The Festival of Britain, 1951." *Albion* 27 (fall 1995): 448–53.

Logan, W. P. D. "Mortality in England and Wales, 1848–1947." *Population Studies* 4 (1951): 132–78.

London County Council General Purposes Committee. *Principal Contracts Completed by the Metropolitan Board of Works*. London: LCC, 1906.

Longford, Elizabeth. *Images of Chelsea*. Vol. 1 of *Images of London,* ed. Ralph Hyde. Richmond-upon-Thames: Saint Helena Press, 1980.

Low, Sidney, and Lloyd C. Sanders. *The History of England During the Reign of Queen Victoria, 1837–1901*. London: Longmans Green, 1907.

Luckin, Bill. *Pollution and Control: A Social History of the Thames in the Nineteenth Century*. Bristol and Boston: Adam Hilger, 1986.

Margetson, Stella. *Fifty Years of Victorian London*. London: Macdonald, 1969.

Mathias, Peter. *The First Industrial Nation: An Economic History of Britain, 1700–1914*. 2d ed. London: Methuen, 1983.

Mayhew, Henry. *London Labour and the London Poor*. London, 1861–62, 1865. Reprint, New York: Dover Publications, 1968.

McCord, Norman. *British History, 1815–1906*. New York: Oxford University Press, 1991.

H. M. Ministry of Works. *History of the King's Works*. 6 volumes. London: Ministry of Works, n.d.

Olsen, Donald. *The Growth of Victorian London*. Harmondsworth: Penguin Books, 1979.

—Review of *The People's Parks,* by Hazel Conway, *Victorian Studies* 36, no. 4 (summer 1993): 491.

Owen, David. *The Government of Victorian London, 1855–1889*. Cambridge: Harvard University Press, 1982.

Passingham, W. J. *The Romance of London's Underground*. New York: Benjamin Bloom, 1972.

Pauley, George. *The Chronicles of a Contractor*. London: Constable, 1926.

Peterson, M. Jeanne. *Family, Love and Work in the Lives of Victorian Gentlewomen*. Bloomington: Indiana University Press, 1989.

Peto, H. *Sir Morton Peto: A Memorial Sketch*. London: privately printed, 1893.

Phillips, Hugh. *The Thames about 1750*. London: Collins, 1951.

Phillips, John A., and Charles Wetherell. "The Great Reform Act of 1832 and the Political Modernization of England." *American Historical Review* 100, no. 2 (April 1995): 411–36.

Port, M. H., ed. *The Houses of Parliament*. New Haven and London: Yale University Press, 1976.

Porter, Dale, and Gloria C. Clifton. "Patronage, Professional Values, and Victorian Public Works: Engineering and Contracting the Thames Embankment." *Victorian Studies* 31, no. 3 (spring 1988): 319–49.

Post Office London Trades Directory. Annual.

Pudney, John. *Crossing London's River: The Bridges, Ferries, and Tunnels Crossing the Thames Tideway in London*. London: J. M. Dent and Sons, 1972.

Rennie, Sir John. *The Autobiography of Sir John Rennie*. London: E. and F. N. Spon, 1875.

Reyburn, Wallace. *Flushed with Pride*. London: Macdonald and Jane's, 1969.

Richardson, A. E. *Robert Mylne, Architect and Engineer, 1733 to 1811*. London: B. T. Batsford, 1955.

Ridley, Jasper. *Lord Palmerston*. London: Constable, 1970.

Robbins, Michael. *The Railway Age in Britain and its Impact on the World*. Baltimore: Penguin Books, 1962; London: Penguin Books, 1965.

Rolt, L. T. C. *Victorian Engineering*. London: Penguin Books, 1970.

Rosen, Christine Meisner, and Joel Tarr, "The Importance of an Urban Perspective in Environmental History." *The Journal of Urban Studies* 20, no. 3 (May 1994): 299–310.

Salter's Guide to the Thames. 28th ed. Oxford: Salter Bros., n.d. (ca. 1926).

Sanderson, Michael. *The Universities and British Industry, 1850–1970.* London: Routledge and Kegan Paul, 1972.

Sheppard, Francis. *London, 1808–1870: The Infernal Wen.* London: Secker and Warburg, 1971.

Simmons, I. G. *Environmental History: A Concise Introduction.* Oxford: Blackwell, 1993.

—*Interpreting Nature: Cultural Consequences of the Environment.* London and New York: Routledge, 1993.

Smith, Denis. "Sir Joseph William Bazalgette (1819–1891), Engineer to the Metropolitan Board of Works." *Transactions of the Newcomen Society* 58 (1986–87): 89–111.

Smith, J. E. *St. John the Evangelist, Westminster: Parochial Memorials.* London: privately printed, 1892.

Timbs, John. *Curiosities of London.* London: John Camden Hotten, 1867.

Trevelyan, G. M. *British History in the Nineteenth Century and After, 1782–1919.* 1937. Reprint, New York: Harper, 1966.

Trueman, A. E. *Geology and Scenery in England and Wales.* Revised by J. B. Whittow and J. B. Hardy. London: Pelican Books, 1971.

Walker, Charles. *Thomas Brassey: Railway Builder.* London: Frederick Muller, 1969.

Walker, G. E., ed. *The Thames Conservancy, 1857–1957.* London: Thames Conservancy, 1957.

Watson, J. G. *The Institution of Civil Engineers: A Short History.* London: Institution of Civil Engineers, 1982.

Webb, R. K. *Modern England From the Eighteenth Century to the Present.* New York: Dodd, Mead, 1968.

Weisburd, Stefi. "Hard Science." *Science News* 134 (1988): 24.

Wiener, Martin. *English Culture and the Decline of the Industrial Spirit.* New York: Cambridge University Press, 1981.

Wohl, Anthony S. *Endangered Lives: Public Health in Victorian Britain.* London: J. M. Dent and Sons, 1983.

Young, G. M. *Victorian England, Portrait of an Age.* London: Oxford University Press, 1936.

Index

63, 144–145; project histories, 31; as private
enterprise, 68–70, 228; financial review,
70; impact of cholera and climate, 76; as
finished projects, 219. *See also*
Metropolitan Board of Works; Office of
Works; *and project types, e.g.,* Docks;
Thoroughfares
Public Works Loan Commission, 156

Railways: underground, 4, 6, 20, 133*f*; as
escape from cities, 18, 115–117; vs
riverways, 52, 127–129, 154; vs
thoroughfares, 118–119; pneumatic,
199–200, 293*n29*; vs steam ferries, 226. *See
also* Metropolitan District Railway
Rats, 12, 221, 268*n10*
Reading (city), **22,** 24, 39
Real estate. *See* Property
Recycling. *See* Sewage disposal and
reclamation
Regent's Park, 18
Rennie, Sir John: London Bridge consultant,
46–47; new London Bridge funding, 154;
Waterloo Bridge, 225
Repton, Humphrey, 17–18
Respiratory ailments, 12
Richmond (city), **22,** 26
Riparian ecology: Thames Embankment, 4;
Embankment history, 5, 8; Thames valley,
39
River Authorities, 248–249
River Cherwell, **22,** 24
River pollution. *See* Water pollution
Riverfront (Thames): conversion from
commerce, 3–4, 6–7, 250; as cultural
microcosm, 8, 218; developments, 9–10, 21,
29, 33–37, 225, 240–241, 247; at
Twickenham, 26; pre-Embankment at
Lambeth, 53*f*, 216*f*; under Office of Woods
and Forests, 80; proprietary interests, 126,
136, 147, 160–165, 203
Rivers Pollution Prevention Act (1876):
passed but unenforced, 222, 274–275*n49*;
amended (1951), 248
Riverside Walk, 240
Royal Commission on the Thames
Embankment. *See* Thames Embankment,
Royal Commission
Royal Commissions on Metropolitan
Improvement. *See* Improvements,
commissions
Royal Festival Hall, 3, 37, 238–239, 250

St. James Park, 18
St. Paul's Cathedral: seen from Festival Hall,
3, 250; location, **27;** sewer excavation effect
on, 75; views pre- and post-Victoria
Embankment, 214*f*

St. Thomas Hospital, **27,** 162, 203, 216*f*, 238
Salisbury, Lord (2nd Marquis of Salisbury).
See Cecil, Robert
Sanitation: in Victorian society, 5–7, 54–55,
220–221, 224; as social construction, 14;
administrative units, 59–60, 62–63, 65;
improvement loans, 156; Embankment
impact on, 219, 224. *See also* Public health
Science, 65, 76, 274*n41*
Scotland: cattle, 11
Sewage: flow and diversion, 11, 118, 220, 245;
cholera from, 19; in Thames River, 21, 25,
52, 56, 66, 71, 74, 222; on Hungerford
Reach, 32; treatment, 50–52, 54–56, 59,
64–67, 221, 224; definitions, 51, 54; diluted
by rainwater, 63, 246; as problem, 76,
117–118, 248, 273*n27*
Sewage disposal and reclamation: recycling
concept, 67–68, 221, 274*n41*; treated
effluents, 68–70, 221, 224; as problem, 76,
221, 223–224
Sewage sludge: disposal as problem, 224
Sewer gases, 56–57, 70
Sewers: as public works, 4–5, 62; London
Main Drainage, 4–8, 21, 50–76, 118, 133*f*,
145, 220, 274*n37*; modernization, 6; as
water pollution control, 6, 39, 58, 66, 76;
rats in, 12, 221; trenches, 15; within
Embankment, 20, 75, 133*f*; lack of, above
London, 25; repairs, 29*n*; prefigured
Thames Embankment, 50; definitions, 51,
54; hook-ups, 56; design, plans, and
technology, 60–64, 70, 72–**73,** 118;
construction, 74–75, 274*n37*; human
ecology and, 75; construction materials,
186, 221
Shaw-Lefevre, George J. (Baron Eversley),
234, 299*n37*
Simmons, I. G.: "cultural ecosystem," 11
Simon, Dr. John, 62–63, 65
Skilled tradesmen: aspirations, 174; union
organizations, 174–176
Snow, John: waterborne nature of cholera, 55,
65, 272*n10*
Social construction: Thames Embankment,
8–10, 133, 249–250; of eighteenth century
landscape, 14; sewers as interactive
process, 51; through public discourse, 108,
134
Somerset House: seen from Festival Hall, 3,
250; location, **27;** views from, 215*f*;
Embankment effect on, 225; as a building,
281*n5*
Steam ferries: docks and landings, 110, 163,
197, 200, 225–226; commuters, 119, 164;
operators, 161, 163–164; Embankment
effect on, 225–226
Suburbs: as artificial countryside, 18, 53–54,

Traffic: as motive for Embankment, 118–119; Embankment effect on, 224, 226–227; roadway modernization, 269n6

Trench, Sir Frederick: embankment plan, 111–113, 207; sewer outfall plan for tidal locks, 118

Trinity House (mariners): high-water mark, 140, 192n, 270–271n40; as navigation agency, 140

Tunnels (animal burrows), 221

Tunnels (engineering objects): for utilities, 20; Thames River, 91, 199; railway, 199; timber framing, 201

Turnpikes. See Thoroughfares

Twickenham (city), 26

Unions, 174–176

Urban ecology: Thames Embankment and, 4, 106; Embankment history and, 5, 8, 14, 218–250; affected by physical construction, 6; cultural ecosystems as, 11–12; development as science, 13; cities as pollution agents, 19, 222

Urban History Group, 13–14

Utilities: Embankment tunnels for, 20; Embankment lighting, 193–194, 196, 200

Vauxhall Bridge, **27,** 29–30; view from, 216f

Vested interest: vs public interest, 8, 16, 122, 126–127, 147–148, 160–190, 208, 222, 233–234, 247; contractors, 172

Vestries: parochialism, 58–59, 62, 152; Metropolis Local Management Act (1855), 63, 229; Metropolitan Board of Works, 84–85, 210

Victoria Embankment: location, 32–35; William Cowper, 88–89; views, 105f, 214f; at Charing Cross Bridge, 133f; plans reviewed and authorized, 141, 163, 192–193; finance, 151–152; brick walls, 186; Temple-Blackfriars section, 195, 204–210; contracts, 195–197, 204, 206, 210–211, 291n4, 292n21; Bazalgette bronze relief, 212; opening ceremonies, 212, 219; claimed as flood cause, 228–229; twentieth century extensions, 239–240. See also Embankments (general)

Victoria Tower Gardens, 232–237

Victorian society: environment, 4–6, 8, 14, 20, 52, 54–56, 76, 217, 220–221; vs French, 10, 14–15, 168, 211, 240; morality of, 15, 18–19, 55, 221; parks in, 18, 209, 269n25; banking and finance, 148–151; institutional fragmentation, 249

Wages and earnings: engineers, 94–95, 171; skilled labor, 175–176; Embankment

specifications, 176, 193, 196; laborers, 176–177

Wales, Prince of. See Albert Edward, Prince of Wales

Walker, James: London Bridge consultant, 47–49; embankment plans, 48–49, 75, 138, 271n48; survey line, 48–49, 75, 271n48

Walker and Burgess (firm), 138

Walkways: Embankment construction and, 4, 193, 217f; morality and, 18; disfavored by some, 28; Festival of Britain (1951), 37; paving contracts and materials, 196, 210–211, 295–296n57; Riverside Walk, 240. See also Cheyne Walk

Water Authorities, 249

Water closets: Victorian attitudes, 5, 54–55, 220–221; patented by Bramah, 55; vs dry closets, 272n12

Water pollution: diseases, 12, 55, 201, 222, 272n10; in Victorian ecology, 14, 222, 274n41; protected by property law, 19, 222; sewage as agent, 21, 25, 51–53, 56, 117–118; Thames River and tributaries, 21, 24–25, 52–53, 57, 221; fish kills and fisheries, 71, 300n56; Embankment effect on, 220; as motive for Embankment, 248–249

Water pollution controls: sewers, 6, 39, 58, 66, 76, 220; regional cost-sharing, 221–222, 248; public opinion vs administrative agencies, 224; Tory values, 274–275n49

Water quality: opinions on, 65–66, 220, 274n41; human ecology and, 75; Embankment effect on, 220

Water supply: law and legislation, 58, 66, 248; effect on public health, 220, 222, 273n27; privatization, 249

Waterfront. See Riverfront (Thames)

Waterloo Bridge: location, **27,** 34, 37; rebuilding, 47, 225

Webster, William: Embankment contractor, 151, 185f, 187, 211, 292n21; collaborated with John Grant, 186–190, 203–204; Abingdon Street Embankment, 232–233

Wembley Stadium, 15

Westminster Bridge: eighteenth century opening, 11; location, **27;** relocated roadway, 30; starts Victoria Embankment, 32; rebuilding (1860), 47; construction supervised by Thomas Page, 78, 198; Boadicea statue, 195; during Albert Embankment construction, 216f

Westminster (city), **22,** 30–31; wharfingers challenged Embankment plans, 146, 160–161; Embankment walkway, 217f; clock tower, 250

Wharfingers: challenged Embankment plans, 126, 146, 160–161, 202

Whig-Liberal views, 17

ABOUT THE AUTHOR

Dale H. Porter is a Professor of History and Humanities at Western Michigan University. He received his BA at Western Michigan University, his MA at Stanford University, and his PhD at the University of Oregon. He has published two books, *The Abolition of the Slave Trade in England, 1784–1807* and *The Emergence of the Past: A Theory of Historical Explanation,* and contributed to many journals, including *History of Technology, Victorian Studies,* and *History and Theory.*